# ANIMAL TEST ALTERNATIVES

# ANIMAL TEST ALTERNATIVES

## Refinement · Reduction · Replacement

edited by
Harry Salem
*U.S. Army Edgewood Research, Development, and Engineering Center*
*Aberdeen Proving Ground, Maryland*

Marcel Dekker, Inc.    New York • Basel • Hong Kong

**Library of Congress Cataloging-in-Publication Data**

Animal test alternatives : refinement, reduction, replacement / edited by Harry Salem.

p. cm.

Includes bibliographical references and index.

ISBN 0-8247-9284-X

1. Toxicity testing. 2. Animal experimentation. I. Salem, Harry.

RA1199.A533 1995

615.9'07—dc20

94-35420

CIP

*The views of the authors do not purport to reflect the position of the Department of Defense. The use of trade names in this book does not constitute an official endorsement or approval of the use of such commercial hardware or software.*

The publisher offers discounts on this book when ordered in bulk quantities. For more information, write to Special Sales/Professional Marketing at the address below.

This book is printed on acid-free paper.



Marcel Dekker, Inc.
270 Madison Avenue, New York, New York 10016

Current printing (last digit):
10 9 8 7 6 5 4 3 2 1

PRINTED IN THE UNITED STATES OF AMERICA

# Preface

The objectives of this book are to present the current concepts and approaches on alternatives to animal research and testing, as well as to review their stage of development and validation, focusing on the three Rs (refinement, reduction, and replacement). The fourth R—responsibility—is demonstrated by the willingness of the many committed scientists to contribute to this book. Indeed, all chapters have been written by experts from the scientific communities (regulatory, government, industry, and academia) who are actively making contributions in the field of alternatives to the use of animals in research and testing.

Because of the diverse efforts ongoing in alternative testing, and because of slow regulatory acceptance, I felt it was important to cover not only eye and skin irritation alternatives, but also many other areas of activity in alternatives in research and testing, including the hepatic, renal, respiratory, cardiovascular, and environmental systems, as well as computer models, cell and tissue cultures, chemical models, and the use of animals lower on the phylogenetic tree. In addition, progress to date and future direction are addressed, as is the regulatory climate for using alternatives.

The contributors describe the state of the art, the techniques used, and the results obtained, as well as the progress toward validation. Many also point out the necessity for mechanistically based tests in the regulatory process, as opposed to simple correlations. In addition, this book provides an insight into regulatory thinking and harmonization.

This book will be of interest to all scientists and students in the life sciences, in particular to those in the fields of pharmacology, toxicology, and related medical sciences. Those working in the pharmaceutical, cosmetic, and agricultural chemical industries, as well as academicians, government regulators, and environmental and risk-assessment scientists will also find this book of great value. In addition, researchers active in seeking and validating alternatives to animals in research and testing through refinement, reduction, and replacement will find this book very useful.

This book is dedicated to Flo, Jerry, Amy, and Joel, for their practice of compassion and respect for all living things, as well as to all investigators who are seeking and applying alternatives to animal testing without compromising human and animal health and welfare.

Harry Salem

# Introduction

**Harry Salem**

*U.S. Army Edgewood Research, Development, and Engineering Center*
*Aberdeen Proving Ground, Maryland*

This book has been prepared by members of the scientific community who are sensitive to the issues of animal experimentation as well as to human health concerns, and who are searching to provide validated substitutions for in vivo animal models.

The objectives are to present new concepts and approaches on alternatives to animal testing, as well as to review their current stage of development and validation, focusing on the three Rs: refinement, reduction, and replacement.

Currently, tests in animals are the primary means of assessing potential hazards of chemical exposure in humans, other than direct testing. Until such time that animals can be completely and confidently replaced in testing by valid and meaningful models, attempts should be made to obtain more information from each animal. This constitutes *refinement*. *Reduction* in the number of animals should result from refinement, and should be considered in an expeditious manner when designing efficacy and safety studies.

The reasons experimentation and testing are conducted are to study the efficacy and safety of chemicals and drugs on a living system that is a surrogate for humans. These data are then extrapolated to humans and, for drug development, are then directly tested. Historically, laws were passed in response to catastrophic events that impinged on the health and safety of the American people. The first Pure Food and Drug Act was passed in 1906, prompted by the adulteration of foods with chemical preservatives, "quack" remedies being promoted and sold to the consumer, and the filth and contamination rampant in the meat-packing industry, conditions that were exposed in Upton Sinclair's book, *The Jungle*. This Pure Food and Drug Act, also known as the Wiley Act, prohibited the manufacture and interstate shipment of adulterated and misbranded food and drugs. The next major relevant law passed was the Federal Food, Drug, and Cosmetic Act of 1938. This was prompted, in part, by the elixir of sulfanilamide episode that resulted in over 100 deaths. Although sulfanilamide tablets and powder have previously been used safely for the treatment of streptococcal infections, the elixir, formulated with ethylene glycol as the vehicle, had not been tested for toxicity, nor was it a requirement of law until that time. This new law of 1938 also regulated cosmetics. Thus the Draize eye irritation test was developed to prevent the sale of dangerous products such as Lash-Lure, a mascara that caused permanent blindness in a woman after its use in the early 1930s. A subsequent catastrophic event was that of thalidomide, which caused birth defects when it was taken during pregnancy. This event resulted in the 1962 Kefauver–Harris Amendment to the Food, Drug, and Cosmetic Act which added the requirement for additional animal testing, not only for safety, but also for efficacy.

Today, with advancing knowledge and technology, more testing may be mandated to rule out long-term adverse effects, other than cancer, following both short- and long-term exposure. Some of these potential adverse effects may even be unknown to us today.

One hopes that much of the testing may be done adequately in vitro, but ultimately, the testing may require whole-animal studies to understand the complex, dynamic interactions and functions of defense and repair mechanisms programmed in the intact animal to maintain homeostasis. This concept of homeostasis was first postulated by Claude Bernard almost 150 years ago.

As the methodology in whole-animal studies is refined, in concert with in vitro tests, the number of animals required for testing may be reduced without compromising the generation of an adequate and useful database to ensure efficacy and safety.

Animal research has contributed significantly to many medical breakthroughs as well as to the database of biological knowledge. Since 1901, 54 Nobel Prizes in physiology and medicine have been awarded for discoveries based on the use of animals. Dr. Joseph Murray, the winner of the 1990 Nobel Prize for Medicine, has stated, "there would not be a single person alive today as a result of an organ transplant without animal experimentation." There is much evidence that remarkable improvement of health and well-being, such as the quality of life, has been dependent on animal research. Diabetes, once a killing disease, took its toll with tragic rapidity until the summer of 1921, when two young Canadian scientists, Frederick Banting and Charles Best, succeeded in isolating insulin, which now enables millions of diabetics to lead normal lives. It was in the laboratories at the University of Toronto that Banting and Best kept the depancreatized dog Marjorie alive with injections of pancreatic extracts that contained what we now know as insulin.

The development and safety testing of vaccines, including the vaccine that led to the eradication of polio, were also dependent on animal experimentation and safety testing. Animal research has resulted in many medical advances, from drug development for the treatment of life-threatening allergies, psychiatric disorders, heart disease, and infections to transplantation and transplant therapy, as well as surgery, including open-heart and coronary bypass procedures.

Recently, when Dr. Steven Rosenberg of the National Institutes of Health received the Research and Development Achievement Award for the development of a most promising breakthrough in the treatment of cancer using immunotherapy, he stated that none of this would have been possible without the use of animals.

Animal experimentation has played a very important role in improving human health and well-being. Without such experimentation, will significant medical breakthroughs continue to be made?

As long as animal experimentation continues, responsible and sensitive scientists will be concerned about the use, care, and welfare of laboratory animals. The scientific community accepts and subscribes to rules and guidelines within which to conduct research and testing on animals. These include accreditation by the American Association for Accreditation of Laboratory Animal Care (AAALAC), review by the Laboratory Animal Use Committee, compliance with the Animal Welfare Act and the Society of Toxicology Code of Ethics, and Good Laboratory Practices.

This book is dedicated to those who pursue this approach and are committed to the principles of the three Rs plus one, responsibility, without compromising the health and safety of the human population. It is hoped that by the use of in vitro systems, such as cell lines, isolated cells, microorganisms, tissue cultures, invertebrates, quantitative structure–activity relations and physiologically based pharmacokinetic modeling, it may be possible to predict toxicity, or at least gain insight about the possible target organs and mechanisms of action. These techniques should at least result in the reduction in the number of animals used in experimentation.

In this book, investigators active in the field present the methods implemented and their findings, consistent with the *four* Rs for efficacy and safety testing of chemicals and drugs.

# Contents

# Contributors

**José Acevedo** In Vitro International, Irvine, California

**Kenneth B. Adler** Department of Anatomy, Physiological Sciences, and Radiology, College of Veterinary Medicine, North Carolina State University, Raleigh, North Carolina

**Melvin E. Andersen*** Center for Extrapolation Modeling, Duke University, Durham, North Carolina

**Robert S. Anderson** Chesapeake Biological Laboratory, University of Maryland, Solomons, Maryland

**John A. Bantle** Department of Zoology, Oklahoma State University, Stillwater, Oklahoma

**Steven I. Baskin** Pharmacology Division, U.S. Army Medical Research Institute of Chemical Defense, Aberdeen Proving Ground, Maryland

**Sonia Sherard Braa** Advanced Tissue Sciences, Inc., La Jolla, California

---

**Present affiliation*: Health Effects Research Laboratory, U.S. Environmental Protection Agency, Research Triangle Park, North Carolina

**Stephen R. Channel** Toxicology Division, Armstrong Laboratory, Wright-Patterson Air Force Base, Ohio

**Nancy A. Chester** Life Sciences Department, U.S. Army Edgewood Research, Development, and Engineering Center, Aberdeen Proving Ground, Maryland

**Paul K. Chien** Department of Biology, University of San Francisco, San Francisco, California

**Leah A. Cohn** Department of Anatomy, Physiological Sciences, and Radiology, College of Veterinary Medicine, North Carolina State University, Raleigh, North Carolina

**Jeffrey R. Cook*** Department of Biological Sciences, State University of New York at Binghamton, Binghamton, New York

**Joseph A. Cotruvo** Office of Pollution Prevention and Toxics, U.S. Environmental Protection Agency, Washington, D.C.

**Benjamin R. Crenshaw, Jr.**[†] Addiction Research Center, National Institute on Drug Abuse, National Institutes of Health, Baltimore, Maryland

**L. M. Dasko-Vincent** Health Effects Research Division, U.S. Army Biomedical Research and Development Laboratory, Fort Detrick, Frederick, Maryland

**Nicholas J. DelRaso** Toxicology Division, Armstrong Laboratory, Wright-Patterson Air Force Base, Ohio

**Tracy Donnelly** Advanced Tissue Sciences, Inc., La Jolla, California

**Kurt Enslein** Health Designs, Inc., Rochester, New York

**George R. Famini** U.S. Army Edgewood Research, Development, and Engineering Center, Aberdeen Proving Ground, Maryland

**Robert A. Finch** U.S. Army Biomedical Research and Development Laboratory, Fort Detrick, Frederick, Maryland

**Thomas J. Flynn** Division of Toxicological Research, Center for Food Safety and Applied Nutrition, U.S. Food and Drug Administration, Laurel, Maryland

**Arthur Furst** Institute of Chemical Biology, University of San Francisco, San Francisco, California

*Present affiliations*:
*Department of Ophthalmology, Anheuser-Busch Eye Institute, St. Louis Medical Center, St. Louis, Missouri
[†]Department of Anesthesiology, Johns Hopkins School of Medicine, Baltimore, Maryland

**Shayne C. Gad** Gad Consulting Services, Raleigh, North Carolina

**Henry S. Gardner, Jr.** U.S. Army Biomedical Research and Development Laboratory, Fort Detrick, Frederick, Maryland

**Alan M. Goldberg** Department of Environmental Health Sciences, Center for Alternatives to Animal Testing, School of Hygiene and Public Health, Johns Hopkins University, Baltimore, Maryland

**Virginia C. Gordon** In Vitro International, Irvine, California

**Carol E. Green** Toxicology Laboratory, SRI International, Menlo Park, California

**Sidney Green** Division of Toxicological Research, Department of Health and Human Services, U.S. Food and Drug Administration, Laurel, Maryland

**Tami M. Greenwalt** In Vitro Technologies, Inc., Baltimore, Maryland

**Kailash C. Gupta** Department of Health Science and Health Effects, U.S. Consumer Product Safety Commission, Bethesda, Maryland

**Mark V. Haley** Life Sciences Department, U.S. Army Edgewood Research, Development, and Engineering Center, Aberdeen Proving Ground, Maryland

**Barry L. Hancock** Toxicology Division, Armstrong Laboratory, Wright-Patterson Air Force Base, Ohio

**Georgia A. Hart** Mountain Technical Center, Schuller International, Inc., Littleton, Colorado

**Jeff D. Harvell** Department of Dermatology, University of California School of Medicine, San Francisco, California

**Thomas W. Hesterberg** Department of Health, Safety, and Environment, Mountain Technical Center, Schuller International, Inc., Littleton, Colorado

**David W. Hobson**† Battelle Memorial Institute, Columbus, Ohio

**Jurij J. Hostynck**‡ University of California School of Medicine, San Francisco, California

**John E. Johnson, Jr.*** Addiction Research Center, National Institute on Drug Abuse, National Institutes of Health, Baltimore, Maryland

**Inger Kidd** Advanced Tissue Sciences, Inc., La Jolla, California

---

*Retired
*Present affiliations*:
†Department of Pharmaceutical Sciences, Dermatological Products of Texas, San Antonio, Texas
‡Euromerican Technology Resources, Inc., Lafayette, California

**Stephen Drew Kirby** Pathophysiology Division, Physiology Branch, U.S. Army Medical Research Institute of Chemical Defense, Aberdeen Proving Ground, Maryland

**Kannan Krishnan** Department of Occupational and Environmental Health, University of Montreal, Montreal, Quebec, Canada

**Wayne G. Landis*** U.S. Army Edgewood Research, Development, and Engineering Center, Aberdeen Proving Ground, Maryland

**Robert L. Lipnick** Health and Environmental Review Division, Environmental Effects Branch, Office of Pollution Prevention and Toxics, U.S. Environmental Protection Agency, Washington, D.C.

**Janna S. Madren-Whalley** Pathophysiology Division, Physiology Branch, U.S. Army Medical Research Institute of Chemical Defense, Aberdeen Proving Ground, Maryland

**Philip S. Magee** BIOSAR Research Project, Vallejo, and Department of Dermatology, University of California School of Medicine, San Francisco, California

**Howard Maibach** Department of Dermatology, University of California School of Medicine, San Francisco, California

**Francis N. Marzulli** Consultant in Toxicology, Bethesda, Maryland

**Millard M. Mershon** Toxicology Division, U.S. Army Medical Research Institute of Chemical Defense, Aberdeen Proving Ground, Maryland

**Charles B. Millard** Toxicology Division, U.S. Army Medical Research Institute of Chemical Defense, Aberdeen Proving Ground, Maryland

**W. R. Mitchell** Health Effects Research Division, U.S. Army Biomedical Research and Development Laboratory, Fort Detrick, Frederick, Maryland

**Laurie M. Mora** Chesapeake Biological Laboratory, University of Maryland, Solomons, Maryland

**Mildred M. Newman** Mountain Technical Center, Schuller International, Inc., Littleton, Colorado

**Sharon J. Northup** I.V. Systems Division, Toxicology and Materials Testing, Baxter Healthcare Corporation, Round Lake, Illinois

**Laura M. Patrone**† Department of Biological Sciences, State University of New York at Binghamton, Binghamton, New York

*Present affiliations*:

*Institute of Environmental Toxicology and Chemistry, Huxley College of Environmental Studies, Western Washington University, Bellingham, Washington

†Mayo Clinic, Scottsdale, Arizona

**Laura S. Rhoads*** Department of Biological Sciences, State University of New York at Binghamton, Binghamton, New York

**Charles E. Ruegg** In Vitro Technologies, Inc., Baltimore, Maryland

**Lemuel T. Russell** Division of Experimental Therapeutics, Department of Pharmacology, Walter Reed Army Institute of Research, Washington, D.C.

**Harry Salem** Life Sciences Department, U.S. Army Edgewood Research, Development, and Engineering Center, Aberdeen Proving Ground, Maryland

**Van M. Seabaugh** U.S. Environmental Protection Agency, Washington, D.C.

**William J. Schmidt** Toxicology Division, Armstrong Laboratory, Wright-Patterson Air Force Base, Ohio

**Paul M. Silber** In Vitro Technologies, Inc., Baltimore, Maryland

**E. Tiffany Spence** Thomas J. Stephens & Associates, Inc., Carrollton, Texas

**Thomas J. Stephens** Thomas J. Stephens & Associates, Inc., Carrollton, Texas

**Richard D. Thomas** National Research Council, National Academy of Sciences, Washington, D.C.

**David G. Thomassen**† Inhalation Toxicology Research Institute, Albuquerque, New Mexico

**Sandra A. Thomson** Life Sciences Department, U.S. Army Edgewood Research, Development, and Engineering Center, Aberdeen Proving Ground, Maryland

**Dennis Triglia**‡ Advanced Tissue Sciences, La Jolla, California

**Charles A. Tyson** Toxicology Laboratory, SRI International, Menlo Park, California

**Robert G. Van Buskirk** Department of Biological Sciences, State University of New York at Binghamton, Binghamton, New York

**James A. Vick** Department of Health and Human Services, U.S. Food and Drug Administration, Washington, D.C.

**Jurgen D. von Bredow**§ Walter Reed Army Institute of Research, Washington, D.C.

---

*Present affiliations*:

*U.S. Army Medical Research Institute of Chemical Defense, Aberdeen Proving Ground, Maryland

†U.S. Department of Energy, Washington, D.C.

‡Houghten Pharmaceuticals, San Diego, California

§Division of Residue Chemistry, Metabolism and Diagnostics Branch, U.S. Food and Drug Administration, Beltsville, Maryland

**Merry Jane Walsh** Toxicology Division, Armstrong Laboratory, Wright-Patterson Air Force Base, Ohio

**David B. Warheit** Haskell Laboratory, E. I. du Pont de Nemours & Company, Newark, Delaware

**Arthur D. Weissman*** Addiction Research Center, National Institute on Drug Abuse, National Institutes of Health, Baltimore, Maryland

**D. R. Wellington** Health Effects Research Division, U.S. Army Medical Research and Development Laboratory, Fort Detrick, Frederick, Maryland

**Robert J. Werrlein** Pathophysiology Division, Physiology Branch, U.S. Army Medical Research Institute of Chemical Defense, Aberdeen Proving Ground, Maryland

**Patricia D. Williams**† Investigative Toxicology Department, American Cyanamid Company, Pearl River, New York

**Leland Y. Wilson** Chemistry Department, La Sierra University, Riverside, California

**Maurice Zeeman** Health and Environmental Review Division, Environmental Effects Branch, Office of Pollution Prevention and Toxics, U.S. Environmental Protection Agency, Washington, D.C.

---

*Present affiliations*:

*Department of Anesthesiology, Seton Hall University School of Graduate Medical Education, South Orange, New Jersey

†Glycomed, Inc., Alameda, California

# 1

# The Evolution of In Vitro Toxicology in the Pharmaceutical Industry

**Patricia D. Williams***

*American Cyanamid Company*
*Pearl River, New York*

In vitro toxicology describes a field of endeavor that simply applies technologies, including isolated organs, isolated tissues, cell culture, biochemistry, and chemistry, to the study of toxic or adverse reactions of xenobiotics. The objective of this chapter is to provide a perspective of the evolution of in vitro toxicology, particularly as it applies to the pharmaceutical industry.

## Historical Perspectives

The decade of the 1980s was marked by prominent interest and activity in the applications of in vitro techniques in toxicology, both in the academic and industrial sectors. This activity was manifested by the appearance of several journals devoted to in vitro, biochemical, or molecular approaches to toxicology; the establishment of laboratories focused on the search for alternatives; and the conduct of numerous symposia and workshops devoted to the topic of in vitro toxicology. Most importantly, the past decade was associated with a growing commitment of industry to in vitro test development. This commitment was reflected by the formation of groups within many industrial safety assessment components that were focused on investigational research, often

*Present affiliation*: Glycomed, Inc., Alameda, California

involving in vitro techniques. Several factors have impinged on these parallel developments in an additive fashion. First, the availability of the applied in vitro technologies themselves (e.g., cell culture, receptor pharmacology), and the personnel trained to perform them, blossomed in the 1970s and 1980s. Second, the contribution of the public sector in the developments of the 1980s must also be recognized. Toxicology, particularly as an industrial activity, has always been strongly influenced by public opinion and pressure. The impetus for the formation of the Food and Drug Administration (FDA), and the issuance of regulatory laws commencing with the Food, Drug, and Cosmetic Act of 1938, can largely be attributed to public reaction to tragedies, such as sulfanilamide in 1937 and thalidomide in 1961. So, too, has public pressure for animal welfare concerns in the last decade played a significant role in solidifying the commitment to investigate and employ in vitro techniques whenever possible.

Lastly, in viewing the progression of in vitro toxicology in the pharmaceutical industry, it is relevant that the advent of toxicology itself, as a profession, is a recent development in this century. This development can be reflected by the formation of the Society of Toxicology in 1961, and the establishment of drug safety units distinct from pharmacology departments within the industry in the 1950s and 1960s. Toxicology, therefore, can be considered as a relatively young profession when compared with pharmacology, whose professional society, American Society for Pharmacology and Experimental Therapeutics (ASPET), was established in 1908. Industrial laboratories, therefore, have been in the forefront in the incorporation of in vitro techniques in toxicology.

## INDUSTRIAL APPLICATIONS

The number and types of in vitro toxicological models employed in the pharmaceutical industry encompass virtually every major target organ of toxicological interest (e.g., liver, kidney, heart, brain, skin, eye). The breadth of the systems available is impressive and, again, signifies a relatively rapid progression of in vitro test development in toxicology.

### Philosophical and Scientific Considerations

The fact that many of the in vitro test systems are employed in toxicology underscores the commitment of industry to the principles of reduction, refinement, and replacement of whole-animal in vivo tests whenever possible. However, industrial toxicologists must appropriately balance this commitment with a fourth R, responsibility. The ethical and legal responsibility of the toxicologist is to assess the safety of new products, and to protect, to the best of his or her ability, the public from harm. Thus, current test procedures cannot be abandoned unless the new tools can be adopted with the assurance that adverse properties will be reliably detected. A key factor in the application of in vitro techniques in toxicology involves the degree of correlation between events occurring in vitro and those that the toxicologist evaluates in the intact animal. This correlation determines the ultimate scientific value of the techniques and of the level of confidence associated with a particular test in terms of its predictability from a safety perspective.

The criteria that determine the degree of correlation or level of confidence in a given test are summarized in Table 1. The first of these is predictability, both qualitatively and quantitatively. Qualitatively, do the rankings or order of toxicities in vitro correlate with the order of toxicity in vivo? From a quantitative aspect, what are the dose–response characteristics from which potency estimates and comparisons can be made? Finally, how do drug concentrations in vitro compare with those achieved in vivo? The second criteria to examine is that of test system identity. To what degree does the in vitro system structurally and functionally mimic the in vivo organ? Third, the area of mechanisms of cellular injury is a key criteria to consider in the utilization of in vitro models for toxicological evaluation. The variety of mechanisms that can play a central role in cell injury underscores the need to recognize specifically how the response of the in vitro test system correlates with the response of the intact

**Table 1** Criteria for Establishing In Vitro–In Vivo Correlations

| | |
|---|---|
| Predictability | |
| Qualitative: rankings (% maximum response) | |
| Quantitative: dose response; relative drug concentrations in vitro and in vivo | |
| Identity | |
| Structure: morphological correlates | |
| Function: tissue specificity (e.g., transport characteristics, metabolic pathways) | |
| Mechanisms of injury | |
| Membrane damage (structural, functional) | |
| Synthetic activity (protein, RNA, DNA) | |
| Metabolic poisoning ($O_2$ utilization/consumption, glycolysis, gluconeogenesis) | |
| Compensatory factors | |
| Biochemical scavengers (glutathione, metalloproteins) | |
| Detoxification pathways | oxidation/reduction/hydrolysis/conjugation |
| Toxification pathways | oxidation/reduction/hydrolysis/conjugation |

organ or organism. Lastly, the topic of compensatory factors needs to be considered. How does the ability of the in vitro system to scavenge toxic reaction products compare with in vivo abilities? What detoxification or toxification pathways relevant to the in vivo fate of toxins or chemicals are present in the in vitro test system? These questions must be addressed before one uses in vitro procedures to establish a clear understanding of the assumptions or limitations of the data to be generated. Although in vitro systems need not fulfill all the correlative criteria outlined to be useful, the successful achievement of these criteria will be required to replace current whole-animal tests.

## The Role of In Vitro Tests in Toxicological Testing

An examination of the role that in vitro tests play in toxicology requires a preliminary discussion of the safety assessment process itself. Preclinical toxicological assessment involves a series of decisions at the preproject and product safety stages of development. Although the vast majority of in vitro techniques have not been fully validated to meet all of the in vitro–in vivo criteria discussed in the preceding section, many can still play an important role within the toxicological tier structure. At the preproject evaluation stage, several chemical analogues of a particular pharmacological series may be under consideration for development. In vitro techniques are particularly powerful at this level of decision-making as product candidates (or project team compounds) are identified. Specifically, in vitro tests facilitate the evaluation of agents for specific target organ effects (e.g., hepatotoxicity, nephrotoxicity), contributing directly to the product selection process. Such preliminary evaluation and screening may greatly enhance the probability of success of new agents in subsequent safety testing in animals and humans. Following selection of a compound for further development, a safety assessment "package" is conducted before clinical introduction of the new agent. At this stage, in vitro techniques are also applied in performing specialized evaluations (e.g., genotoxicity, neuromuscular transmission) as well as in resolving issues that may have arisen in in vivo safety studies. These in vitro data are submitted to regulatory agencies as part of the overall safety assessment of a new chemical entity. In vitro techniques are also useful in problem-solving activities, such as exploring the mechanisms of target organ toxicity, or performing species comparisons in vitro. Similarly, at the clinical stages of drug testing, safety issues may be most appropriately explored through the use of in vitro techniques to study mechanisms of an adverse finding (i.e., to compare species, including humans).

In vitro techniques permit the toxicologist to participate in the drug discovery process itself. Since the development of safe, as well as effective, products, is the ultimate goal of the pharmaceutical industry, the role that in vitro models can play in the design of new agents and improvement of existing ones is particularly exciting. Specifically, in vitro systems permit toxicologists, as well as biochemists and chemists, to explore structure–activity relations involved in toxic reactions.

The potential to define structural features associated with toxic reactions, and to determine if those characteristics are distinct from those required for the desired therapeutic activity, represents an integral role the toxicologist can play in the drug discovery process. For existing products, the modeling of toxic reactions in vitro, particularly at the mechanistic level, provides avenues for designing strategies to ameliorate or prevent injury at the target organ level.

An example of a simple in vitro model that has contributed in multiple ways to the safety assessment process involves the use of isolated renal membranes to study the nephrotoxicity of aminoglycoside antibiotics. By using isolated renal brush border membranes, mechanistic information and structure–activity relations were obtained by examining the binding of radiolabeled aminoglycosides (e.g., gentamicin) to the membranes in vitro [1,2]. In addition, inhibitors of aminoglycoside nephrotoxicity in vivo were identified using the isolated renal membrane model [3], contributing directly to product improvement and design.

## THE FUTURE

The future of in vitro techniques in toxicological assessment takes us back to our earlier discussion of the philosophical and scientific considerations operating in the evolution of alternative methods.

Scientifically, the future will depend on the level of confidence achieved that in vitro systems provide information that is representative of the in vivo processes the toxicologist seeks to model and predict. The degree to which this level of confidence will evolve is directly proportional to the fulfillment of the scientific criteria (see Table 1) that determine the strength of the in vitro–in vivo correlations obtained. Because gaps remain in our knowledge relative to these important criteria, the future will also depend on advancements in available knowledge and technology concerning biochemical or in vitro toxicological events. Specifically, challenges for the future involve the continued development of current in vitro models to provide physiological and morphological characteristics that correlate more closely with target organs in vivo. These challenges will include defining media constituents that will facilitate the

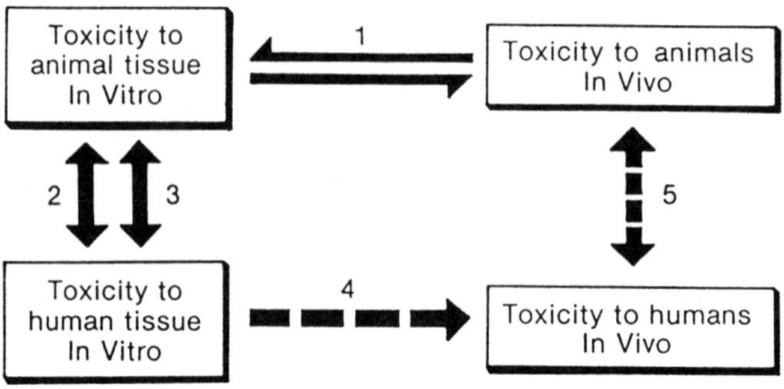


**Figure 1** The use of in vitro techniques in performing interspecies comparisons. 1, correlation of in vitro with in vivo observations; 2, development of parallel in vitro systems in humans; 3, comparison of in vitro sensitivity of different species; 4, potential toxicity in vivo; 5, comparison of in vivo sensitivity of different species.

retention of normal morphology and metabolizing capabilities in vitro, as well as the establishment of systems that reproduce the dynamic features of organs in vivo. With in vitro–in vivo correlations playing a key role in in vitro test development, the growth of technologies that focus on differentiated functions, cellular relations, human models, and other in vivo properties, such as fluid dynamics, will be critical to the future applications of in vitro techniques in toxicology. As predictability and correlation of in vitro models with specific in vivo events is achieved, toxicologists can look forward to performing meaningful species comparisons in vitro, involving direct preclinical assessments in humans (Fig. 1).

## CONCLUSION

In summary, an examination of the current state of the art in the pharmaceutical industry illustrates exciting and important role(s) that in vitro systems can play in toxicological assessments. With the development of in vitro models more closely representing organ function and dynamics, the ability to accurately predict toxic reactions in vivo will be enhanced. As time and experience build scientific confidence and with validation of these in vitro models, one can envision the toxicologist utilizing in vitro models more extensively in the assessment of toxicity. However, to advance alternative methodology to this stage, it is clear that the toxicologist must continue to take an active role in the research of toxic mechanisms and the development of the tools to study them.

Without such a commitment, the field of in vitro toxicology might be expected to remain static. Thus, although the current uses of in vitro techniques have significantly reduced animal use and advanced our understanding of toxic mechanisms, the full potential of in vitro models to further reduce, refine, and replace whole-animal tests has certainly not been realized. The future, in fact, is only limited by the amount of energy and creativity we, as scientists, can apply to in vitro test development.

For the present, the utilization of in vivo toxicological models is imperative for responsible risk assessment of new chemical entities. At the same time, the use of the many in vitro models currently available can serve as valuable adjuncts to these in vivo assessments, not only reducing the number of animals used in risk assessment, but providing unique information and possibilities for scientists involved in the drug discovery and development process.

## REFERENCES


1. Williams, P. D., Bennett, D. B., Gleason, C.R., and Hottendorf, G. H., Correlation between renal membrane binding and nephrotoxicity of aminoglycosides, *Antimicrob. Agents Chemother.* *31*:570–574 (1987).
2. Williams, P. D., and Hottendorf, G. H., $^3$H-Gentamicin uptake in brush border and basolateral membrane vesicles from rat kidney cortex, *Biochem. Pharmacol.* *35*: 2253–2256 (1986).
3. Williams, P. D., Hottendorf, G. H. and Bennett, D. B., Inhibition of renal membrane binding and nephrotoxicity of aminoglycosides, *J. Pharmacol. Exp. Ther.* *237*:919–925 (1986).

# 2

# Relating In Vitro to In Vivo Exposures with Physiologically Based Tissue Dosimetry and Tissue Response Models

**Melvin E. Andersen***

*Duke University*
*Durham, North Carolina*

**Kannan Krishnan**

*University of Montreal*
*Montreal, Quebec, Canada*

The assessment of potential health risks associated with human exposure to chemicals is performed in four steps: (a) hazard identification, (b) exposure assessment, (c) dose–response assessment, and (d) risk characterization [1]. Toxicology studies in animals and in animal tissue preparations provide qualitative and quantitative information useful for risk assessment. Qualitative information is usually more useful for hazard identification (i.e., estimating what kinds of adverse responses may be caused by a particular chemical). Quantitative results provide information for the dose–response assessment (i.e., deciding under which exposure conditions these effects are likely to occur in animals and in exposed persons). In vitro systems are often applied to study (a) metabolic rates and pathways of metabolism, (b) mechanisms of action of the toxicant, and (c) tissue responses, using tissue from various species, including humans. These in vitro test systems have been more frequently applied for hazard identification than for dose–response assessment, even

*Present affiliation*: U.S. Environmental Protection Agency, Research Triangle Park, North Carolina

though these systems are amenable to detailed characterization of quantitative dose–response relations.

To utilize this rich in vitro data source in the dose–response assessment portion of contemporary risk assessment, it is necessary to know how the observations in vitro relate to in vivo exposure situations. Two important questions are: What types of in vivo exposures give rise to tissue and cellular concentrations equivalent to those achieved in the in vitro studies, and how accurately do in vitro dose–response curves predict dose–response behavior in the intact organism? Physiologically based dosimetry and physiologically based tissue response models permit a wide range of extrapolations [2,3], including extrapolation from results with multiple in vitro test systems to predict toxicity and dose–response characteristics under realistic exposure conditions in the living animal [4].

The interested reader is directed to several reviews related to aspects of physiologically based modeling [5–7]. This chapter discusses the integration of in vitro results with these physiologically based models and the prospects for wider application of these integrated techniques in chemical risk assessment, including extrapolation based on in vitro results with human tissue samples.

## INTEGRATED EXPOSURE–DOSE–RESPONSE RELATIONSHIPS

Toxicity in intact animals represents the culmination of a series of interrelated steps: exposure, delivery of active chemical from the external environment to target tissues, initial interaction of toxic chemical with cellular constituents, and progression from these initial interactions to cellular alterations and, eventually, to overt deleterious responses at the cellular, tissue, or organism level. Schematically,

Exposure→Target Tissue Dose→Initial Cellular Interactions

and then,

Initial Cellular Interactions→Altered Cell Function→Toxicity

Tissue dose of the toxic moiety, the unchanged parent chemical, or reactive metabolites, is not always proportional to exposure concentration of the parent chemical, especially at the higher-exposure levels frequently used in toxicological studies (e.g., cancer bioassays). In these cases, tissue disposition of a chemical changes from one exposure level to another, and the basis of this exposure level-dependent behavior has to be ascertained to adequately predict response incidence across exposure levels. To distinguish the concepts of exposure level and dose to critical tissues, the use of the term *exposure–dose–response assessment* has been proposed [8]. The expression, exposure–dose–response assessment, refers to the quantitative relationship between exposure levels and target tissue dose and, additionally, the relation between tissue dose and the observed or predicted responses in animals and humans.

In vitro toxicology studies typically produce concentration–response relationships relative to particular endpoints. Ideally, the goal of these in vitro studies should be to predict accurately the exposure–dose–response relationship for the intact animal. Little attention has yet been paid to the ability of these in vitro results to quantitatively predict outcome in vivo. Progress in relating the exposures and toxicity observed in vitro to in vivo conditions has been hampered by lack of application of quantitative tools. An understanding of the critical biological determinants that constitute the in vitro and in vivo systems should make it feasible to determine the in vivo conditions that produce tissue concentrations and toxic effects comparable with those seen in vitro. This comparison can be effectively accomplished when in vitro data are incorporated within a physiologically based modeling framework.

## In Vitro and In Vivo Exposures

The concentration–response relationship obtained in vitro may well be observed in the intact test organism, provided (a) the exposure concentrations are comparable and (b) the critical tissue response or repair elements are adequately expressed in the in vitro system. The exposure concentration in in vitro experiments refers to the concentration of chemical in the incubation medium (i.e., in proximity to the target cells or tissues). In vitro concentrations can generally be easily estimated by appropriate chemical assays. The in vivo exposure concentration refers to the concentration in the ambient environment. To compare in vitro and in vivo concentrations, the concentrations in proximity to target cells in vivo have to be compared with the in vitro exposure conditions. Concentrations, here, may relate to parent chemical or reactive metabolite(s). In vitro and in vivo exposures would be considered similar when the appropriate tissue dose metric is similar for the in vivo exposure and the in vitro test situation. To be of practical value then, it is necessary to determine, or to reliably estimate, the external exposure conditions that would give rise in vivo to the instantaneous or cumulative tissue dose achieved by the in vitro exposure situation. The ability to compare these situations requires a quantitative description that relates tissue dose to exposure concentration within a pharmacokinetic framework. Physiologically based pharmacokinetic models provide this framework.

## PHYSIOLOGICAL DOSIMETRY MODELS

The steps from exposure, to tissue dose, to initial cellular effects are frequently included in descriptions of the pharmacokinetics or dosimetry of toxic chemicals. The accumulation and elimination of precursor chemical and reactive metabolites from each tissue within a physiologically based pharmacokinetic (PB-PK) model (Fig. 1) are described by appropriate mass–balance differential equations (MBDE). Terms and abbreviations used frequently in the equations are given in Table 1. A representative MBDE for the liver, expressing the rate of change of the amount of chemical in the liver ($A_l$), usually contains terms for parent chemical entering and exiting the tissue by the blood flow ($Q_l$) with arterial ($C_a$) and venous blood ($C_{vl}$) concentrations and for loss (or possibly gain) of chemical in the tissue by metabolism. The following MBDE for the liver [Eq. (1)] includes two pathways of metabolism, a saturable component and a first-order component.

$$\frac{dA_l}{dt} = Q_l(C_a - C_{vl}) - \frac{V_{max} \cdot C_{vl}}{K_m + C_{vl}} - K_f \cdot C_{vl} \cdot V_l \tag{1}$$

In most PB–PK models, amount of chemical in the tissue ($A_t$) is determined by integrating Eq. (1) from an appropriate starting condition ($A_{l,0}$) by numerical techniques.

$$A_l = \int \left(\frac{dA_l}{dt}\right) + A_{l,o} \tag{2}$$

The concentration in the tissue ($C_t$) is estimated by dividing the amount in the tissue ($A_t$) by the tissue volume ($V_t$); the free concentration, in well-mixed, flow-limited model structures [9], is the tissue concentration divided by the tissue blood partition coefficient ($P_t$). Thus, for the liver,

$$C_l = \frac{A_l}{V_l} \tag{3}$$

and

$$C_{vl} = \frac{C_l}{P_l} \tag{4}$$

The full model contains a MBDE for each tissue.

In other cases, such as methotrexate binding to dihydrofolate reductase [10], or dioxin binding to cytochrome P-4501A2 in the liver [11], there may be specific tissue-binding interactions. Now, the mass conservation equation contains free, partitioned, and specifically bound forms of the chemical.

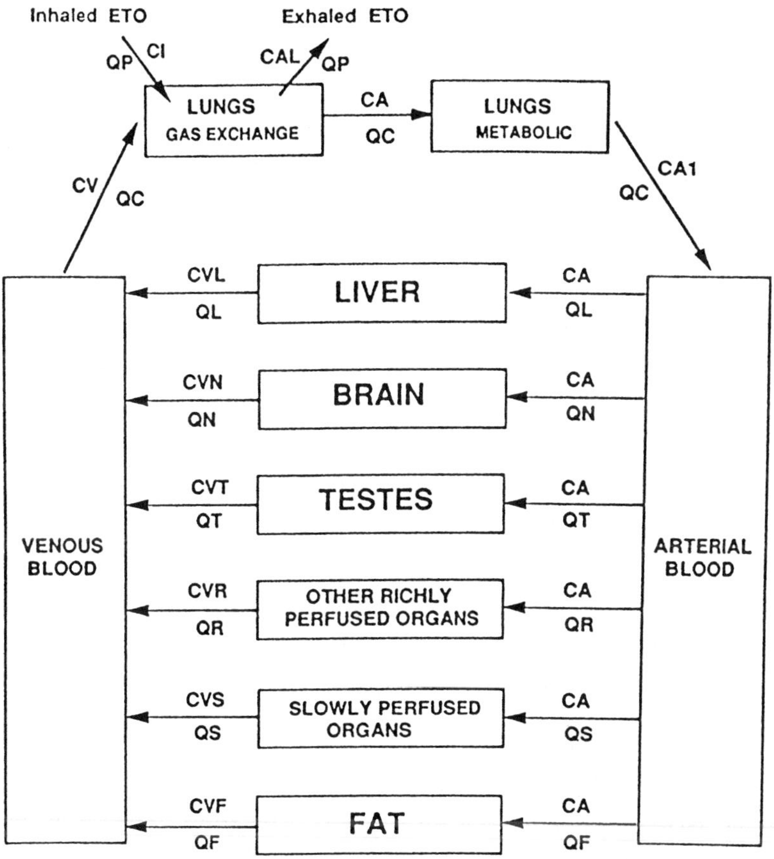


**Figure 1** A schematic of a physiologically based pharmacokinetic (i.e., dosimetry) model for ethylene oxide (EtO). Q terms refer to blood and airflow rates; C terms refer to concentrations. The concentrations each represent the concentration of chemical coming from the previous compartment; $CV_i$ refers to the venous blood concentration exiting tissue, i. QC, QP, CI, and CAL are, respectively, cardiac output, alveolar ventilation rate, inspired concentration, and alveolar concentration of EtO.

**Table 1** Definitions of Some Terms

| Calculated variables | Units |
| --- | --- |
| $A_t$ amount in tissue | mg |
| $C_t$ concentration in tissue | mg/L |
| $C_a$ arterial blood concentration | mg/L |
| $C_{vt}$ venous blood from tissue | mg/L |
| Tissue compartments | |
| $V_t$ tissue volume | L |
| $Q_t$ tissue blood flow | L/hr |
| $P_t$ tissue partition coefficient | L blood/L tissue |
| Biochemical parameters | |
| $V_{max}$ maximum velocity | mg/hr |
| $K_m$ Michaelis dissociation constant | mg/L |
| $B_{max}$ binding maximum | mg/tissue |
| $K_b$ binding dissociation constant | mg/L |
| $k_f$ first-order metabolic rate constant | /hr |
| $k_{mm}$ second-order MM binding constant | /M/hr |
| $k_{dt}$ first-order detoxification rate constant | /hr |
| $k_{repair}$ first-order rate constant for adduct repair | /hr |

$$A_t = C_{vt} \cdot P_t \cdot V_t + \frac{B_{max} \cdot C_{vt}}{K_b + C_{vt}} \tag{5}$$

Metabolism of certain xenobiotics produces metabolites that can react with cellular constituents to elicit toxic effects. The rate of change of the amount of reactive metabolites ($A_{rm}$) available for reaction with critical macromolecules (MM) in a tissue depends on the rate of formation in the tissue, the rate of reaction with the target macromolecule(s), and the rate of detoxifying reactions in the tissue.

$$\frac{dA_{rm}}{dt} = \text{formation} - \text{reaction with target molecules} - \text{detoxification}$$

For the case in which the saturable metabolism of parent chemical produces a reactive intermediate, the MBDE for the amount of reactive metabolite ($A_{rm}$) and the steady-state concentration of the reactive metabolite ($C_{rm,ss}$) are

$$\frac{dA_{rm}}{dt} = \frac{V_{max} \cdot C_{vt}}{K_m + C_{vt}} - k_{mm} \cdot V_t \cdot C_{rm} \cdot \text{MM} - k_{dt} \cdot C_{rm} \cdot V_t \tag{6}$$

and,

$$C_{\mathrm{rm,ss}} = \frac{V_{\max}\cdot C_{\mathrm{vt}}/(K_{\mathrm{m}} + C_{\mathrm{vt}})}{(k_{\mathrm{mm}}\cdot \mathrm{MM} + k_{dt}\cdot R)} \tag{7}$$

When these metabolites are highly reactive and never circulate beyond the tissue in which they are produced, these simple equations serve to predict the average target tissue exposure to $C_{\mathrm{rm}}$. In other cases, the metabolites are reactive, but stable, and are eliminated by metabolism after distribution by the systemic circulation. The PB–dosimetry model then has to permit circulation and access of these metabolites to other tissues. Separate PB models then are provided for the important metabolites [12–14].

The next process to be included in these dosimetry models is the initial cellular interactions of parent chemical or reactive metabolite with cellular target molecules. These cellular interactions correspond to the necessary initiating steps linking tissue dose to subsequent toxicity. A few examples may be useful here. A reactive, mutagenic parent compound, such as ethylene oxide, reacts with DNA to form DNA–adducts that, when present during cell division, increase the probability of mutation and malignant transformation [15]. Other reactive species, such as the trichloromethyl free radical formed by carbon tetrachloride metabolism, react with proteins and lipids, altering the properties of these macromolecules [16]. Sufficient degradation of the pool of functional macromolecule leads to cell damage and cell death.With dioxin, interactions between dioxin and a cellular protein, the *Ah* receptor, eventually lead to changes in cell concentrations of growth factors that appear, in a complex fashion, to alter cell growth and differentiation rates [17,18].

The accumulation of protein or DNA adducts is expected to be related to the average tissue concentration of the reactive chemical species (either $C_{\mathrm{vt}}$ or $C_{\mathrm{rm}}$ or a combination of the two). The time-average concentration of an adduct formed from a reactive metabolite is dependent on rates of formation and tissue repair of the adduct. The equation for the change in concentration of the adduct and the steady-state concentration can be expressed in a generic form, where formation is by a second-order reaction and repair is first-order relative to the adduct concentration.

$$\frac{d\,(\mathrm{adduct})}{dt} = k_{\mathrm{mm}}\cdot C_{\mathrm{rm}}\cdot \mathrm{MM} - k_{\mathrm{repair}}\cdot(\mathrm{adduct}) \tag{8}$$

and,

$$(\mathrm{adduct})_{\mathrm{ss}} = \frac{k_{\mathrm{mm}}\cdot C_{\mathrm{rm}}\cdot \mathrm{MM}}{k_{\mathrm{repair}}} \tag{9}$$

With dioxin [18] or peroxisomal proliferators [19], the initial cellular interactions are believed to occur through specific protein receptors; the chemical and receptor form a complex that alters gene transcription. Protein induction by dioxin, for instance, has been modeled by assuming an interaction between dioxin, the *Ah* receptor, and sites on DNA [20]. Dioxin modified the zero-order rate of synthesis ($k_{syn}$), but not the first-order rate constant of degradation ($k_{degrad}$). Thus, the steady-state concentration of protein or growth factor becomes:

$$\text{Growth factor}_{ss} = k_{syn}\left[1 + \frac{\text{Synmax}\cdot(L\text{–}Ah)^n}{K_d^{\,n} + (L\text{–}Ah)^n}\right] - k_{degrad}\cdot(\text{Growth factor}) \tag{10}$$

where *L–Ah* is the concentration of dioxin–*Ah* receptor complexes within the cell. The relationship depends on the affinity of the dioxin–*Ah* complex for the sites on DNA, given by the dissociation constant $K_d$. Synmax is the maximum rate of synthesis that can be sustained. The exponent, $n$, allows cooperativity in binding to multiple sites. These physiological models for initial cellular interactions (i.e., DNA binding, protein or lipid binding, and growth factor expression) bridge the gap between tissue dosimetry and tissue response models for specific toxic effects.

## Physiological Response Models

The initial cellular interactions lead to an alteration in cellular function and, eventually, to toxic effects. These response relations can also be described by quantitative physiological-modeling approaches, and such relations have been developed for a limited number of endpoints [21]. Probably, the most extensive work has focused on models for chemical carcinogenesis that, themselves, derive from more general attempts to develop biologically realistic two-stage models of cancer [22] that include mutation rates and growth rates of the normal and intermediate cell types involved in the progression to cancer [23].

In the two-stage, Moolgavkar–Venzon–Knudson (MVK) cancer model, tumors arise from the sequential mutation of both alleles of a critical gene. The tumor incidence function depends on the number of susceptible cells, $N_0$, the mutation rate for conversion of normal cells to intermediate cells with one mutation, $\mu_1$, the birth and death rates of these intermediate cells, $\alpha_2$ and $\beta_2$, respectively, and the mutation rate for conversion of intermediate to fully malignant cells, $\mu_2$. For time-independent parameters, the approximate incidence at any time, $I(t)$, is:

$$I(t) = \mu_1\cdot\mu_2\cdot N_0\int_0^t \exp(\alpha_2 - \beta_2)(t - s)ds \tag{11}$$

The complete exposure–dose–response model for chemical carcinogenesis includes the necessary relationships between tissue dose (i.e., chemical concentrations, protein–DNA adducts, or growth factor concentrations) and the mutation, birth, and death rates in these cancer response models.

In creating dose–response models for an individual response, for instance, cell death ($\beta_2$), we generally lack sufficient detail to accurately define all steps at the cellular level and, thus, it is necessary to define mathematical functions that correlate the response with adduct level or with growth factor level. One possible relationship is linearity between the measure of the initial cellular interactions and response.

A linear relationship may exist, for example, between mutation rates ($\mu_1$,$\mu_2$) and DNA–adduct levels, or perhaps for the relationship between cell proliferation rates ($\alpha_2$) and growth factor concentrations. More complex relationships have been suggested for rates of cell death, for which the sensitivity to death has been expressed as a log-normal probability distribution of the amount of protein adducts or the rate of metabolism to reactive intermediates [24,25]. Mutation, cell growth and differentation rates, and cell death rates influence various toxicity endpoints, including immunotoxicity, specific organ toxicity, and teratogenesis, in addition to cancer.

This brief summary outlines the general structure of a few PB–dosimetry and PB–tissue response models. There is a broad diversity of such models, with significant individualization required for specific chemicals [5,6]. The purpose of providing these examples and the equations that are contained in these models is to focus attention on the individual biological determinants that underlie these exposure–dose–response models and to begin to examine the areas in which in vitro studies can contribute to development of these models. The ultimate goal of the integration of physiological-modeling strategies with in vitro results is to show that in vitro results, placed in the context of these predictive physiological models, generate predictions of dosimetry, initial cellular interactions, and toxicity that compare well with observations in test animals.

## In Vitro Studies and Tissue Dosimetry

To apply these models to predict tissue dosimetry, it is necessary to have estimates of the physiological, biochemical, and biological parameters in these various equations. The anatomical and physiological parameters in a test species are usually available from the literature and require little additional research studies on whole animals. Many of the biochemical and physiochemical parameters can potentially be obtained from in vitro systems (Table 2). Among the important areas to which in vitro studies contribute are skin or gut

**Table 2** Some In Vitro Systems Used to Estimate Parameters for Physiologically Based Dosimetry Model

| Parameters | Techniques | Tissue preparations |
|---|---|---|
| Partition coefficients | Vial-equilibration | Tissue homogenates, cells, minced tissue |
| | Equilibrium dialysis, ultrafiltration | Tissue homogenates, cells |
| Binding constants | Scatchard analysis, Equilibrium dialysis | Purified preparations, homogenates, cells |
| Metabolic constants | Spectrophotometric, spectrofluorimetric, HPLC assays, others | Homogenates, subcellular fractions, intact cells, isolated perfused organs |

preparations, to determine the rates, extent, and mechanisms of absorption of xenobiotics; tissue, cellular, and subcellular preparations, to assess rates and affinities of metabolizing enzyme pathways; and various intact tissue preparations, to assess the initial interactions between toxic chemicals and target tissues.

Chemicals gain entry to tissues by various routes of administration, such as inhalation, skin application, or ingestion, where the chemicals have to diffuse through cellular barriers before reaching the circulation. The use of appropriate in vitro models for absorption with skin and gut could permit characterization, on a quantitative level, of the determinants of absorption and inclusion of this information in the models [26,27]. Currently, the only way to accurately examine the rate of absorption with most chemicals is to expose the animals and measure blood concentrations at various times after administration.

In the body some clearance mechanisms remove chemical directly from the circulation. Organic acids, such as penicillin, are removed by filtration and secretion into the kidney filtrate. Organic dyes are eliminated by biliary excretion after clearance from the hepatic blood flow. The kinetics of these chemicals depend on their solubilities in tissue and the blood flow to these specific organs. By far the most common mode of elimination of chemicals from the body is by metabolism. In vitro methods have been widely used to examine the nature of metabolites formed and rates of biotransformation of various toxic chemicals. These studies have been performed with isolated perfused tissues, tissue slices, isolated cells from the target tissues, and various subcellular preparations, mitochondria, microsomes, and such. But do these in vitro preparations give results consistent with metabolic clearance in the living animal? Some limited work has focused on the quantitative evaluation of the ability of in vitro kinetic constants to predict clearances and half-lives in vivo. Examples include studies with drugs [28] and with volatile organic compounds of interest because of their presence in the general environment and in occupational settings [29,30]. Frequently, the units employed by investigators using in vitro systems are not conducive to extrapolation. Maximum velocities are expressed per unit microsomal protein, with no indication of the yield of protein per unit liver weight. For PB–PK modeling the $V_{max}$ needs to be expressed per weight of tissue or per total tissue. When metabolic rates are also accurately known for reactions of metabolite(s), this information can be integrated to generate PB–PK models that describe the disposition of both parent chemical and major reaction products.

The amount of chemical in a tissue depends on the partitioning and tissue-binding characteristic of the material. In vitro techniques have been used to estimate partition coefficients for volatile chemicals and partitioning of both charged and neutral, nonvolatile chemicals [31–34]. Binding relations with particular receptors are characterized in vitro with various tissue preparations and analysis with Scatchard plots, or by other suitable techniques. These constants together with blood flow rates and rates of metabolism when integrated in Eqs. (1, 5) permit estimation of tissue concentrations of these test chemicals. Nonlinear processes are accommodated by capacity limited binding ($B_{max}$) and metabolism ($V_{max}$), or sometimes, by changes in physiological parameters, such as altered body composition during aging [35], or by pharmacological effects of exposure on blood flows or breathing rates.

## COMPARING EXPOSURES

In vitro studies sometimes precede detailed in vivo toxicity research. A question that always lurks near the surface when discussing the results of these in vitro toxicity results is whether the concentrations used have any relevance at all to in vivo exposure conditions. It is frequently possible to expose cells in suspension or tissues in culture to concentrations of chemical that would be impossible to achieve, even under the most unrealistic exposure situations. How do we know when this overexposure in vitro is a problem for interpretation? One possibility involves the application of these PB–dosimetry models. These predictive PB–PK models readily serve as tools to assess the relevance of these in vitro exposure concentrations. In these cases, PB–PK models estimate the in vivo exposure conditions necessary to produce the concentrations achieved in vitro. If the conclusion is reached that these tissue concentrations arise in the animal only at completely unreasonable exposure levels, the in vitro system should be regarded as of little or no relevance for predicting human risk.

The reverse application is equally possible. Once all necessary model parameters are known, these models can predict target tissue concentrations of chemical and metabolite achieved during those in vivo exposures that give rise to overt effects in the animal. These concentrations can then be used as the test concentrations for studies with in vitro cell systems. In vitro concentrations probably should not be used if they exceed overtly toxic concentrations in vivo by more than severalfold.

## Tissue Response Modeling

The initial interactions of chemicals and reactive metabolites with tissues are easily evaluated in any number of in vitro preparations. Hepatocytes, renal tubule cells, purified lung Clara cells, among a host of similar tissue preparations, are useful for evaluating specific tissue metabolism, including activation of chemicals to metabolites that react with target macromolecules in the cell. Repair processes for clearance of adducts could be studied in isolated cells in culture, for instance. The elaboration of specific growth factors could also be evaluated in appropriate cell lines or tissue preparations and correlated with cell proliferation or cell differentiation rates. Many other endpoints could also be evaluated, with judicious choice of cell system and exposure conditions. One longer-term advantage of work with cell systems is the possibility of conducting similar work on human tissue as appropriate. Human hepatocytes could be used side by side with animal hepatocytes to estimate rates of metabolism, extent of macromolecular binding, and sensitivity to damage. When the in vitro–in vivo extrapolation procedures for rats were sufficiently well developed, they could readily be applied to the results from human tissues. Successful in vitro–in vivo extrapolation could reduce our reliance on animal test for estimating expected human health risks. Instead of relying on extrapolations from animal test results, in vitro toxicity tests with human tissues could be used together with physiological modeling to conduct dose–response evaluations. Animal studies would still be a mainstay of hazard identification studies, along with appropriate in vitro response studies.

## Dose–Response Relationships

These in vitro cell and tissue systems bridge the gap between dosimetry and tissue response at the level of the intact animal. Some examples for the three cases discussed in the foregoing include in vitro hepatotoxicity of chlorinated aliphatic chemicals [36,37], in vitro analysis of DNA repair and replication

[38], and in vitro correlation between DNA adduct formation and chromosomal breakage [39]. These systems also begin to permit examination of the quantitative dose–response behavior relating initial cellular interactions and toxicity. The dose–response relationships give the information on the intensity of response at different dose levels. For instance, they describe the probability of mutation arising owing to a particular concentration of DNA–adducts in a cell, or the probability of cells dying because of a particular concentration of macromolecular adducts in the target cells, or the probability of cell proliferation owing to a particular increase in growth factor concentration. Integrating these in vitro dose–response curves with the PB–PK models permits prediction of the dose–response behavior in the intact animal.

Although the in vitro–in vivo extrapolation of dosimetry and some of these initial cellular interactions appears very possible, more advanced pathological states undoubtedly represent the interplay of complex biological processes mediated by multiple organ systems in the body and may be very difficult to predict based simply on in vitro results from cells from a single tissue examined in vitro. Cell death may follow macromolecular binding, but aspects of cell–cell communication in the intact tissue or elaboration of growth factors in other tissues may moderate the response, compared with that observed in isolated cells. Mutation rates depend on the cell cycle times in vivo, dynamic repair processes for DNA–adducts, and immune surveillance of initiated cells. Growth factors may be mitogenic in a cell suspension, but elicit homeostatic responses in the intact tissue that subsequently depress overall tissue proliferation [40]. This paradoxical behavior, mitostimulatory factors producing inhibition of cell proliferation in an intact organ, may occur frequently in mature tissues on which little net proliferation can occur without increasing the tissue size well beyond normal. Tissue response modeling is in its infancy and must be more fully developed before it can be used to integrate in vitro observations for predicting full scale toxic effects in animals.

## RISK ASSESSMENT

In vitro approaches tantalize with their promise for reducing reliance on toxicity tests in animals and improving risk assessment. However, their main use remains qualitative; they establish the potential of a chemical to elicit an effect. How will their more complete potential be realized? The more pervasive application of in vitro toxicity results in risk assessment will only be possible as these in vitro results are combined with quantitative mathematical descriptions that integrate in vitro findings into the context of the living animal and relate the in vitro to the in vivo exposure conditions. Dosimetry models already are at a stage of development where they can be applied with in vitro results to predict behavior in animals and in humans. Extrapolation based on the initial cellular interactions examined in vitro also appears possible, but more validation is necessary in these areas to provide confidence in these approaches. Predicting dose–response characteristics of toxicity directly based on in vitro results is in its infancy and will move forward only if there is commitment to coordinated development of in vitro technologies and physiological modeling of these more complex events in vivo. Without a clear strategy describing the manner in which these in vitro studies will influence risk assessment, in vitro toxicology will continue to contribute primarily to hazard identification. This strategy is provided by the detailed biological framework provided by these physiological dosimetry and tissue response models. The partnership of quantitative modeling strategies and in vitro research promises to open an immense body of in vitro results for more direct consideration in the risk assessment process.

## ACKNOWLEDGMENT

The work presented in this chapter (CR813113) was supported in part by the EPA Cooperative Agreement, and has been reviewed and approved for publication by the Health Effects Research Laboratory, U. S. Environmental Protection Agency.

## REFERENCES

1. National Academy of Sciences, *Risk assessment in the Federal Government: Managing the Process*, National Academy Press, Washington, DC, 1983.
2. Clewell, H. J. III, and Andersen, M. E., Risk assessment extrapolations and physiological modeling, *Toxicol. Ind. Health 1*:111–131 (1985).
3. Leung, H. W., Development and utilization of physiologically based pharmacokinetic models for toxicological applications, *J. Toxicol. Environ. Health 32*:247–267 (1991).
4. Krishnan, K., Gargas, M. L., and Andersen, M. E., In vitro toxicology and risk assessment. *Alternative Methods in Toxicology*, Vol. 9 (A. M. Goldberg, ed.), Mary Ann Liebert, New York, 1992.
5. D'Souza, R. W., and Boxenbaum, H., Physiological pharmacokinetic models: Some aspects of theory, practice and potential, *Toxicol. Ind. Health 4*:151–171 (1988).
6. Gerlowski, L. E., and Jain, R. K., Physiologically based pharmacokinetic modeling: Principles and applications, *J. Pharm. Sci. 72*:1103–1126 (1983).

7. Himmelstein, K. J., and Lutz, R. J., A review of the application of physiologically based pharmacokinetic modeling, *J. Pharmacokinet. Biopharm.* *7*:127–137 (1979).
8. Andersen, M. E., Krishnan, K., Conolly, R. B., and McClellan, R. O., Mechanistic toxicology research and biologically-based modeling: Partners for improving quantitative risk assessments, *CIIT Activities* *12*:1–7.
9. Andersen, M. E., Physiological modeling of organic chemicals, *Ann. Occup. Med.* *35*:305–321 (1991).
10. Bischoff, K. B., Dedrick, R. L., Zaharko, D. S., and Longstretch, J. A., Methotrexate pharmacokinetics, *J. Pharm. Sci.* *60*:1128–1133 (1971).
11. Leung, H. W., Paustenbach, D. J., Murray, F. J., and Andersen, M. E., A physiologically based pharmacokinetic description of the tissue distribution and enzyme inducing properties of 2,3,7,8-tetrachlorodibenzo-*p*-dioxin in the rat, *Toxicol. Appl. Pharmacol.* *103*:399–410 (1990).
12. Deitz, F. K., Rodriguez-Giaxola, M., Traiger, G. J., Stella, V. J., and Himmelstein, K. J., Pharmacokinetics of 2-butanol and its metabolites in the rat, *J. Pharmacokinet. Biopharm.* *9*:553–576 (1981).
13. Fisher, J. W., Gargas, M. L., Jepson, G. W., Allen, B., and Andersen, M. E., Physiologically-based pharmacokinetic modeling with trichloroethylene and its metabolite, trichloroacetic acid in the rat and mouse, *Toxicol. Appl. Pharmacol.* *109*:183–195 (1991).
14. Gargas, M. L., Clewell, H. J. III, and Andersen, M. E., Metabolism of inhaled dihalomethanes in vivo: Differentiation of kinetic constants for two independent pathways, *Toxicol. Appl. Pharmacol.* *82*:211–223 (1986).
15. Krishnan, K., Gargas, M. L., Fennell, T. R., and Andersen, M. E., A physiologically based description of ethylene oxide dosimetry in the rat, *J. Toxicol. Ind. Health* *8*:121–140 (1992).
16. Recknagel, R. O., Carbon tetrachloride hepatotoxicity, *Pharmacol. Rev.* *19*:145–208 (1967).
17. Sutter, T. R., Guzman, K., Dold, K. M., and Greenlee, W. F., Targets for dioxin: Genes for plasminogen activator inhibitor-2 and interleukin-1β, *Science* *254*:415–418 (1991).
18. Whitlock, J. P., Genetic and molecular aspects of 2,3,7,8-tetrachlorodibenzo-*p*-dioxin action, *Annu. Rev. Pharmacol. Toxicol.* *30*:251–277 (1991).
19. Isseman, I., and Green, S., Activation of a member of the steroid hormone receptor superfamily by peroxisome proliferators, *Nature* *347*:645–650 (1990).
20. Andersen, M. E., Mills, J. J., Gargas, M. L., Kedderis, L. B., Birnbaum, L. S., Neubert, D., and Greenlee, W. F., Modeling receptor-mediated processes with dioxin: Implications for pharmacokinetics and risk assessment, *J. Risk Anal.* *13*:25–34 (1993).
21. Conolly, R. B., and Andersen, M. E., Biologically-based pharmacodynamic models: Tools for toxicological research and risk assessment, *Annu. Rev. Pharmacol. Toxicol.* *31*:503–523 (1991).

22. Armitage, P., and Doll, R., The age distribution of cancer and a multistage theory of carcinogenesis, *Br. J. Cancer 8*:1–12 (1954).
23. Moolgavkar, S. H., Carcinogenesis modeling: From molecular biology to epidemiology, *Annu. Rev. Public Health 7*:151–169 (1986).
24. Bogen, K. T., Risk extrapolations for chlorinated methanes as promoters vs initiators of multistage carcinogenesis, *Fundam. Appl. Toxicol. 15*:536–557 (1990).
25. Reitz, R. H., Medrala, A. L., Corley, R. A., Quast, J. F., Gargas, M. L., Andersen, M. E., Staats, D. A., and Conolly, R. B., Estimating the risk of the liver cancer associated with human exposures to chloroform, *Toxicol. Appl. Pharmacol. 105*:443–459 (1990).
26. Hotchkiss, S. A., Chidgey, M. A. J., Rose, S., and Caldwell, J., Percutaneous absorption of benzlacetate through the rat skin in vitro: Validation of an in vitro model against in vivo data, *Food. Chem. Toxicol. 28*:443–447 (1990).
27. Dykes, P. J., Edwards, M. J., O'Donovan, M. R., Merrett, V., Morgan, H. E., and Marks, R., In vitro reconstruction of human skin: The use of skin equivalent as potential indicators of cutaneous toxicity, *Toxicol. In Vitro 5*:1–8 (1991).
28. Rane, A., Wilkinson, G. R., and Shand, G., Prediction of hepatic extraction ratio from in vitro measurements of intrinsic clearance, *J. Pharmacol. Exp. Ther. 200*:420–424 (1977).
29. Hilderbrand, R. L., Andersen, M. E., and Jenkins, L. J., Jr. (1981). Prediction of in vivo kinetic constants for metabolism of inhaled vapors from kinetic constants measured in vivo, *Fund. Appl. Toxicol. 1*:403–409.
30. Sato, A., and Nakajima, T., A vial-equilibration method to evaluate the drug metabolizing enzyme activity for volatile hydrocarbons, *Toxicol. Appl. Pharmacol. 47*:41–46 (1979).
31. Gargas, M. L., Burgess, R. J., Voisard, D. E., Cason, G. H., and Andersen, M. E., Partition coefficients of low molecular weight volatile chemicals in various liquids and tissues, *Toxicol. Appl. Pharmacol. 98*:87–99 (1989).
32. Fiserova-Bergerova, V., and Diaz, M. L., Determination and prediction of tissue–gas partition coefficients, *Int. Arch. Occup. Environ. Health 58*:75–87.
33. Sato, A., and Nakajima, T., Partition coefficients of some aromatic hydrocarbons and ketones in water, blood, and oil, *Br. J. Ind. Med. 36*:231–234 (1979).
34. Lin, J. H., Sugiyama, Y., Awaza, S., and Hanano, M., In vitro and in vivo evaluation of the tissue-to-blood partition coefficient for physiological pharmacokinetic models, *J. Pharmacokinet. Biopharm. 10*:637–647 (1982).
35. Birnbaum, L. S., Pharmacokinetic basis of age-related changes in sensitivity to toxicants, *Annu. Rev. Pharmacol. Toxicol. 31*:101–128 (1991).
36. Tyson, C. A., Correspondence of results from hepatocyte studies with in vivo response, *Toxicol. Ind. Health 3*:459–478 (1987).
37. Tyson, C. A., Gee, S. J., Hawn-Prather, K., and Story, D. L. Correlation between in vitro and in vivo toxicity of some chlorinated aliphatics, *Toxicol. In Vitro 3*:145–150 (1989).

38. Doolittle, D. J., Muller, G., and Scribner, H. E., The in vivo–in vitro hepatocyte assay for assessing DNA repair and DNA replication: Studies in the CD-1 mouse, *Food Chem. Toxicol. 25*:399–401 (1987).
39. Walk, R. A., Muller, T., Yeats, S., and Linz, U., DNA confirmation assay: Determination of in vitro DNA adduct formation and strand breaks. *In Vitro Toxicol. 2*:59–65 (1989).
40. Jirtle, R. J., Meyer, S. A., and Brockenbrough, J. S., Liver tumor promoter phenobarbital: A biphasic modulator of hetpatocyte proliferation, *Chemically-Induced Cell Proliferation: Implications for Risk Assessment* (B. Butterworth, T. J. Slaga, W. Farland, and M. McClain, eds.), Wiley-Liss, New York, 1991, pp. 209–216.

# 3

# The Use of Theoretical Descriptors in Quantitative Structure–Activity Relationships

**George R. Famini**

*U.S. Army Edgewood Research, Development, and Engineering Center*
*Aberdeen Proving Ground, Maryland*

**Leland Y. Wilson**

*La Sierra University*
*Riverside, California*

Quantitative structure–activity relationships (QSARs) have been used extensively in correlating structural features to physical, chemical, and toxicological properties. The basic tenet of a QSAR is that there is a connection between the microscopic (molecular structure) and the macroscopic (empirical) properties. Furthermore, this connection can be used to predict empirical properties directly from the molecular structure. This relationship was first quantitated by Hammett, who developed the linear free energy relationship (LFER).

## Linear Solvation Energy Relationships

From the concept of Hammett, Kamlet and Taft developed a methodology for developing free energy relationships based on solute–solvent interactions [1–4]. This relationship, defined as the *general linear solvation energy relationship*, is shown in Eq. (1).

$$\log \text{Property} = \text{Steric} + \text{Polarizability} + \text{Hydrogen Bonding} \qquad (1)$$

In this way, a given property can be described as a linear expression consisting of contributions by steric factors, polarization, and polarizability factors, and hydrogen-bonding factors. In the multiple-solute–single-solvent system, the specific terms are as follows: steric: molar volume ($V_m$); polarizability: spectroscopically determined polarizability ($\pi^*$); hydrogen bonding: spectroscopically determined acidity and basicity terms ($\alpha$ and $\beta$). In practice, not all four terms are required for every relationship. The coefficients and associated statistical *t*-scores can be used to gauge the importance of each descriptor for every property correlation. In this way the LSER can be used to infer insight to solute–solvent interactions.

One major difficulty of this approach has been the nature of the descriptors. All are empirically determined; therefore, reducing the usefulness of this approach for a priori predictions. Some attempts have been made correlating more fundamental structural and electronic descriptors with the Kamlet–Taft solvatochromic parameters, with moderate degrees of success [5].

### Applications of Theoretical Chemistry

Theoretical chemistry has been used in the past to supply structural and electronic descriptors for QSAR and QSAR-like equations. In this way, empirically derived descriptors can be replaced in relationships, such as the LSER, with descriptors that will make the relationships more generally applicable. The theoretical linear solvation energy relationship (TLSER) is such a derivation, using the LSER philosophy and general structure, but replacing the empirically derived descriptors with computationally derived descriptors.

## CALCULATIONAL PROCEDURES

All geometries were optimized using the MNDO algorithm within MOPAC [6,7]. The molecular volumes were generated from the optimized geometries using the method of Hopfinger, as incorporated in the U.S. Army-developed molecular-modeling package MMADS [8,9]. All experimental data and LSER parameters were taken from works of the original authors.

## THE THEORETICAL LINEAR SOLVATION ENERGY RELATIONSHIP DESCRIPTORS

The TLSER descriptors have been developed with two main goals in mind. First, the TLSER descriptors should correlate optimally with the LSER parameters. Second, the property correlation equations with the TLSER should yield correlation coefficients and standard deviations as accurate as the LSER. Furthermore, the TLSER descriptors should be as generally applicable to any solute–solvent interaction as are the LSER descriptors.

The TLSER descriptors that have been developed follow that of the LSER, and fit into one of the categories listed in Eq. (1). The steric term for the TLSER is the molecular van der Waals volume ($V_{mc}$). Volume calculations of this type are standard in most molecular-modeling packages available today. As would be expected, correlation between $V_m$ and $V_{mc}$ is high, with a correlation coefficient of 0.979 and a standard error of the estimate of 5 ml/mol.

The polarizability term used in the TLSER is determined from the method of Stewart and Dewar incorporated in MOPAC [10]. Dividing the resulting polarizability by the $V_{mc}$ results in a size-independent polarizability index ($\pi_I$). This term defines the cases in which the electron cloud can be moved or polarized. Aromatics would rank high on the scale, and alkanes low. In addition, the $\pi_I$ is inversely proportional to the electronegativity.

Similar to the LSER, the hydrogen-bonding term is separated into acid (acceptor) and base (donor) terms. Furthermore, because the hydrogen bond (or the Lewis bond, or in essence any bond) can be separated into covalent and electrostatic parts, individual descriptors are needed for the TLSER to describe this. The energies of the highest occupied molecular orbital (representing the basicity, ($\epsilon_b$) and the lowest unoccupied molecular orbital (representing the acidity, $\epsilon_a$) are used to describe the covalent interactions. Similarly, the electrostatic portions of the basicity and acidity are represented by selected atomic formal charges in the molecule, either the most negative (for the basicity, $q_-$), or the most positive hydrogen (for the acidity, $q_+$). The basicity terms ($\epsilon_I$ and $q_-$) correlate highly with the LSER β-term, with a correlation coefficient of 0.9518 and a standard error of the estimate of 0.09.

The final generalized TLSER equation, is shown below:

$$\log P = mV_{mc} + p\pi_I + a\epsilon_a + a'qH_+ + b\epsilon_b + b'q$$

The relationship is similar in appearance to the LSER, and contains all of the key points that have made the LSER successful.

## REPRESENTATIVE CORRELATIONS

Similar to the LSER, all six TLSER descriptors are seldom required or are statistically significant for a given solute–solvent property. Generally, the equations reduce to a three- or four-descriptor correlation. Table 1 shows a representative set of correlations using the TLSER descriptors [11–16]. In each case, the descriptors that were significant at the 95% confidence level were

**Table 1** TLSER Correlations

| Property | $V_{mm}$ | $\pi_1$ | $\epsilon_b$ | $\epsilon_a$ | $q_-$ | $qH_+$ | $C$ | $N$ | $R$ | SD |
|---|---|---|---|---|---|---|---|---|---|---|
| $K_{ow}$ | 2.995 | −0.847 | 1.730 | n/a | −5.415 | n/a | −3.960 | 64 | 0.957 | 0.357 |
| $LC_{50}$ | −0.928 | −10.557 | −1.442 | n/a | 0.443 | n/a | 18.082 | 32 | 0.943 | 0.574 |
| $EC_{50}$ | −4.067 | −4.147 | 3.902 | n/a | −2.777 | n/a | 11.356 | 25 | 0.982 | 0.332 |
| EA | n/a | −2444 | n/a | 3540 | 1430 | −383 | 20545 | 23 | 0.951 | 308 |
| $k_m$ | 1.185 | −0.610 | n/a | n/a | −2.290 | −1.077 | −0.083 | 19 | 0.986 | 0.073 |
| $k_{oe}$ | −97.7 | 874 | 1256 | n/a | n/a | n/a | −2184 | 10 | 0.986 | 26.9 |
| $C$(tadpole) | −2.13 | −4.891 | −4.980 | n/a | 5.00 | n/a | 11.99 | 39 | 0.979 | 0.237 |
| $C$(gold orfe) | 2–2.64 | 94.885 | n/a | n/a | 4.15 | −1.77 | 7.55 | 30 | 0.975 | 0.237 |

included, and all others were dropped from the correlations. An in-depth treatment of this technique is given in a recent journal article, and several technical reports [17,18]. The next paper will describe some correlations completed in depth pertaining to selected physical properties.

## CONCLUSIONS

The diversity of the properties successfully correlated with the TLSER show the general applicability of both the LSER and the application of theoretical techniques to the LSER. In addition, the TLSER descriptors, based solely on theoretically derived and determined parameters, result in as good a correlation as the LSER descriptors. By using these equations, then, solute–solvent interactions can be related to fundamental descriptions (structural and electronic) of the molecule. Furthermore, by using the TLSER descriptors, a priori prediction of solute–solvent properties can be made using these equations.

## REFERENCES

1. Kamlet, M. J., Taft, R. W., and Abboud, J.-L. M., *J. Am. Chem. Soc. 91*:8325 (1977).
2. Kamlet, M. J., Taft, R. W., Abboud, J.-L. M., and Abraham, M. J., *J. Prog. Org. Chem. 48*:2877 (1983).
3. Kamlet, M. J., and Taft, R. W., *Acta Chem. Scand. B39*:616 (1985).
4. Kamlet, M. J., Taft, R. W., Famini, G. R., and Doherty, R. M., *Acta Chem. Scand. B41*:589 (1987).
5. Lewis, D. F. V., *J. Comp. Chem. 8*:1084 (1988).
6. Thiel, W., and Dewar, M. J. S., *J. Am. Chem. Soc. 99*:4899 (1977).
7. Stewart, J. J. P., *MOPAC: A General Molecular Orbital Package* FJSRL-TR-86-0003, Frank J. Seiler Research Laboratory, U. S. Air Force Academy, Colorado Springs, CO, June 1988.
8. Hopfinger, A. J., *J. Am. Chem. Soc. 102*:7126 (1980).
9. Leonard, J. M., and Famini, G. R., *A User's Guide to the Molecular Modeling Analysis and Display System*, CRDEC-TR-030, U. S. Army Chemical Research, Development and Engineering Center, Aberdeen Proving Ground, MD, January 1989.
10. Stewart, J. J. P., and Dewar, M. J. S., *Chem. Phys. Lett. 111*:416 (1984).
11. Kamlet, M. J., Taft, R. W., Abboud, J.-L. M., Famini, G. R., and Doherty, R. M., *J. Pharm. Sci. 74*:807 (1985).
12. De Zwart, D., Slooff, W., *Aquat. Toxicol 4*:129 (1983).
13. Hafkensheid, T. L., and Tomlinson, E., *Int. J. Pharm. 17*:1 (1983).
14. Larsson, L., *Acta Chem. Scand. 11*:1131 (1953).
15. Dorohio, D., and Iancu, D., *Anal. Stüntiche Univ 20*:59 (1981).

16. Giusti, D. M., Conway, R. J., and Lawson, C. T., *J. Water Poll. Control Fed.*, M. Randic, *J. Am. Chem. Soc. 97*:6609 (1975).
17. Wilson, L. Y., and Famini, G. R., *J. Med. Chem. 34*:1668 (1991).
18. For a more complete treatment of the TLSER descriptors, see either reference 9, or the following U. S. Army Technical Reports, available through NTIS: G. R. Famini, *Using Theoretical Descriptors in Structure Activity Relationships I. Molecular Volume*, CRDEC-TR-88031, U. S. Army Chemical Research, Development and Engineering Center, Aberdeen Proving Ground, MD, January 1988, unclassified report, NTIS ADA191522; G. R. Famini, *Using Theoretical Descriptors in Structure Activity Relationships II. Polarizability INDEX*, CRDEC-TR-88137, U. S. Army Chemical Research, Development and Engineering Center, Aberdeen Proving Ground, MD, September 1988, unclassified report, NTIS ADA199594; L. Y. Wilson and G. R. Famini, *Using Theoretical Descriptors in Structure Activity Relationships III. Electronic Descriptors*, CRDEC-CR-88083, U. S. Army Chemical Research, Development and Engineering Center, Aberdeen Proving Ground, MD, August 1988, unclassified report, NTIS ADA200482; G. R. Famini, *Using Theoretical Descriptors in Structure Activity Relationships IV. Molecular Orbital Basicity and Electrostatic Basicity*, CRDEC-TR-013, U. S. Army Chemical Research, Development and Engineering Center, Aberdeen Proving Ground, MD, November 1988, unclassified report, NTIS ADA202132; G. R. Famini, *Using Theoretical Descriptors in Structure Activity Relationships V. A Review of the Theoretical Parameters*, CRDEC-TR-085, U. S. Army Chemical Research, Development and Engineering Center, Aberdeen Proving Ground, MD, July 1989, unclassified report, NTIS ADA213580.

# 4

# Structure–Activity Predictions as Alternatives to Animal Tests

**Kurt Enslein**

*Health Designs, Inc.*
*Rochester, New York*

Animal tests that have been performed in the past can form the basis for equations modeling specific toxicity endpoints. These equations can then be used to predict the same endpoints for chemicals for which the bioassays have not been performed. These principles of structure–activity relationships (SARs) [1,2] have been implemented in the TOPKAT program. The SAR models exist for the following mammalian toxicity endpoints:

Rat oral $LD_{50}$ [3]
Rat oral maximum-tolerated dose [4]
Rabbit skin irritation (Draize) [5]
Rabbit eye irritation (Draize) [6]
Carcinogenicity (two-species, two-sex) [7]
Teratogenicity (frank malformations) [8]
Rat oral chronic LOAEL [9]
Mouse inhalation $LC_{50}$ [10]

Data used for the development of these models are carefully screened and, often, "harmonized" to make them suitable for the stringent requirements of predictive equations.

Estimates produced by the TOPKAT program can be internally validated through technology provided within it. This is a crucial requirement, and differentiates the TOPKAT program from other predictive systems. The TOPKAT SAR models have found use in both regulatory as well as industrial applications. Regulatory applications have included submissions as part of Premanufacturing notifications (PMN) to the U.S. Environmental Protection Agency (EPA), the examination of impurities in marketed drugs, the complementing of Material Safety Data Sheets (MSDS) with estimated values in place of actual assay data, the evaluation of intermediates of chemical manufacture, the estimation of rat oral median lethal dose ($LD_{50}$) values for experimental chemicals, and the partial fulfillment of the Food and Drug Administration (FDA) environmental assessment requirements for Phase I Investigational New Drug (IND) studies.

In addition to several of the foregoing examples that have also found application in industry, TOPKAT has been used for the prioritization of discovery candidates, the reduction of toxic effects for developmental compounds, the cross-validation of bioassays, dose-ranging for bioassays, and performing toxicity experiments in calculo.

## REFERENCES

1. Purcell, W. P., Bass, G. E., and Clayton, J. M., *Strategy of Drug Design: A Guide to Biological Activity*, John Wiley & Sons, New York, 1973.
2. Martin, Y. C., Advances in the methodology of quantitative drug design, *Drug Design* (E. J. Ariens, ed.), Academic Press, New York, 1979, pp. 1–72.
3. Gombar, V. K., Enslein, K., and Blake, B. W., Prediction of rat oral $LD_{50}$ values from multiple QSAR equations, in preparation.
4. Gombar, V. K., Enslein, K., Hart, J. B., Blake, B. W., and Borgstedt, H. H., Estimation of maximum tolerated dose for long-term bioassays from acute lethal doses and structure by QSAR, *Risk Anal. 11*:509 (1991).
5. Enslein, K., Borgstedt, H. H., Blake, B. W., and Hart, J. B., Prediction of skin irritation severity by structure–activity relationships. *In Vitro Toxicol. 1*:129 (1987).
6. Enslein, K., Blake, B. W., Tuzzeo, T. M., Borgstedt, H. H., Hart, J. B., and Salem, H., Estimation of rabbit eye irritation scores by structure–activity equations, *In Vitro Toxicol. 2*:1 (1988).
7. Enslein, K., Borgstedt, H. H., Tomb, M. E., Blake, B. W., and Hart, J. B., A structure–activity model of carcinogenicity based on NCI/NTP bioassays and food additives, *Toxicol. Ind. Health 33*:267 (1987).
8. Gombar, V. K., Borgstedt, H. H., Enslein, K., Hart, J. B., and Blake, B. W., A QSAR model of teratogenesis, *Quant. Struct. Activity Relation. 10*:306 (1991).

9. Gombar, V. K., Enslein, K., Carter, J. R., Blake, B. W., Huque, K. I., Ramanujam, V. M. S., Mumtaz, M. M., Reisman, D. J., Peirano, W. B., Knauf, L., and DeRosa, C., Prediction of effects levels of chemicals from quantitative structure–activity relationships (QSAR) models. I. Chronic lowest observed adverse effect level (LOAEL), (in press).
10. Prediction of inhalation $LC_{50}$ in the mouse, *HDi Toxicol. Newslett. 13* Feb. (1991).

# 5

# The Aquatic Toxicology of Isopropylamine: Comparison of Experimentally Derived Values with Structure–Activity Predictions

**Nancy A. Chester, Mark V. Haley, and Wayne G. Landis***

*U.S. Army Edgewood Research, Development, and Engineering Center*
*Aberdeen Proving Ground, Maryland*

Research that derives quantitative structure–activity relationships (QSARs) has produced a variety of equations with predictive capabilities. Properly applied and validated, these equations should have the capability to enable the scientist to make better decisions about environmental toxicity. Potential applications include participation in the risk assessments of contaminated sites, containing multiple toxicants with limited datasets, and in the exploration of residual toxicity associated with the breakdown of a potentially new product. Unfortunately, it is usually difficult to examine the power of the model or equation with an unknown. As one of the goals of this chapter, the prediction of two environmental QSARs are compared with experimentally derived results.

As part of a continuing series of environmental toxicity and hazard assessment studies, the toxicity of isopropylamine (IPA), a potential industrial pollutant, for two representative freshwater species, *Daphnia magna* and *Selenastrum capricornutum*, was examined. In an effort to investigate the efficacy of alternative-screening and range-finding methods for the *D. magna* toxicity tests, a QSAR equation that is part of the program, TOPKAT, was used

**Present affiliation*: Huxley College of Environmental Studies, Western Washington University, Bellingham, Washington

to generate a model-specific toxicity estimate, which was compared with results of toxicity tests.

Isopropylamine, also known as 2-propamine and 2-aminopropane, is a colorless, flammable liquid, with a strong ammonialike odor. This strong base is miscible with water, alcohol, and ether, and is currently used in the synthesis of pharmaceuticals, pesticides, rubber accelerators, dyes, and surface-active agents that may serve as potential sources of exposure. The chemical may enter the body by inhalation, ingestion, skin absorption, or eye and skin contact. Harmful effects and symptoms include irritation of the eyes, nose, throat, and skin, pulmonary edema, visual disturbance, skin and eye burns, and dermatitis. Details of chemical and hazard or toxicity data are further described in Table 1 [1,2].

TOPKAT, developed by Health Designs, Inc. (HDi) [3], is an integrated program developed for the estimation of toxic effects of chemical structures using statistical QSAR techniques, such as multivariate regression and discriminate analysis. Various structure–activity models have been developed by HDi for various toxic endpoints, including equations for a *D. magna* median effective concentration ($EC_{50}$) and a rat oral median lethal dose ($LD_{50}$) to *D. magna* $EC_{50}$, for which the derivations have previously been described [4,5]. Although log *P* (log of octanol/water partition coefficient) is the parameter most often used for QSAR studies of aquatic species, the models used in this study were constructed using molecular connectivity indices (MCI), topological descriptors that quantify salient structural features, and substructural keys as the parameters descriptive of the compounds in the model's database that contribute most to the explanation of the *D. magna* $EC_{50}$. For example, the following are 6 examples of the 17 parameters used for the model: pri-

**Table 1** Chemical Properties and Toxicity Data for Isopropylamine

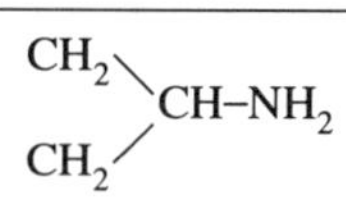


CAS Number 75-31-0

| Physical properties | | Toxicity data | |
|---|---|---|---|
| Molecular weight | 59.11 | Rat oral $LD_{50}$ | 820 mg/kg |
| Boiling point (1 atm) | 32.4°C | Rat inhalation LCLo | 800 ppm/8 hr |
| Vapor pressure (20°C) | 478 mm Hg | Rabbit, skin (severe irritation) | 10 mg/24 hr |
| Specific gravity (15/4°C) | 0.694 | | |
| pH (0.1 *N* aqueous solution) | 11.5 | | |

mary amine bound to an aromatic ring atom, aliphatic alcohol, oxygen-substituted aryl ester, benzene, valence path MCI, order 2, valence path-cluster, order 4.

Initial performance tests that validate the model's regression equation used the cross-validation method. Briefly, one chemical is removed from the dataset, the equation is recalculated, and the omitted compound is estimated with the equation. The process is repeated for all compounds, and the statistics are calculated on the estimates. These results can be expressed as the fraction of compounds predicted within certain factors. This model predicted 85.6% of the compounds from the dataset within a factor of 10; however, the TOPKAT program includes specifications for the validation of an estimate. Users must incorporate toxicological experience and understanding of both the compound in question and those on which the model was based, to provide confidence in the prediction. This communication addresses the aquatic toxicity of IPA and reports the TOPKAT-generated *D. magna* toxicity estimates, and their validation.

## MATERIALS AND METHODS

### Toxicity Tests

The 48-hr *D. magna* acute toxicity and 96-hr *S. capricornutum* growth inhibition tests conformed to applicable American Society for Testing and Materials (ASTM) and U.S. Environmental Protection Agency (EPA) standards [6–8], and have been described previously [9,10]. Daphnia cultures were originally obtained from Dr. Freida Taub at the University of Washington (Seattle, WA) and were reared in the laboratory as described by Goulden et al. [11] using hardened reconstituted water [7]. The 48-hr tests were conducted using a 16:8 light/dark (315 ft-candles) cycles, with temperatures maintained at 20±1°C. Because of increases in pH as the percentage of IPA in solution increased, two separate daphnid assays were conducted to determine the role of pH in IPA toxicity. The pH solution in the first assay ranged from 7.3 to 10.3. The second assay used a stock of IPA adjusted to pH 7.0 with 10% HCl, resulting in solution pH ranging from 7.1 to 7.4. The $EC_{50}$s were calculated using the probit analysis [12] and were verified with graphically tabulated $EC_{50}$s derived from mortality and concentration values [13]. Additionally, a parallel set of controls was run, in which no IPA was added, and pH was adjusted from pH 8.3 to 11.0 with 0.5 M NaOH. The percentage mortality was plotted against pH to observe the effect of pH on the survivorship of daphnia. Water for this test and for the culture of the *D. magna* used for the test, was obtained from a well source with a treatment system that provides water at starting pHs of approximately 8.3. All other test and culture conditions remained the same as those just described.

The *S. capricornutum* was grown in a semiflow-through culture apparatus [14] on T82MV [15], and test conditions were the same as those for daphnia tests. Test vessels were inoculated with approximately $8.0 \times 10^4$ cells per milliliter of the algal stock. Two separate assays were again conducted to assess the effect of pH on toxicity. Buffered test pH was 7.0, whereas unadjusted test solutions ranged in pH from 7.5 to above 11.0 (the pH probe was not accurate above pH 11.0). The percentage growth inhibition was calculated from the area under the growth curve [16] and was plotted against concentration. Subsequent $IC_{50}$s (concentration that inhibits 50% growth compared with control) were derived graphically from a least-square regression line [17,18] and equation, calculated by using a function from the Macintosh-compatible Cricket Graph scatterplot program.

## TOPKAT Quantitative Structure–Activity Relationship Estimation and Validation

The structure of isopropylamine was entered into TOPKAT using the simplified molecular input line entry system (SMILES) [19]. SMILES is a linear notation scheme for entry of chemical structures into computer databases. The language was developed to overcome complexities associated with the Wiswesser line notation method [20], and is much simpler and more user-friendly. The *D. magna* $EC_{50}$ and the rat oral $LD_{50}$ to *D. magna* $EC_{50}$ models were selected to obtain an $EC_{50}$ estimation [4]. A two-step validation process was employed to ascertain the confidence for the toxicity estimate: (a) a determination was made for whether the major substructures of the compound being estimated were present in the database of the model selected, and (b) the database was examined for structurally related compounds for which assay results and also TOPKAT-predicted estimates may either support or refute the estimate in question. This information, together with the user's own experience and knowledge of the compound, provides a measure of confidence that can be placed in the prediction. The resulting estimate was compared with the actual pH-unadjusted $EC_{50}$.

## RESULTS

Results of assays appear in Table 2. The dose–response curves for the two types of toxicity tests are presented in Figures 1 and 2. The unadjusted 48-hr $EC_{50}$ of

**Table 2** Summary of Toxicity Tests Conducted with Isopropylamine

| | pH uncontrolled | pH adjusted |
|---|---|---|
| *Daphnia magna* | | |
| $EC_{50}$ (mg/L) | | |
| Probit | 91.6; 95% confidence limits = 77.1–104.2 | 197.6; 95% confidence limits = 167.4–230.6 |
| Graphic | 89.4 | 195.3 |
| *Selenastrum capricornutum* | | |
| ($IC_{50}$ mg/L) | | |
| Probit | <62.5 | 118.4; 95% confidence limits = 72.5–166.0 |
| Graphic | | 120.3 |

IPA to daphnia is 91.6 mg/L, whereas the pH-adjusted $EC_{50}$ is 197.6 mg/L. Figure 1b shows the fatal effects of increasing pH on *D. magna*. At pH 11.0, 35% of the daphnia population are immobile after 48 hr. The pH-adjusted 96-hr algal growth toxicity test resulted in an $EC_{50}$ of 118.4 mg/L. The unadjusted toxicity test would result in an $EC_{50}$ below 62.5 mg/l, the lowest concentration tested, and one which inhibited over 96% of the algal growth.

The TOPKAT *D. magna* $EC_{50}$ model estimated the $EC_{50}$ for unadjusted IPA to be 12.2 mg/L. The characteristics that contributed to the estimate are presented in Table 3. A search for similar structures in the database used to compile the daphnia QSAR revealed six compounds, 2-ethylhexylamine, allylamine, ethanolamine, ethylamine, *n*-butylamine, and cyclohexylamine, that had a primary amine. In each case but one, the *D. magna* $EC_{50}$ QSAR overestimates the actual toxicity, just as it did with IPA. Comparisons of the actual $EC_{50}$s with the predicted values are also presented in Table 3.

The TOPKAT rat oral $LD_{50}$ to *D. magna* $EC_{50}$ predicted an $EC_{50}$ of 16.5mg/L, similar to the estimate for toxicity predicted by the *D. magna* $EC_{50}$ without the inclusion of rat toxicity data (see Table 3). Five compounds were found in the database that were similar enough in structure to IPA for comparison: 2-ethylhexylamine, allylamine, *n*-butylamine, ethanolamine and ethylamine. The model overestimated the toxicity of all the compounds, except ethanolamine.

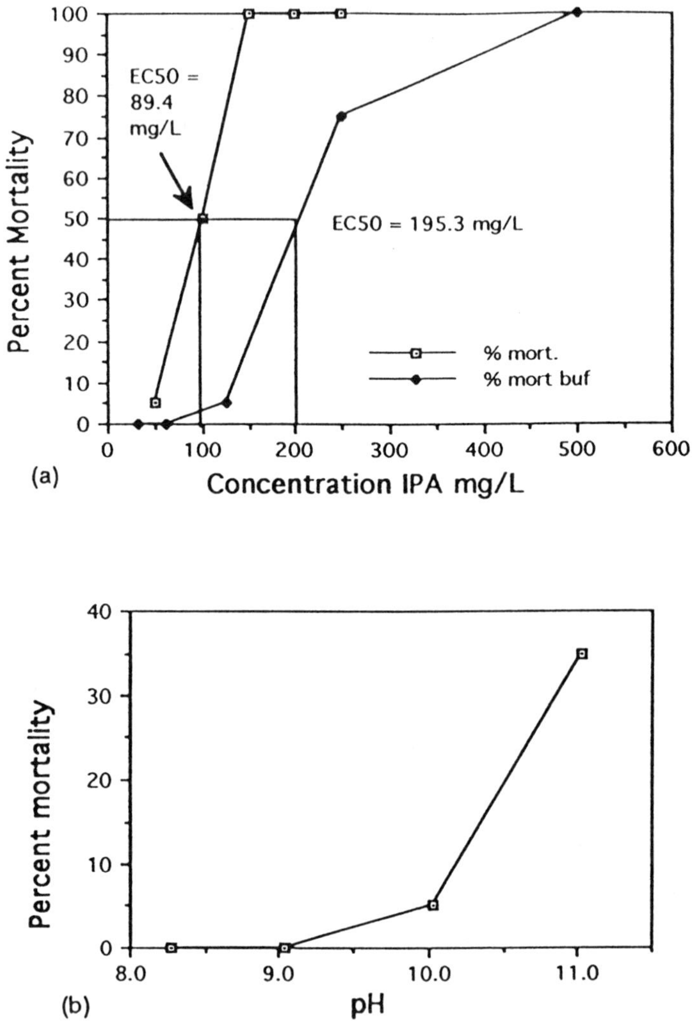


**Figure 1** Daphnia toxicity tests: (a) The dose–responses curves for the toxicity tests using the pH-adjusted and pH-unadjusted testing solutions. Both curves are steep, with approximately the same slopes. (b) Survivability of *D. magna* at different pHs under the conditions of the toxicity tests.

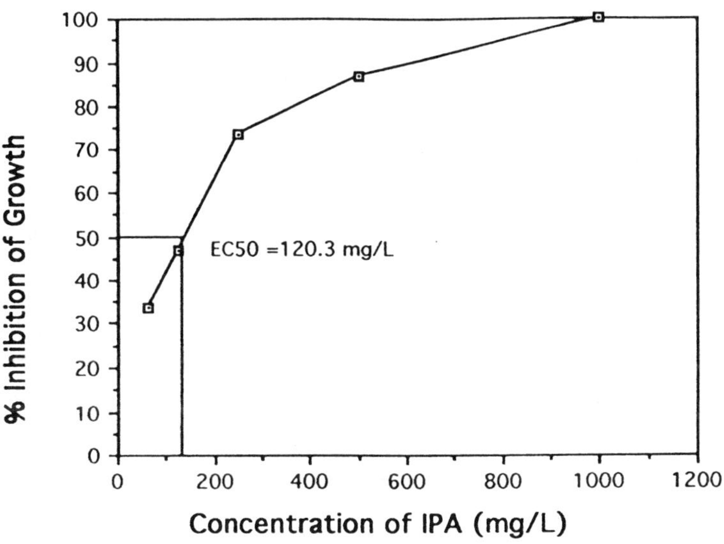


**Figure 2** The pH-adjusted 96-hr algal growth test: The curve exhibits partial inhibition of algal growth over a wide range of concentrations.

## DISCUSSION

It is apparent from the reductions of toxicity in the pH-adjusted tests, and the decrease in survivorship of *Daphnia* at higher pHs, that pH contributes to the overall toxicity of IPA for both *D. magna* and *S. capricornutum*. Yet, pH is not the sole cause of immobility or mortality and growth inhibition. The effect of IPA on an aquatic ecosystem into which it has been introduced will depend, in part, on the system's buffering capacity with the addition of a base.

After running the TOPKAT program, it was found that the *D. magna* $EC_{50}$ model overestimates the toxicity of IPA. Following the two-step validation process, the question of full representation of IPA in the model's database was satisfied. No major substructural features were outside the descriptor–design set on which the equation was based, which can lead to a poor TOPKAT performance. To address the second step of the validation, the model's database was searched, and several very similar compounds were found, the actual assay results, and TOPKAT estimates, of which refuted the estimate generated for IPA. The daphnia model overestimates the toxicity of these similar compounds also; however, an estimate for IPA can be considered valid by reporting a range of $EC_{50}$s. This range, 12.2–110.0 mg/L, is bracketed by the estimate, which is

**Table 3** Estimate of $EC_{50}$ Using the TOPKAT *D. magna* 48-hr $EC_{50}$ Model and Rat Oral $LD_{50}$ to *D. magna* $EC_{50}$

| | | |
|---|---|---|
| Name: Isopropylamine | | |
| *Daphnia magna* $EC_{50}$ model | | |
| Key | Cross product | |
| Primary amine (noncylcic) RNH2 (R = alkyl) | 0.961 | |
| Valence-adjusted path MCI order 1 | 0.437 | |
| Constant term | 2.287 | |
| Total | 3.685 | |
| Estimate of $EC_{50}$ as log (1000/M) = 3.685 or 12.2 mg/L | | |
| Compounds used in validation of estimate | Actual $EC_{50}$ | Predicted $EC_{50}$ |
| 2-Ethylhexylamine CAS:104-75-6 | 2.2 | 4.44 |
| Allylamine CAS: 107-11-9 | 110.0 | 14.1 |
| Cyclohexylamine CAS:108-91-8 | 80.0 | 6.9 |
| *n*-Butylamine CAS:109-73-9 | 75.0 | 30.8 |
| Ethanolamine CAS:141-43-5 | 140.0 | 49.6 |
| Ethylamine CAS:75-04-7 | 110.0 | 12.0 |
| Rat oral $LD_{50}$ to *Daphnia magna* $EC_{50}$ | | |
| Key | Cross product | |
| Rat oral $LD_{50}$ | 0.808 | |
| Primary or secondary aliphatic amine | 0.868 | |
| Valence-adjusted path MCI order 2 | 0.077 | |
| Constant term | 1.802 | |
| Total | 3.555 | |
| Estimate of $EC_{50}$ as log (1000/M) = 3.555 or 16.5 mg/L | | |
| Compounds used in validation of estimate | Actual $EC_{50}$ | Predicted $EC_{50}$ |
| 2-Ethylhexylamine CAS:104-75-6 | 2.2 | 1.8 |
| Allylamine CAS: 107-11-9 | 110.0 | 13.3 |
| *n*-Butylamine CAS:109-73-9 | 75.0 | 8.6 |
| Ethanolamine CAS:141-43-5 | 140.0 | 213.3 |
| Ethylamine CAS:75-04-7 | 110.0 | 25.5 |

the lowest $EC_{50}$ value, and the highest assay-derived $EC_{50}$ available from the most similar database compound. In reporting a range of $EC_{50}$s, confidence in this estimate is considered moderate. That the actual $EC_{50}$ for IPA falls within this reported range confirms the usefulness of the predicted results.

A similar performance was observed in the estimation of daphnid toxicity by the rat oral $LD_{50}$ to *D. magna* $EC_{50}$ model. The inclusion of the rat toxicity data apparently improves the performance significantly when the basic models are compared for predictive performance using a larger dataset [4].

These results indicate that had TOPKAT been available to run before conducting the actual daphnia tests, it would have proved a valuable source for the determination of starting concentrations, thereby eliminating initial range-finding tests. TOPKAT also consistently overestimated the toxicity of the compound, although the estimates were within an order of magnitude of the experimentally derived values. This consistent overestimation of toxicity may be a characteristic of the model in dealing with amine-type compounds.

In the future, it will be advantageous to continually test estimates and experimentally derived values to examine the ability of the environmental toxicity models included in TOPKAT. This could be done in a post hoc fashion, as was done here, or by consulting the literature for appropriate data. Verification of the usefulness of QSAR models with new data and compounds has the potential of providing estimates of hazard for situations not directly amenable to experimental manipulation.

## REFERENCES


1. Genium Publishing Corporation, Material safety data sheet: Isopropylamine, no. 483, August 1982, *Material Safety Data Sheets Collection Plus Update Service, Organic Materials*; Genium Publishing Corporation, Schenectady, NY, 1979.
2. Sittig, M., Isopropyl amine, *Handbook of Toxic and Hazardous Chemicals and Carcinogens*, 2nd ed., Noyes Publications, Park Ridge, NJ, 1985, pp. 534–535.
3. Health Designs, Incorporated, *Topkat User Guide*, Health Designs, Inc., Rochester, NY, 1987.
4. Enslein, K., Tuzzeo, T. M., Blake, B. W., Hart, J. B., and Landis, W. G., Prediction of *Daphnia magna* $EC_{50}$ values from rat oral $LD_{50}$ and structural parameters, *Aquatic Toxicology and Environmental Fate, Vol. 11, ASTM STP 1007* (G. W. Suter II and M. A. Lewis, eds.), American Society for Testing and Materials, Philadelphia, 1989.
5. Enslein, K., Estimation of toxicological endpoints by structure–activity relationships, *Pharmacol. Rev. 36*:131S–135S (1984).
6. American Society for Testing and Materials, Guide for Conducting Acute Toxicity Tests With Fishes, Macroinvertebrates, and Amphibians, Standard E729, American Society for Testing and Materials, Philadelphia, 1986.
7. Environmental Protection Agency, Part 797, Environmental effects testing guidelines, subpart 3, aquatic guidelines, *Code of Federal Regulations*, 40, part 700 to end. U. S. Government Printing Office, Washington, DC, 1986.
8. Environmental Protection Agency, *User's Guide: Procedures for Conducting*

Daphnia magna *Toxicity Bioassays*, EPA-600/8-87/011. U. S. Environmental Protection Agency, Las Vegas, NV 1987.

9. Haley, M. V., Johnson, D. W., Hart, G. S., Muse, W. T., and Landis, W. G., The toxicity of brass particles to the microalgae *Ankistrodesmus falcatus* and *Selenastrum capricornutum*, *J. Appl. Toxicol. 6*:281–285 (1988).
10. Johnson, D. W., Haley, M. V., Hart, G. S., Muse, W. T., and Landis, W. G., Acute toxicity of brass particles to *Daphnia magna*, *J. Appl. Toxicol. 6*:225–228 (1986).
11. Goulden, C. E., Comotto, R. M., Hendrickson, J. A. Jr., Hornig, L. L., and Johnson, K. L., Procedures and recommendations for the culture and use of *Daphnia* in bioassay studies, *Aquatic Toxicology and Hazard Assessment: Fifth Conference, ASTM STP-766* (J. G. Pearson, R. B. Foster, and W. E. Bishop, eds.), American Society for Testing and Materials, Philadelphia, 1982.
12. Finney, D. J., *Statistical Method in Biological Assay*, Charles Griffin & Co., London, 1978.
13. Finney, D. J., The median lethal dose and its estimation, *Arch. Toxicol. 56*:215–218 (1985).
14. Goulden, C. E., and Henry, L., Daphnia bioassay workshop, The Academy of Natural Sciences of Philadelphia, Philadelphia, 1983.
15. Taub, F. B., and Read, P. L., Standardized aquatic microcosm protocol. Final Report, Contract 223-80-2352, Vol. II, U. S. Food and Drug Administration, Washington, DC, 1983.
16. The Organization for Economic Co-operation and Development, Test guideline 201: Alga, growth inhibition test, *OECD Guidelines for Testing of Chemicals*. OECD, Paris, France, 1981, p. 8.
17. Stephen, C. E., and Rogers, J. W., Advantages of using regression analysis to calculate results of chronic toxicity tests, *Aquatic Toxicity and Hazard Assessment, Vol. 8, ASTM STP 891* (R. C. Bahner and D. J. Hansen, eds.), American Society for Testing and Materials, Philadelphia, 1985, pp. 328–338.
18. Stephan, C. E., Topics on expressing and predicting results of life cycle tests, *Aquatic Toxicology and Environmental Fate, Vol. 11, ASTM STP 1007* (G. W. Suter II and M. A. Lewis, eds.), American Society for Testing and Materials, Philadelphia, 1989, pp. 263–272.
19. Weininger, D., SMILES, a chemical language and information system. 1. Introduction to methodology and encoding rules, *J. Chem. Inf. Comput. Sci. 28*:31–36 (1988).
20. Wiswesser, W. J., *A Line-Formula Chemical Notation*, W. Y. Crowell, New York, 1954.

# 6

# Structure–Activity Relationships and the Validation of In Vitro Toxicology Tests

**Robert L. Lipnick, Maurice Zeeman, and Joseph A. Cotruvo**

*U.S. Environmental Protection Agency*
*Washington, D.C.*

Ideally, an in vitro toxicology test should be able to provide qualitative information on both the potency and type of biological response of the whole organism. Many tests under development are designed to predict a single potency endpoint, for which validation consists of a comparison of experimental potency in the whole-animal study relative to this single in vitro endpoint.

Toxicologists are trained not only to report defined endpoints in whole-animal experiments, but also to make a variety of qualitative observations that can provide important clues to the ability of a chemical to elicit additional adverse effects. The inability of most in vitro tests to model the spectrum of effects observed in whole-animal experiments represents a limitation in their use as animal replacements. A potential bridge between in vitro and in vivo tests could be the use of decision tests that provide guidance in selecting appropriate in vitro endpoints for testing. We can define *decision tests* as triggers for in vitro hazard tests, but not replacements. Such decision tests need not require biological cells or tissues, but could be based on purely chemical and biochemical measurements, as for example, measuring a chemical's intrinsic reactivity or susceptibility to metabolic transformation. Ideally, such tests would be able to categorize a chemical's mode and molecular mechanism of action. Knowledge of the molecular structure correlates for such mechanisms that are based on existing data for similar substances is also of great value in this endeavor.

## MODE AND MOLECULAR MECHANISM OF ACTION

The mode and molecular mechanism of action of a chemical should be distinguished. The former may be considered as a pattern or cluster of common biological responses exhibited by a class of chemicals, such as irritants, whereas the latter requires an understanding of the relation between this group of biological responses and what actually takes place at the molecular level. Irritancy can arise by various molecular mechanisms. Knowledge of the relation between mode and molecular mechanism can provide the link for the selection of appropriate in vitro tests and their relation to the corresponding whole-animal tests [1].

### Molecular Mechanism Domain

Molecular mechanisms (in ascending order of complexity) can be divided into the following general categories: narcosis (baseline toxicity), electrophile, proelectrophile, pharmacophore, and propharmacophore [1]. With increasing complexity, the number of parameters required to interpret the effect at the molecular level is expected to increase, whereas the domain of applicability of an in vitro model is expected to decrease.

#### Narcosis Mechanism (Baseline Toxicity)

Narcosis represents the most fundamental toxicological mechanism and corresponds to baseline toxicity [2]. In the presence of more specific mechanisms, the narcosis response will be masked by more specific effects elicited at lower toxicant concentrations. Nonelectrolyte chemicals of low reactivity, such as saturated aliphatic monohydric alcohols, ethers, amides, hydrocarbons, and chlorinated hydrocarbons, are associated with a narcosis-type response [3].

A variety of validation test studies now exist within the in vitro literature, based on chemicals known to act by this mechanism. It has been known since the discoveries of Meyer [4] and Overton [5,6], at the turn of the century, that the potencies of such chemicals in a wide variety of whole-animal and cellular systems are all correlated with their partition coefficients, which provide a means of calculating a minimum biophase active concentration. Since all of these data correlate with partition coefficient, they also correlate with one another. Validation studies that are based solely on chemicals acting by narcosis can provide an unrealistic appraisal of the applicability of the test if it is proposed for use for chemicals acting by more specific mechanisms. In addition, chemicals acting by narcosis yield excellent correlations with in vitro data from virtually any such system and, therefore, provide very unrealistic assessments of the validity of such in vitro methods for other chemicals. This condition is similar to the problem of intercorrelation of properties related to such potencies, despite the fact that the partition coefficient is the one of fundamental significance [7].

In fact, the partition coefficient has been used in numerous quantitative structure–activity relationship (QSAR) studies. Thus, a baseline toxicity QSAR model,

$$\log(1/\mathrm{LD}_{50}) = 0.805X - 0.971 \log(0.0807\ 10^{x} + 1) + 0.924$$
$$(r = 0.824;\ s = 0.208)$$

for the rat oral median lethal dose ($LD_{50}$; in mol/kg) has been derived from data on 54 saturated monohydric alcohols and monoketones, where $X$ is calculated $\log P$, and $P$ is the $n$-octanol/water partition coefficient [8]. A comparison of predicted and observed $LD_{50}$ values for various chemicals was useful in categorizing their mechanism relative to narcosis or more specific mechanisms [1].

Electrophile Mechanism

Chemicals capable of undergoing covalent bond formation with nucleophilic sites (e.g., sulfhydryl groups) within the organism, such as benzyl chloride, allyl bromide, acrolein, chloroacetonitrile, ethylene oxide, α-nitrostyrene, and acetaldehyde [1], can act by an electrophile mechanism. The potency of such action is controlled by the intrinsic reactivity of such substances and their rates of uptake, transport, metabolism, and excretion. The toxicity of electrophiles can be modeled by QSAR by using a reactivity parameter (pseudo-first-order reaction rate with a model nucleophile) and log $P$, respectively [9]. Electrophiles have been implicated in lachrymatory and vesicant effects [10], and $p$-nitrobenzyl halides have been used as model nucleophiles in assessing skin sensitization of electrophile toxicants [11].

An acceptable in vitro system for testing electrophiles would be able to simulate in the whole animal the effects of reactivity on the rate of covalent bond formation with toxicant target sites and loss by glutathione and other detoxicant nucleophiles, as well as the effect of lipophilicity on uptake, transport, and excretion.

Proelectrophile Mechanism

Proelectrophiles are chemicals requiring metabolic activation to serve as electrophile toxicants. In principle, at least one other parameter relating to this metabolic transformation (e.g., redox potential) would be required in a QSAR model. Examples of proelectrophile toxicants are propargyl alcohol (alcohol dehydrogenase activation), pentaerythritol trially ether (monooxygenase activation), and 1,3-dibromopropane (glutathione transferase activation) [1]. The toxicity to fish of nitroaromatic compounds has been related to a proelectrophile QSAR model [12],

$$\log(1/LC_{50}) = 0.96 \log P - 8.81E_{1/2} - 0.68$$
$$(n = 20;\ r = 0.964;\ s = 0.18)$$

where $E_{1/2}$ is the polarographic half-wave potential (a measure of rate of metabolic reduction to the electrophile toxicant).

### Pharmacophore and Propharmacophore Mechanisms

Chemicals acting by pharmacophore and propharmacophore mechanisms (more common for drugs and agricultural chemicals than industrial chemicals) are able to undergo receptor-mediated binding, either with or without metabolic activation, respectively. Because of the high degree of stereoelectronic specificity required for receptor binding, QSAR models for chemicals acting by these mechanisms are generally restricted to closely related structures. Useful molecular descriptors for such studies can be derived through simulated interactions with the putative receptor, using molecular-modeling methods [13]. Such chemicals would appear to require specialized in vitro tests, such as those involving receptor binding now employed in exploratory drug research.

## CHEMICAL SELECTION

The selection of chemicals for evaluation of in vitro tests seems to be an area that has largely been neglected. In most studies, such selection seems to be done without benefit of a formal systematic procedure. As a result, such selection is made by (a) relying on what chemicals of interest have already been tested with the corresponding whole-animal test, (b) choosing readily available chemicals belonging to the class of interest, or (c) attempting to validate the test for all possible chemicals by choosing sample compounds from different categories of chemicals, such as pesticides, industrial chemicals, metals, surfactants, and so on. As used here, the term *category* refers to ad hoc classifications used for such studies, and it should not be considered as having any fundamental significance.

The basing of such a choice on what chemicals have already been tested in the whole animal may seem pragmatic, but may, in fact, provide a false sense of validity for the larger universe of chemicals or even lead to an unduly pessimistic conclusion about the predictive value of the test relative to a more limited special category of interest. Most members of the larger group may be useful for exploring only a small number of mechanisms, whereas chemicals belonging to a specific class of interest may all act by a common mechanism.

## Application of Structure–Activity Relationships and Quantitative Structure–Activity Relationships to Chemical Selection

Structure–activity relationships (SARs) and quantitative structure–activity relationships (QSARs) can provide useful paradigms for systematizing the selection of test chemicals [13]. Although the use of a finite set of chemicals for validation can never assure the predictability of a test for all chemicals, a thoughtful selection process can more efficiently achieve this objective and provide greater confidence in the applicability of the test.

The greatest confidence in such validation studies can be gained if the application of an in vitro test can be limited, at least initially, to a very well-defined, discrete class of chemicals. A major difficulty in choosing such a set of chemicals is that the scope of such a category is in itself difficult to define. Within the QSAR paradigm, molecules are commonly compared in terms of their properties, rather than in relation to the skeletal connections in the whole molecules. The use of such chemical parameters or molecular descriptors may seem like a simple solution to defining what chemicals should be tested, but a more difficult question arises concerning the choice and rationale for selecting properties and the extent to which these properties may contain redundant information. For example, partition coefficients between water and various immiscible organic solvents, such as chloroform, will be collinear with those of octanol/water in many, but not all chemicals. These partition coefficients, in fact, depend on more fundamental properties, including molar volume and the ability to serve as a hydrogen-bonding donor and acceptor.

In the Hansch approach to QSAR, hydrophobic (e.g., $\log P$), electronic (e.g., $pK_a$) and steric (measure of fit to site of action) properties are used as parameters or molecular descriptors in analysis as well as in chemical selection. For example, suppose an investigator was testing a series of *para*-substituted benzoic acid derivatives. Chemicals would need to be selected to represent the range of all three of these parameters likely to be encountered in future reliance on the in vitro test. The chemicals should be chosen so that they are evenly distributed within this three-dimensional space. There is an implicit assumption of lack of change in molecular mechanism of action within this group of chemicals.

Hansch has characterized the domain of such a testing space as *spanned substituent space* [14]. Compounds outside this domain may be predicted, but such predictions should be considered as extrapolations, rather than interpolations.

An additional complication arises because of the inability to uniquely define the dimensionality of chemical space encompassed and the nature of these dimensions. This difficulty is related to the even more fundamental question of the relation between the dimensionality of the test set and the larger universe of chemicals for which such predictions may be desired.

## Water Solubility-Related Limitations

Many in vitro tests consist of cells suspended in an aqueous medium to which the test substance is added. Potency is judged based on the relative concentration required for a substance added to this phase to produce a defined biological response. Additional insight can be gained about the physicochemical limitations of such tests by their analogy to experience in testing chemicals with fish and other aquatic organisms in which uptake also occurs from a solution in water.

Testing substances with a reasonable degree of water solubility should not pose special experimental problems in an in vitro test, but this is not true when testing highly insoluble compounds. In aquatic toxicology, it is not uncommon to artificially solubilize such compounds by using a small amount of cosolvent that is miscible in water.

Nevertheless, it is essential to exclude the possibility of additional toxicity being contributed by such a cosolvent, and to ensure that the cosolvent is used to produce more rapid dissolution and does not lead to testing above actual water solubility. Thus, for some substances, toxicity is intrinsically limited by water solubility, and the use of such supersaturated solutions may provide misleading results. Saturated solutions for highly insoluble substances can also be prepared efficiently with the use of generator columns. It is recommended that the actual test concentrations be determined analytically when testing such insoluble substances, and that the use of nominal concentrations be avoided, as has been the current practice in aquatic toxicology. Use of measured concentrations also avoids problems of loss owing to adsorption to the glass walls of the test chamber.

For liquid solutes, there is a high correlation between water solubility and the octanol/water partition coefficient [15]. However, for solutes that melt above room temperature, the correlation needs to be corrected for the transition from the solid to the liquid phase as part of the dissolution process [16], and chemicals with similar partition coefficients acting by the same mechanism would be expected to show equivalent cutoffs. Solubility decreases with increasing melting point for solutes that have the same partition coefficient, and melting temperature is a function of intermolecular hydrogen bonding, symmetry, and conformational flexibility. These factors should be considered in identifying such highly insoluble compounds and defining appropriate procedures before in vitro testing.

### Problems From Testing Volatile Chemicals

Chemicals that are volatile from water require special experimental considerations to ensure that the actual concentration is known and does not decrease with time. This can be accomplished, as in aquatic toxicology testing, by the use of a controlled closed system, continuous flow, and analytical monitoring of the actual concentration over time. In designing studies for potentially volatile compounds, it is useful to obtain a prior estimate of volatility (air/water partition coefficient), which can be approximated directly from the chemical structure [17].

### Nonlinear Relationships

According to Richardson's rule, the toxicity of chemicals within a homologous series increases with increasing chain length and molecular weight [18]. In the rat oral $LD_{50}$ QSAR equation cited earlier for baseline narcosis toxicity, the form of the model gives rise to a nonlinear equation, with maximum toxicity corresponding to a log $P$ value of close to 2.

In fact, Richardson's rule applies only if pseudo-steady-state thermodynamic equilibrium can be achieved between the site of administration and the site of action. This is not true in whole-animal studies; except with gaseous or vapor administration. Richardson found such apparent relationships only because he did not investigate higher members of a series [18]. In aquatic toxicology and corresponding in vitro tests, linear relationships will be observed over a larger range of compounds within a series and the in vitro test could thus predict greater potency than would be observed in the whole-animal test.

## CONCLUSIONS

Chemical and biochemical tests, along with structure–activity relationships, could be used to select a battery of single-endpoint in vitro tests to disclose the spectrum of biological responses observable in the whole-animal test. The QSAR models already exist for predicting the potency of some chemicals relative to certain whole-animal endpoints. In selecting chemicals for the validation of in vitro toxicology tests, careful appraisal needs to be made of the class or classes of chemicals for which the test is to be applied, along with corresponding molecular mechanisms of toxicity and domain of spanned substituent space parameters. Special procedures used in aquatic toxicology for testing very insoluble or volatile chemicals should be employed in corresponding in vitro tests. Nonlinear relationships are expected between potency measured in whole-animal and in vitro tests owing to differences in uptake, transport, metabolism, and excretion.

## REFERENCES

1. Lipnick, R. L., Outliers: Their origin and use in the classification of molecular mechanisms of toxicity, *Sci. Total Environ. 109/110*:131–153 (1991).
2. Lipnick, R. L., Narcosis: Fundamental and baseline toxicity mechanism for non-electrolyte organic chemicals. *Practical Applications of Quantitative Structure–Activity Relationships (QSAR) in Environmental Chemistry and Toxicology* (W. Karcher and J. Devillers, eds.), Kluwer Academic Publishers, Dordrecht, The Netherlands, 1990, pp. 129–144.
3. Lipnick, R. L., A QSAR study of Overton's data on the narcosis and toxicity of organic compounds to the tadpole, *Rana temporaria*, *Aquatic Toxicology and Hazard Assessment:* 11th Symposium (G. W. Suter II and M. Lewis, eds.), American Society for Testing and Materials, STP 1007 Philadelphia, 1989, pp. 468–489.
4. Lipnick, R. L., Hans Horst Meyer and the lipoid theory of narcosis, *Trends Pharmacol. Sci. 10*:265–269 (1989).
5. Lipnick, R. L., Charles Ernest Overton: Narcosis studies and a contribution to general pharmacology, *Trends Pharmacol. Sci. 7*:161–164 (1986).
6. Lipnick, R. L., ed, *Charles Ernest Overton: Studies of Narcosis and a Contribution to General Pharmacology*, Chapman & Hall, London, and Wood Library-Museum of Anesthesiology, 1991.
7. Lipnick, R. L., and Filov, V. A., Nikolai Vasilyevich Lazarev, toxicologist and pharmacologist, comes in from the cold, *Trends Pharmacol. Sci. 13*:56–60 (1992).
8. Lipnick, R. L., Pritzker, C. S., and Bentley, D. L., Application of QSAR to model the toxicology of industrial organic chemicals to aquatic organisms and mammals, *QSAR in Drug Design and Toxicology* (D. Hadži and Jerman-Blažiž, eds.), Proceedings of the 6th European Symposium on Quantitative Structure–Activity Relationships, Portorose, Yugoslavia, September 22–26, 1986, Elsevier, New York, 1987, pp. 301–306.
9. Hermens, J., Quantitative correlation studies between the acute lethal toxicity of 15 organic halides to the guppy (*Poecilia reticulata*) and chemical reactivity towards 4-nitrobenzylpyridine, *Toxicol. Environ. Chem. 9*:219–236 (1985).

10. Dixon, M., Biochemical research on chemical warfare agents, *Nature 158*:432–438 (1946).
11. Roberts, D. W., Goodwin, B. F. J., Williams, D. L., Jones, K., Johnson, A. W., and Alderson, J. C. E., Correlations between skin sensitization potential and chemical reactivity for *p*-nitrobenzyl compounds, *Food Chem. Toxicol. 21*:811–813 (1983).
12. Deneer, J. W., Sinnige, T. L., and Hermens, J. L. M., A QSAR study of fish toxicity of nitroaromatic compounds, *QSAR in Drug Design and Toxicology* (D. Hadži and Jerman-Blažič, eds.), Proceedings of the 6th European Symposium on Quantitative Structure–Activity Relationships, Portorose, Yugoslavia, September 22–26, 1986, Elsevier, New York, 1987, pp. 352–354.
13. Franke, R., *Theoretical Drug Design Methods*, Elsevier, Amsterdam, 1984, 412 pp.
14. Hansch, C., On the predictive value of QSAR, *Biological Activity and Chemical Structure* (J. A. Keverling Buisman, ed.), Elsevier, Amsterdam, 1977, pp. 47–61.
15. Hansch, C., Quinlan, J. E., and Lawrence, G. L., The linear free-energy relationship between partition coefficients and the aqueous solubility of organic liquids, *J. Org. Chem. 33*:347–350 (1968).
16. Banerjee, S., Yalkowsky, S. H., and Valvani, S. S., Water solubility and octanol/water partition coefficients of organics. Limitations of the solubility–partition coefficient correlations, *Environ. Sci. Technol. 14*:1227–1229 (1980).
17. Hine, J., and Mookerjee, P. K., Structural effects on rates and equilibriums. XIX. Intrinsic hydrophilic character of organic compounds. Correlations in terms of structural contributions, *J. Org. Chem. 40*:292–298 (1975).
18. Lipnick, R. L., Narcosis, electrophile and proelectrophile toxicity mechanisms: Application of SAR and QSAR, *Environ. Toxicol. Chem. 8*:1–12 (1989).

# 7

# Applications of Liver and Kidney Cell Systems That Can Reduce Animal Usage

**Charles A. Tyson and Carol E. Green**

*SRI International*
*Menlo Park, California*

Interest in the development and use of primary cell systems for toxicological and pharmacological applications has increased dramatically over the last 15 years [1,2]. Abstracts from past Society of Toxicology annual meetings have been surveyed for reports on purely in vitro cell or tissue culture studies, which were found to increase from 11% of the total presentations in 1980 to 24% in 1990; other studies with subcellular preparations or cells from animals treated first with the toxicant added a few more percentage points. Many of these studies were developmental or involved basic research on toxic mechanisms, but technology has progressed to the point where some of these systems are being applied to a variety of toxicological and pharmacological problems.

The purpose of this chapter is to provide examples of the kinds of information obtainable with primary cell and tissue systems and their potential applications. Most of the work described here has been done with isolated hepatocytes (HEP) and renal proximal tubule suspensions (RPT) because they are common targets of drug- and chemical-related toxicity and because they play central roles in the disposition and excretion of foreign compounds after absorption.

## METHODS AND MATERIALS

### Isolation Procedures

Detailed methods of isolating HEP from either whole liver or biopsy specimens by collagenase perfusion are described elsewhere [3–5]. The viability of the freshly isolated cells was assessed by trypan blue exclusion; only preparations with greater than 80% viable cells were used. Erythrocytes were harvested from heparinized or EDTA-treated whole blood by centrifugation and stored in culture medium at 4°C overnight; they were recentrifuged and resuspended in culture medium at approximately 1.6 g hemoglobin per deciliter before use. The RPT were isolated in situ by a modified collagenase perfusion method in which deferoxamine was added to the perfusate to reduce oxidative damage to cells during isolation [6]. When a section of a baboon kidney was received at the laboratory for use, the perfusion was conducted through a cannula inserted into a vessel on the outer surface, and the procedures thereafter were the same [7]. Viability of RPT preparations was determined primarily by measuring nystatin-stimulated oxygen consumption [6,7]. Only preparations with greater than 35% stimulation retained enough viability throughout the experiment (4 hr or longer) to be used.

### Culture Conditions

Both HEP suspension ($1 \times 10^6$ viable cells per milliliter of culture medium) and monolayer ($\sim 0.25 \times 10^6$ viable cells per milliliter) cultures were used, depending on the experimental objectives. A coculture system—HEP attached to culture dishes with erythrocytes suspended in the medium—was also used [8]. The culture medium was hormone-supplemented Waymouth's 752/1 containing either 0.2 or 2.0% bovine serum albumin (BSA) [5,7–9]. The RPTs were incubated in suspension in the same medium containing 2% BSA at a protein level of 0.5 mg/ml, with shaking, under essentially the same experimental conditions [7].

All incubations were conducted at 37°C under an atmosphere containing 20% or higher $O_2$, 5% $CO_2$, and the balance $N_2$. The HEP and RPT suspensions were shaken at 70 osc/min in airtight Erlenmeyer flasks, with sidearms for serial sampling.

### Assays

Amphetamine (AMP) metabolism was monitored by using D-[7-$^{14}$C]AMP sulfate (Research Products Intl., Mt. Prospect, IL), and high-pressure liquid chromatography (HPLC) was used for metabolite separation [5]. Acetaminophen

(APAP)–protein adducts were determined immunochemically after electrophoretic separation of the proteins by using an affinity-purified anti-APAP antibody developed in Dr. E. A. Khairallah's laboratory at the University of Connecticut (Storrs) [10,11]. Trichloroethylene (TCY; [U-$^{14}$C] radiolabel from Amersham, Arlington Heights, IL) and its metabolites were assayed by gas chromatography [9]. The 3-(4,5-dimethylthiazol-2-yl)-2,5-diphenyltetrazolium bromide (MTT) assay was conducted spectrophotometrically, using a Dynatech microplate reader [12]. In cyanide (CN) antidote experiments, cytotoxicity was monitored by determining cell adenosine triphosphate (ATP) levels using the lucerifin–luceriferase assay after aspiration of the erythrocyte-containing culture medium and solubilization of HEP [8]. All chemicals and other reagents were purchased from established commercial suppliers and used without further purification.

## RESULTS AND DISCUSSION

### Use for Comparative Metabolism

Early work focused on the value of HEP systems as in vitro models for interspecies comparisons of drug and chemical metabolism. Studies on AMP and TCY metabolism provide clear examples of the ways in vitro data can be used both prospectively to select the most representative animal models for absorption, distribution, metabolism, and excretion (ADME) studies and pharmacokinetic and toxicity studies, and retrospectively, to help clarify the significance for humans of findings in laboratory animals [4,9].

Metabolism of AMP occurs in the liver by two principal pathways: (a) ring hydroxylation, which forms *p*-hydroxy-AMP; and (b) oxidative deamination, which forms phenylacetone, benzoic acid, and hippuric acid as principal metabolites. In vivo, the rat metabolizes the drug primarily by ring hydroxylation, whereas in rabbits, dogs, squirrel monkeys, and humans, oxidative deamination is the principal pathway (Fig. 1a). These findings were reproduced in vitro (see Fig. 1b). Also, a good inverse rank correlation was observed between the rates of hepatic AMP metabolism and the amount of AMP excreted unchanged in animal and clinical studies. Each species was unique in the way it handled AMP (or, for that matter, all other drugs and chemicals that have been examined by us), but the rat, the most commonly used species in toxicity and metabolic disposition studies, was clearly not representative of human exposure in this case.

These results are instructive from another standpoint: If a test strategy is designed to support an Investigational New Drug Application (IND) for

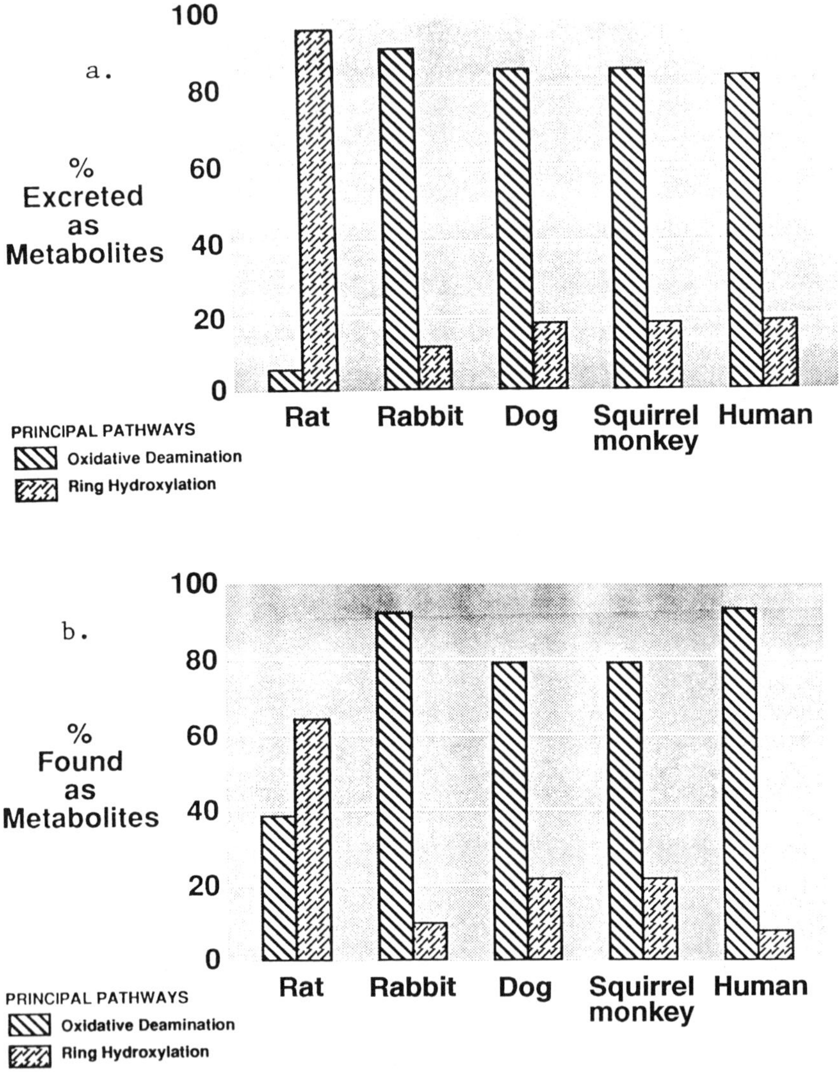


**Figure 1** Amphetamine (AMP) metabolism. (a) In vivo 24- or 48-hr urine collection. (b) In vitro 4-hr incubation with HEP.

amphetamine without the benefit of in vivo data, and ADME and pharmacokinetic studies were conducted in rats, the wrong species would have been chosen as a model for the way humans handle this drug. If the same experiments were conducted with dogs, the results would be entirely different. Without human studies, it would not be known which species is more representative. Preliminary testing in hepatocytes from both species and comparison with human in vitro data allows determination of the more appropriate species for pharmacokinetic and ADME studies and weighing of the toxicological data accordingly before human exposure.

In vivo studies in several laboratories have demonstrated that TCY is hepatocarcinogenic in the mouse but not in the rat. Because this compound is a proved hepatocarcinogen in an experimental model, regulatory agencies are highly concerned about the consequences of human exposure in the workplace and the environment. A metabolite, trichloroacetic acid (TCA), appears to be primarily responsible. The mouse metabolizes TCY four times faster than the rat at comparable doses and produces correspondingly larger amounts of TCA.

Comparative metabolism studies in rat and human hepatocytes under identical experimental conditions were proposed as a way of determining whether humans are more or less at risk from internally formed TCA. Incubation of TCY with these cells produced trichloroethanol, its glucuronide conjugate, and TCA as the principal metabolites in roughly the same proportions as found in vivo in these species. Much lower levels of TCA were present in the human cell incubations compared with those in rat HEP, both in total amount and relative to trichloroethanol plus conjugate (Fig. 2). Because the same principal metabolites were formed as in vivo, it may be concluded that, all other factors being equal, humans are less prone than rats, a known resistant species, to develop hepatocarcinogenicity from TCY exposures.

## Use for Cytotoxicity Screening

Several laboratories are exploring the potential value of primary cell cultures as cytotoxicity screens to predict the general or organ-specific toxicity of previously untested compounds. A series of ongoing studies were conducted with the aim of defining the context within which these screens can be used (i.e., what useful information they can provide). In early work, it was demonstrated that the cytotoxic potential of chlorinated aliphatic compounds in HEP suspensions were closely correlated with in vivo toxicity for these chemicals when corrections were made for differences in expiration and volatility in the in vitro and in vivo systems [13]. The cytotoxicity of dinitrotoluenes in HEP suspensions also correlated well with their hepatotoxic potential in rats, the toxicity of

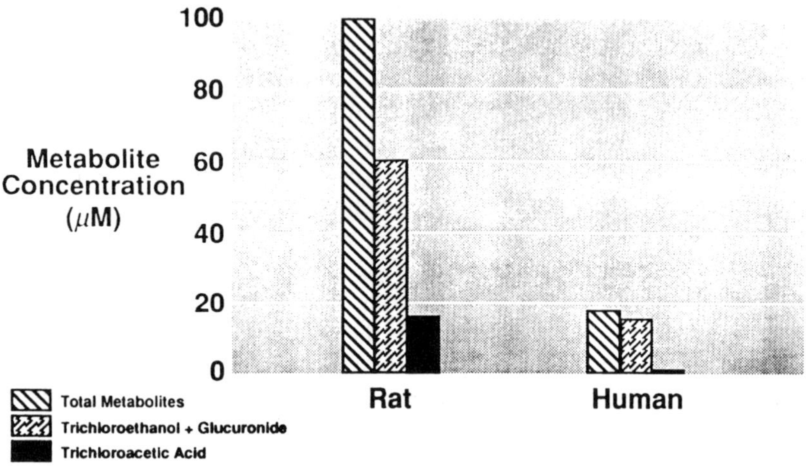


**Figure 2** Trichloroethylene (TCY) metabolism. In vitro 4-hr incubation with HEP.

which, based on molecular orbital calculations, is related to the electronic, rather than the physicochemical, properties of the compounds [14]. Cytotoxic potentials determined for cephalosporins and aminoglycosides in RPT systems show very good rank correspondence with the nephrotoxic potentials of these classes of antibiotics [15,16]. These results and those from other laboratories suggest that HEP and RPT are practical models for estimating the toxic potencies of structurally related compounds to target cells for those chemical classes. Occasional anomalies and exceptions to this "rule" that have appeared in the literature need to be explained, and the investigator must often balance the imprecision of the inverse ranking orders of some individual compounds in the series against the overall correlative value (for which $R^2 > 0.80$ in the cases cited).

If these systems are useful for ranking the cytotoxic potency of untested compounds relative to some reference compounds within the series, they must also have value for detecting interspecies differences in response. Insight was obtained on this point in two separate studies. In one, the cytotoxic potential of nine compounds from the Multicenter Evaluation of In Vitro Cytotoxicity (MEIC) list was determined in a rat HEP assay, using MTT-tetrazolium dye reduction as the indicator, and the results were compared with available in vivo data [12]. The cytotoxic potentials correlated better with rat oral $LD_{50}$ values than did circulating concentrations of the compounds known to cause toxicity

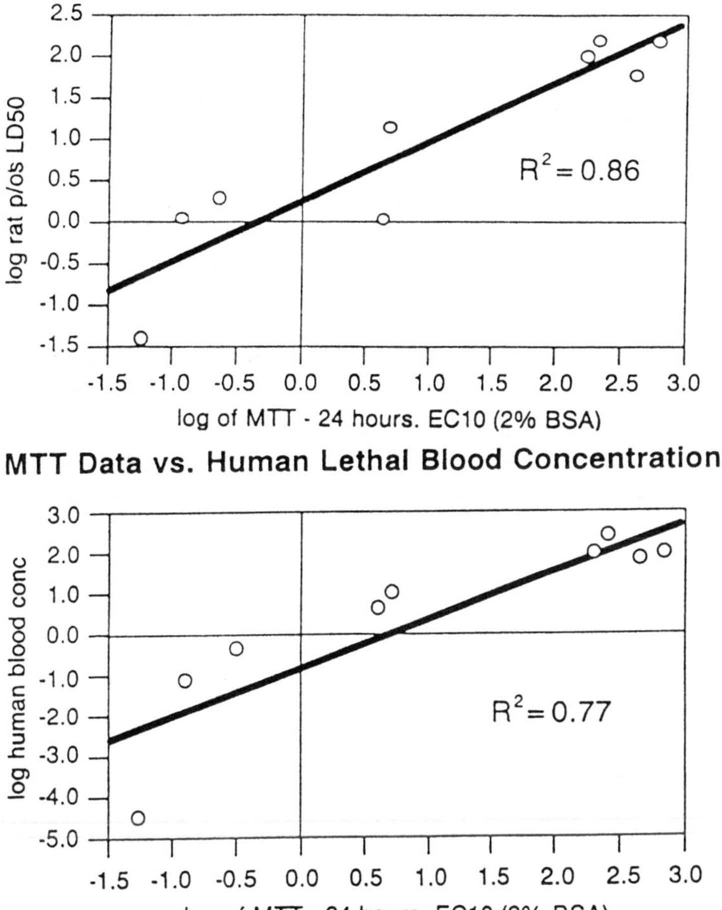


**Figure 3** Effective concentration for 10% change in MTT response for MEIC compounds in 24-hr HEP cultures compared with in vivo response.

in humans (Fig. 3). Although the comparison cannot be made more rigorous in the absence of rat blood or tissue concentrations at lethal doses, a reasonable expectation is that rat HEP should predict rat toxicity more accurately than human toxicity, and in this sense, the results support that expectation. Others have tested ten compounds from the same list in vitro and demonstrated that human hepatocytes predicted acute toxicity in humans more accurately than did either rat hepatocytes or mouse 3T3 cells [17].

With acetaminophen (APAP) as a test compound, many investigators had reported earlier that mouse, hamster, rat, rabbit, and dog HEP show the same relative susceptibility to the drug as is found in vivo [18,19]. These in vitro data also suggest that humans are relatively resistant compared with other species (Fig. 4). Acetaminophen binds to cell proteins at toxic levels, and the same protein-bound adducts were found immunochemically in mouse and human HEP as in vivo [7]. (The latter result was obtained with a liver from an accidental fatality after APAP overdose.) Whereas RPT from CD-1 mice showed a binding pattern similar to that in liver and kidney, no protein adducts were found in incubations with rat tubules (as with intact kidney in in vivo experiments) or with those from a baboon (human kidney being unavailable at the time) [7]. Although the differences in response remain to be clarified, they were profound among these species. Kidney failure from APAP overdose is of more serious concern to some clinicians than liver injury, and the relevance of the mouse, which some investigators use as a model, for evaluating various treatments to reverse APAP toxicity in the human may be questionable.

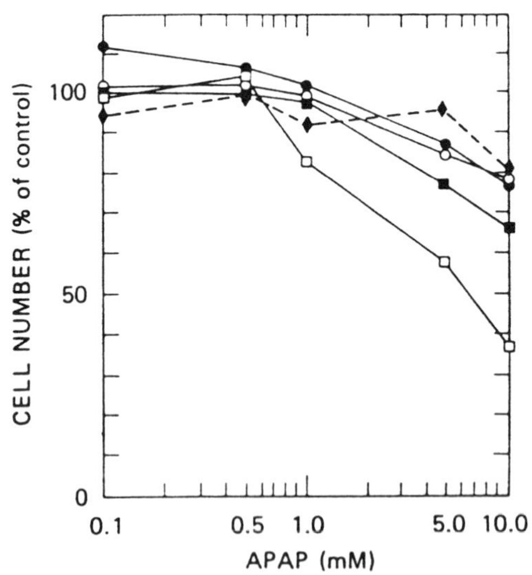


**Figure 4** Acetaminophen (APAP)-induced cytotoxicity in 24-hr HEP cultures. -□-□- hamster; -■-■- dog; -○-○- rat; -●-●- rabbit; -♦-♦- human.

## Use as Screens for Pharmacological Activity

The usefulness of primary cell systems for screening for pharmacological activity in vitro is also becoming increasingly evident. For example, the U.S. Army Medical Research and Development Command was interested in a screening system that would allow selection of the best prospective anticyanide agents for further evaluation from among a large number of candidate compounds. The currently recommended treatment for cyanide (CN) poisoning in this country is a combination of sodium nitrite ($NaNO_2$) and sodium thiosulfate ($Na_2S_2O_3$); cobalt compounds are used in Europe. To provide optimal versatility in the screen, a coincubation system was developed that comprised HEP monolayer cultures (as both target cells for CN action and a source of the enzyme rhodanese, which uses $Na_2S_2O_3$ as a substrate for CN detoxification) and erythrocyte suspensions (which provide hemoglobin for nitrite-induced cyanmethemoglobin formation). Cobalt complexes CN directly and nonenzymatically.

In the coincubation system and under the experimental conditions used, CN-induced ATP depression occurs only in the HEP and is a convenient indicator of cytotoxicity. $NaNO_2$ and $Na_2S_2O_3$ were each partially effective and their combination was most effective in reversing this effect in rat cell cultures (Fig. 5). This result is known to occur in CN-treated animals in vivo.

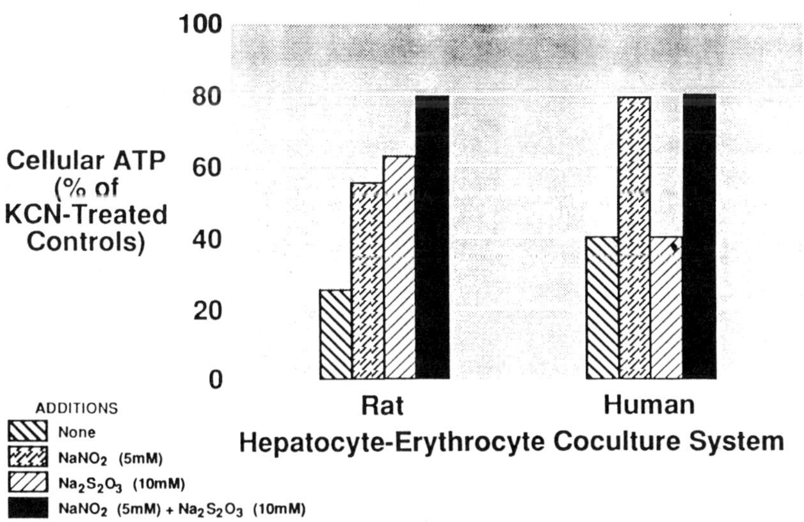


**Figure 5** Antidote reversal of CN-induced cytotoxicity.

In contrast, $NaNO_2$, but not $Na_2S_2O_3$, reversed ATP depression in the human cell culture, and the effectiveness of the combination was totally dependent on the $NaNO_2$ component (see Fig. 5) [20]. The apparent reduced effectiveness of $Na_2S_2O_3$ in the human cell system corresponds to the known lower rhodanese activity in human liver compared with rat liver. These results both verify the value of this approach to screening CN antidotes and encourage a closer examination of the merits of $Na_2S_2O_3$ for human therapy.

Another pharmacological application for HEP cultures has been in screening drugs and drug–ionophore combinations for treating Fe overload in blood-transfused β-thalassemia patients [21]. Our laboratory has also developed a promising lipocyte assay for screening drug candidates with potential effectiveness in treating hepatic fibrosis. Reaggregate brain cell and heart cell cultures from both animal and human tissue are being evaluated for pharmacological and toxicological applications.

## CONCLUSIONS

These examples demonstrate the benefits of using primary cell screens to reduce animal use and to provide data of toxicological and pharmacological significance. One rat liver or a pair of kidneys provides enough cells or tissue for 30–40 separate experimental flasks for screening or mechanistic studies, with a corresponding reduction in the number of animals and test costs required to obtain the necessary information. In addition, a strong mechanistic basis for using these systems as screens for acute toxicity and metabolism has been developed in many laboratories. Toxicologists are beginning to realize that information from these screens is as important as that obtained from mutagenicity and carcinogenicity screens, because primary cell screens provide key data for making decisions at critical junctures in the developmental process, avoiding the wasteful use of animals for testing less than optimal compounds. The use of such data for extrapolations to safe dose or exposure levels for humans is an ultimate, additional goal that will be achieved when in vitro data are routinely incorporated into physiologically based pharmacokinetic models for risk assessments.

Proper interpretation of in vitro results, however, is not always as straightforward as in the examples cited. For instance, renal tubule preparations may be acceptable screens for assessing metal- and antibiotic-induced nephrotoxic potential, but several compounds are known to cause kidney toxicity by metabolites formed by the liver. The ability to estimate the nephrotoxic potentials of such compounds from in vitro data requires the use of coincubation systems and conditions that accurately model in vivo organ specificity and potency. Progress in this direction is occurring in a few laboratories. The hepatocyte–erythrocyte system was the first to show that primary cell coincubation systems could mimic in vivo response and potency with pharmacologically active drugs. Another, classic example is the Ames test, in which liver S-9 postmitochondrial fraction is added to *Salmonella* bacteria to assess whether metabolic activation of the test compound is required for mutagenesis. Similar strategies with kidney RPT and cells from other potential target organs supplemented with S-9 or intact liver cells can be envisioned.

## ACKNOWLEDGMENTS

The authors gratefully acknowledge receipt of human donor livers from the California Donor Transplant Network, San Francisco; the Stanford University Medical Center, Palo Alto; and the National Disease Research Interchange, Philadelphia, which made this research possible. This work was supported by funding from NIGMS and NIEHS/NIH, USAMRDC/Department of Defense, the Environmental Protection Agency, and the Kitasato Research Institute, Tokyo (a fellowship for Dr. K. Otoguro).

## REFERENCES

1. Goldberg, A. M., *In Vitro Toxicology New Directions*, Vol. 7, *Alternative Methods in Toxicology*, Mary Ann Liebert, New York, 1989.
2. Gad, S. C., Recent developments in replacing, reducing, and refining animal use in toxicologic research and testing, *Fundam. Appl. Toxicol. 15*:8–16 (1990).
3. Green, C. E., Dabbs, J. E., and Tyson, C. A., Functional integrity of isolated rat hepatocytes prepared by whole liver vs. biopsy perfusion, *Anal. Biochem. 129*: 269–276 (1983).
4. Green, C. E., Le Valley, S. E., and Tyson, C. A., Comparison of amphetamine metabolism using isolated hepatocytes from 5 species including humans, *J. Pharmacol. Exp. Ther. 237*:931–936 (1986).
5. Allen, K. L., and Green, C. E., Isolation of human hepatocytes by biopsy perfusion. *In Vitro Biological Systems*, Vol. 1, *Methods in Toxicology* (C. A. Tyson and J. M. Frazier, eds.), Academic Press, San Diego, pp. 262–278, 1993.
6. Green, C. E., Dabbs, J. E., Tyson, C. A., and Rauckman, E. J., The effect of oxygen tension and antioxidants on isolated rat renal proximal tubules, *Renal Failure 12*:147–156 (1990).
7. Tyson, C. A., Cohen, S. D., and Khairallah, E. A., Comparative toxicity: Mechanistic studies on acetaminophen action in vitro and in vivo in various species including man. *In Vitro Toxicology Mechanisms and New Technology*, Vol. 8, *Alternative Methods in Toxicology* (A. M. Goldberg, ed.), Mary Ann Liebert, New York, pp. 163–170, 1991.
8. Gee, S. J., Le Valley, S. E., and Tyson, C. A., Application of a hepatocyte–

erythrocyte coincubation system to studies of cyanide antidotal mechanisms, *Toxicol. Appl. Pharmacol. 88*:24–34 (1987).
9. Knadle, S. A., Green, C. E., Baugh, M., Vidensek, M., Short, S. M., Partos, X., and Tyson, C. A., Trichloroethylene biotransformation in human and rat hepatocytes, *Toxicol. In Vitro 4*:537–541 (1990).
10. Bartolone, J. B., Sparks, K., Cohen, S. D., and Khairallah, E. A., Immunochemical detection of acetaminophen-bound liver proteins, *Biochem. Pharmacol. 36*: 1193–1196 (1987).
11. Birge, R.B., Bartolone, J. B., Emeigh Hart, S. G., Nishanian, E. V., Tyson, C. A., Khairallah, E. A., and Cohen, S. D., Acetaminophen hepatotoxicity: Correspondence of selective protein arylation in human and mouse liver in vitro, in culture, and in vivo, *Toxicol. Appl. Pharmacol. 105*:472–482 (1990).
12. Otoguro, K., Komiyama, K., Omura, S., Nourse, L., and Tyson, C., An in vitro cytotoxicity assay using rat hepatocytes and MTT and coomassie blue dye as indicators, *ATLA 19*:352–360 (1991).
13. Tyson, C. A., Gee, S. J., Hawk-Prather, K., Story, D. L., and Milman, H. A., Correlation between in vivo/in vitro toxicity with some chlorinated aliphatics, *Toxicol. In Vitro 3*:145–150 (1989).
14. Spanggord, R. J., Myers, C. J., Le Valley, S. E., Green, C. E., and Tyson, C. A., Structure–activity relationship for the intrinsic hepatotoxicity of dinitrotoluenes, *Res. Chem. Toxicol. 3*:551–558 (1990).
15. Sina, J. F., Noble, C., Bean, C. L., and Bradley, M. O., Renal tubules in vitro as a model for nephrotoxicity, *In Vitro Toxicology Model Systems and Methods* (C. A. McQueen, ed.), Telford Press, Caldwell, NJ, pp. 263–290, 1989.
16. Tyson, C. A., Dabbs, J. E., Cohen, P. M., Green, C. E., and Melnick, R. L., Studies of nephrotoxic agents in an improved renal proximal tubule system, *Toxicol. In Vitro 4*:403–408 (1990).
17. Jover, R., Ponsoda, X., Castell, J. V., and Gómez-Lechón, M. J., Evaluation of the cytotoxicity of ten chemicals on human cultured hepatocytes: Predictability of human toxicity and comparison with rodent cell culture systems, *Toxicol. In Vitro 6*:47–52 (1992).
18. Green, C. E., Dabbs, J. E., and Tyson, C. A., Metabolism and cytotoxicity of acetaminophen in hepatocytes isolated from resistant and susceptible species, *Toxicol. Appl. Pharmacol. 76*:139–149 (1984).
19. Davies, D. S., Tee, L. B. G., Hampden, C., and Boobis, A. R., Acetaminophen toxicity in isolated hepatocytes, *Adv. Exp. Biol. Med. 197*:993–1003 (1986).
20. Tyson, C. A., *Mechanism of Chemical Action and Treatment of Cyanide Poisoning*, Annual Progress Report AD A192091 on Contract DAMD17-82-C-2211, SRI International, Menlo Park, CA, June 1, 1986.
21. Tyson, C. A., Le Valley, S. E., Chan, R., Hobbs, P. D., and Dawson, M. I., Biological evaluation of some ionophore–polymeric chelator combinations for reducing Fe overload, *J. Pharmacol. Exp. Ther. 228*:676–681 (1984).

# 8

# In Vitro Methods for Hepatotoxic Assessment of Halogenated Fatty Acids

**Nicholas J. DelRaso, Stephen R. Channel, Merry Jane Walsh, Barry L. Hancock, and William J. Schmidt**

*Wright–Patterson Air Force Base, Ohio*

The Johns Hopkins Center for Alternatives for Animal Testing has estimated that approximately 17–22 million animals are used annually in United States laboratories, and that testing of a new chemical would cost about 1.5 million dollars. Furthermore, pressures from animal rights groups will result in a need for alternatives to whole-animal testing. Total replacement of whole-animal testing is not now realistic; however, animal reduction and refinement are achievable. It has been recently reported that animal use has been reduced by the employment of in vitro methods [1].

Halogenated hydrocarbons compose an important class of chemicals used by the U.S. Air Force and industry. These compounds possess physical properties that are suited for meeting various Air Force needs. Some of these compounds are ideal candidate hydraulic fluids, lubricants, and extinguishants owing to their low flammability; however, these compounds are both nephrotoxic and hepatotoxic [2,3]. Furthermore, they induce hepatic peroxisome proliferation [4,5], which has been linked to tumorigenesis in rodents [6].

Many qualitative and quantitative methods exist for measuring cellular toxicity in vitro. Endpoints of toxicity for selected target cells can range from cell death [7], to short-term reversible alterations in cellular structure and function [8]. Biological potency can be correlated with chemical structure and ranked by employing cells in bioassay screens. The rationale for the use of cultured cells to assess toxicity is based on the premise that "the actions of chemicals that produce disease and death in animals are ultimately exerted at the cellular level" [9], and is reviewed in detail by Bradlaw [10].

The purpose of this study was to determine if in vitro methods could be used to screen halogenated fatty acids for hepatotoxicity and peroxisomal fatty acid β-oxidation induction. By employing in vitro biochemical and fluorescence assays, critical cellular and subcellular changes can be identified that are not easily assessed in vivo and can give insight into the development of hypotheses for mechanisms of toxicity. This will result in a reduction of animals and of cost in the evaluation of chemical toxicity.

## MATERIALS AND METHODS

### Rat Liver Perfusion

Fischer 344 rat livers were perfused, and hepatocytes were isolated and enriched, as previously described [11], with the following modifications: (a) Perfusion media were supplemented with 15 mM HEPES. (b) Washout perfusion medium was supplemented with 0.5 mM EGTA. (c) Digestion medium was supplemented with 0.26 U/ml of collagenase (type A; Lot CGB-118913, Boehringer-Mannheim Biochemicals, Indianapolis, IN).

### Primate Liver Perfusion

An aseptic recirculating perfusion system was designed to perfuse pieces of rhesus monkey (*Macaca mulatta*) liver. An approximate 10-g liver biopsy section was cut into three pieces (~3.0 g). Digestion medium was the same as that used to perfuse rat livers, except that Sigma (St. Louis, MO) collagenase type IV (lot 109F6804) was used at 317 U/ml. Liver pieces were digested for 45 min.

### Hepatocyte Isolation and Culture

Primary rat and primate hepatocytes were isolated by low-speed isodensity Percoll (Pharmacia, Piscataway, NJ) centrifugation, as described by Kreamer et al. [12]. Rat hepatocytes ($2 \times 10^8$ cells) and primate hepatocytes ($4 \times 10^7$ cells) were adjusted to a cell density of $1 \times 10^6$/ml, respectively. Two milliliters of cell suspension was added to rat tail collagen-precoated 60-mm Falcon tissue culture plates (0.2 mg/plate). After attachment, rat hepatocytes were refed with Williams E culture medium (Gibco, Grand Island, NY), supplemented with HEPES (18 mM), sodium bicarbonate (2.2 mg/ml), bovine serum albumin (0.5 mg/ml), 5-aminoluvelinic acid (0.1 mg/ml), Sigma insulin–transferrin–

sodium selenite solution (10 μg/ml), gentamicin (50 μg/ml), and 5% fetal calf serum (FBS; Gibco, Grand Island, NY). Primate hepatocytes were refed with Williams E culture medium supplemented as above, except that 5% Stimu-CYTE serum-free medium supplement (Upstate Biotechnology, Inc., Lake Placid, NY) was used as a substitute for serum.

## Exposure

Sterile stock solutions of trimerchlorotrifluoroethylene (CTFE) acid, tetramer CTFE acid, and perfluorodecanoic acid (PFDA) were made up in DMSO at concentrations of 10 mg/ml, respectively. The final concentration of test agents and DMSO in tissue culture plates were 50 μg/ml (0.1 mM) and 0.5%, respectively. Plates were dosed daily with change of culture medium. Rat hepatocytes were exposed for 96 hr, whereas primate hepatocytes were exposed for 120 hr. At the termination of exposure, hepatocytes were scraped using a rubber policeman in 150 μl of 154 mM KCl/50 mM Tris-HCl buffer (pH 7.2–7.4). Hepatocyte samples were then homogenized by sonication on ice.

## Biochemical Assays

Medium lactate dehydrogenase (LDH) enzyme levels were determined using a DuPont ACA V discrete clinical analyzer (DuPont, Huffman Estates, IL), as previously described [11]. Acyl CoA oxidase (ACO) [13] and carnitine acetyltransferase (CAT) [14] peroxisomal enzyme activities were assayed, as previously described. Cytochrome P-450 activity, as determined by 7-ethoxycoumarin-*o*-deethylase (7-ECOD) activity, was assayed according to the method of Gray et al. [13]. Reduced glutathione (GSH) was assayed according to the method of Hissin and Hilf [15], as described by Tyson and Green [16]. Glutathione-*S*-transferase (GST) was assayed according to the method of Ahmed et al. [17]. Protein determinations were assayed according to the method of Bradford [18].

## Fluorescence Plate Reader

Dose–response studies were conducted in 24-well Corning plates (Fisher Scientific, Pittsburgh, PA) that were seeded at approximately $2 \times 10^5$ cells per well. After 24 hr, cells were dosed at four dose levels, ranging from 25 (0.05 mM) to 500 μg/ml (2.0 mM), in 0.2 ml (of medium) with tetramer CTFE carboxylic acid and PFDA and incubated an additional 24 hr, respectively. Following exposure, cell monolayers were washed with calcium free–phosphate buffered saline (CF-PBS).

Rhodamine-123 (R-123; 1.0 mg/ml PBS) was added to triplicate wells for each treatment group at final concentrations of 10 μg/ml. After 60 min of incubation at 37°C, the wells were washed three times in CF-PBS, and 0.2 ml of CF-PBS was added to all wells. Negative controls consisted of empty wells that were treated with R-123, washed three times, and refed CF-PBS. Fluorescence determinations were quantified on the Millipore CytoFluor 2300 fluorescence plate reader, using a 485-nm excitation/530-nm emission filter set combination at a sensitivity of 3. Data were expressed as the percentage of control.

## Flow Cytometer

Primary rat hepatocytes were exposed to a low (0.2-mM) and high (2.0-mM) concentration of tetramer CTFE acid for 4 hr in suspension and then analyzed by flow cytometry. Rat hepatocytes from control and treatment groups were suspended in Dulbecco's phosphate-buffered saline (PBS), containing 5% serum to a cell density of 1 × 106/ml. One milliliter of each sample was transferred into a 5-ml cell culture tube, and 10 μl of a 100 μl/ml stock solution of R-123 was added. The tubes were placed into a 37°C incubator, with 5% $CO_2$, atmosphere for 30 min. Immediately following the incubation, the samples were run on a FACScan (Becton Dickinson) flow cytometer, using an argon laser for excitation at 488 nm. Light was collected from the forward scatter (FSC), side scatter (SSC), and fluorescence 1 (FL1, 530/30-nm) detectors. Cell populations were displayed as dot plots of FSC versus SSC. Debris was identified and excluded by establishing a gate to include intact cells, and gated cells were then displayed as FSC versus FL1. Untreated cells established the control values; subsequent samples were compared with this control in the FL1 parameter.

## Statistics

Biochemical measurements for each treatment group were compared by an analysis of variance (ANOVA) using SYSTAT software statistics package (SYSTAT, Inc., Evanston, IL). Means found to be significant by ANOVA were compared with the Tukey post hoc test with type I error level held at $p < 0.05$.

# RESULTS

Rat hepatocytes and primate hepatocytes were exposed to PFDA, trimer CTFE acid, and tetramer CTFE acid (Fig. 1) at a dose of 0.1 mM for 96 and 120 hr, respectively. The dose chosen (0.1 mM) was based on previous in vitro toxicity studies with PFDA that used primary rat hepatocytes. This dose of PFDA

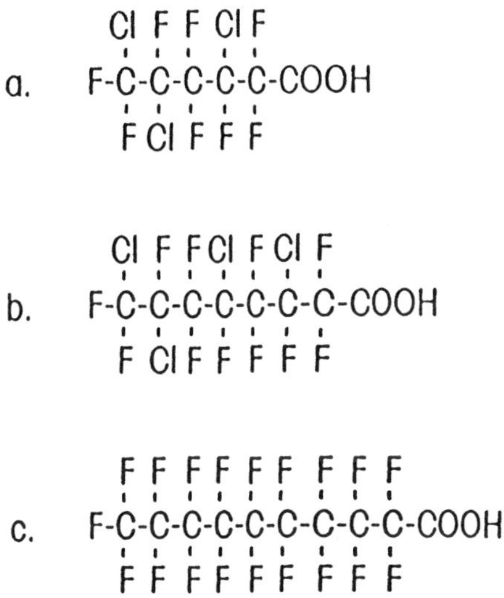


**Figure 1** Chemical structure of halogenated carboxylic acid test agents. (a) Trimer CTFE acid; (b) tetramer CTFE acid; (c) PFDA.

resulted in severe toxicity in hepatocytes and was used for comparative purposes with the CTFE acids at similar doses.

Exposure of rat or primate hepatocytes to PFDA resulted in significant cytotoxicity (Fig. 2). Primate hepatocytes exposed to PFDA in serum-free medium resulted in significant cell loss over the exposure period, such that levels of lactate dehydrogenase (LDH) could not be detected. Rat hepatocytes exposed to PFDA in medium containing 5% serum resulted in a significant fourfold increase in LDH leakage when compared with control cells (see Fig. 2). This fourfold increased LDH leakage represented 40 ± 10% of the total available LDH present in control cells. Control cell LDH leakage was 10% of total LDH.

Only primate hepatocytes exposed to the tetramer CTFE acid resulted in a significant eightfold increase in LDH leakage when compared with control (see Fig. 2). This eightfold increase represented 27 ± 3% of the total LDH. Control LDH leakage from primate hepatocytes was approximately 3% of the total. Exposure of rat hepatocytes to either trimer or tetramer CTFE acids did not result in any significant change in LDH leakage when compared with control (see Fig. 2).

Exposure of rat hepatocytes for 24 hr to either the tetramer CTFE acid or

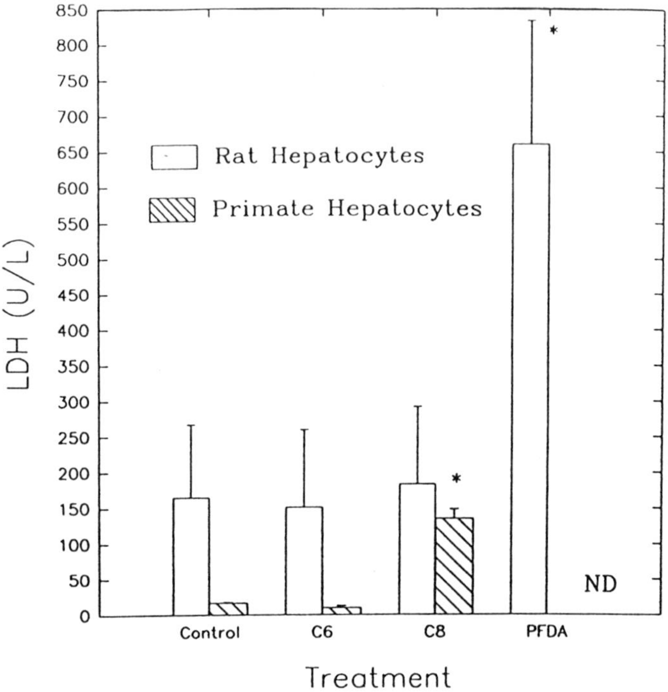


**Figure 2** Medium LDH enzyme levels from rat and primate hepatocyte cultures following exposure to various halogenated fatty acids at 0.1 mM for 96 hr and 120 hr, respectively. Bars represent the mean LDH (international units/L) from three experiments with triplicate plates per treatment group. Control LDH leakage from rat and primate hepatocyte cultures were ~10% and ~3% of total, respectively. (*) Significantly different than control at $p < 0.05$.

PFDA, over a dose range of 0.05–1 mM, indicated similar dose–response effects, as determined by mitochondrial R-123 fluorescence measured in a fluorescence plate reader (Fig. 3). A 300% increase in R-123 fluorescence occurred at the highest dose of PFDA (0.4 mM) and tetramer CTFE acid (1 mM) used. No dose–response effects were found for the trimer CTFE acid in rat hepatocytes exposed at concentrations ranging from 0.2 to 1.4 mM (data not shown). Rat hepatocytes were exposed to a low (0.2 mM) and high (2.0 mM) concentration of tetramer CTFE acid for 4 hr and then analyzed for R-123 fluorescence with a flow cytometer. Flow cytometric analysis indicated a

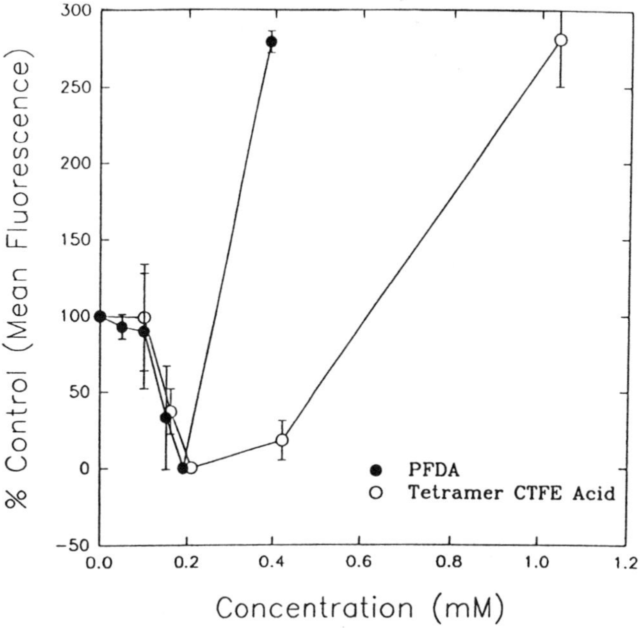


**Figure 3** Dose–response effects of increasing tetramer CTFE acid and PFDA concentrations on rhodamine-123 dye retention by primary rat hepatocytes exposed for 24 hr. Data points represent the average fluorescence from a single experiment with quadruplicate wells per treatment group. Fluorescence was determined in a CytoFluor 2300 fluorescence plate reader. The amount of fluorescence plotted is the percentage of nonexposed control cells.

significant 71% reduction in R-123 fluorescence at the high CTFE acid concentration when compared with control and low CTFE acid concentration (Fig. 4).

Rat and primate hepatocytes were exposed to the halogenated carboxylic acid test agents at 0.1 mM for 96 and 120 hr, respectively, and the activity of peroxisomal ACO and CAT enzymes were then determined. Control ACO activity in rat and primate hepatocytes could not be determined under the experimental conditions used. However, ACO activity was measurable in rat hepatocytes exposed to the test agents (Table 1). In contrast, no ACO activity could be determined in similarly exposed primate hepatocytes. Rat hepatocytes exposed to the trimer and tetramer CTFE acids indicated significant eightfold and sevenfold increases in CAT activity when compared with control, respec-

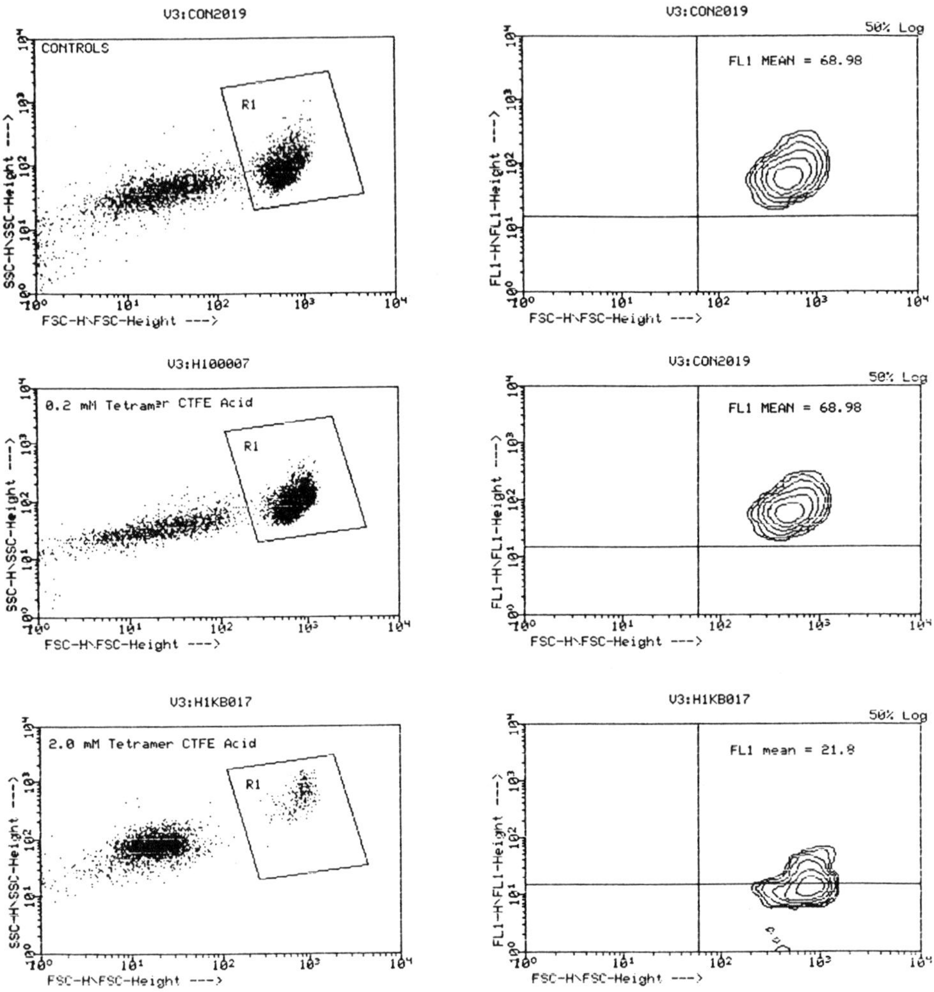


**Figure 4** The effect of tetramer CTFE acid on primary rat hepatocyte size (forward scatter [(FSC)]; *x*-axis) vs. granularity (side scatter [SCC]; *y*-axis) after 4-hr exposure in suspension culture. Contour plots on right represent intact hepatocytes (gate R1 of scatter plot). FL1 represents mean R-123 fluorescence. (*) Significantly different than control at $p < 0.05$.

**Table 1** Peroxisomal Fatty Acid β-Oxidation Activities in Primary Rat and Primate Hepatocytes After Exposure to Various Halogenated Fatty Acids

| | | Assay | |
|---|---|---|---|
| Treatment[a] | Species[b] | Palmitoyl CoA β-oxidation (nmol/min/mg) | Carnitine acetyltransferase (nmol/min/mg) |
| Control | Rat | 0 | 3.5 ± 1.4[c] |
| | Primate | 0 | 5.4 ± 1.3[d] |
| $C_6$ CTFE acid | Rat | 13.2 ± 7.7 | 28.7 ± 13.6* |
| | Primate | 0 | 12.6 ± 1.3* |
| $C_8$ CTFE acid | Rat | 11.3 ± 5.3 | 23.7 ± 11.1* |
| | Primate | 0 | 7.2 ± 3.3 |
| PFDA | Rat | 2.8 ± 2.5 | 11.4 ± 3.9 |
| | Primate | 0 | 0 |

[a]All treatment groups were dosed at 50 μg/ml (~0.1 mM) and 0.5% DMSO. Controls were dosed with DMSO at 0.5%.
[b]Primary rat hepatocytes were exposed for 96 hr. Primary rhesus monkey hepatocytes were exposed for 120 hr.
[c]Values represent the mean and standard deviation from three experiments with triplicate plates for each treatment group.
[d]Values represent the mean and standard deviation from one experiment with triplicate plates for each treatment group.
*Significantly different from control, as determined by a multifactorial ANOVA with Tukey post hoc comparisons of significant means at $p \leq 0.05$.

tively (see Table 1). A significant twofold increase in CAT activity was observed only in primate hepatocytes exposed to the trimer CTFE acid.

Hepatocyte metabolic and detoxification capabilities were assessed in 96- and 120-hr–exposure cultures. Cytochrome P-450 activity, as determined by 7-ECOD activity, was significantly increased only in rat hepatocytes exposed to the trimer CTFE acid (Table 2). Measurements of 7-ECOD activity in primate hepatocytes were not determined. Exposure of rat hepatocytes to either PFDA or tetramer CTFE acid resulted in significantly reduced GSH levels (see Table 2). However, GST activity was not significantly reduced (see Table 2). Exposure of primate hepatocytes to the tetramer CTFE acid resulted in significant reductions in GSH level and GST activity (see Table 2). Because of the cell loss from the culture plates,GSH level in primate hepatocytes exposed to PFDA could not be measured by the method employed here. Although GST

**Table 2** Phase I and Phase II Metabolic Reaction Determinations in Primary Rat and Primate Hepatocytes After Exposure to Various Halogenated Fatty Acids

| | | | Treatment[a] | | |
|---|---|---|---|---|---|
| Assay | Species[b] | Conc | $C_6$ CTFE acid | $C_8$ CTFE acid | PFDA |
| 7-ECOD (nmol/min/mg) | Rat[c] | 0.25 ± 0.04 | 0.52 ± 0.11* | 0.34 ± 0.15 | 0.10 ± 0.03 |
| | Primate[d] | ND[e] | ND | ND | ND |
| GSH (nmol/mg) | Rat | 15.7 ± 3.8 | 22.9 ± 5.0 | 6.0 ± 0.7* | 2.3 ± 1.5* |
| | Primate | 8.0 ± 0.7 | 8.7 ± 0.9 | 2.0 ± 0.5* | NM[f] |
| GST (nmol/min/mg) | Rat | 609.0 ± 204.3 | 676.7 ± 114.2 | 539.2 ± 81.5 | 462.3 ± 133.3 |
| | Primate | 689.6 ± 34.7 | 653.2 ± 119.8 | 248.3 ± 42.0* | 89.7 ± 14.3* |

[a]All treatment groups were dosed at 50 μg/ml and 0.5% DMSO. Controls were dosed with DMSO at 0.5%.
[b]Primary rat hepatocytes were exposed for 96 hr. Primary rhesus monkey hepatocytes were exposed for 120 hr.
[c]Values represent typical average means from three experiments with triplicate plates for each treatment group.
[d]Values represent the mean and standard deviation from one experiment with triplicate plates for each treatment group.
[e]ND, not determined.
[f]NM, not measurable.
*Significantly different than control as determined by a multifactorial ANOVA with Tukey post hoc comparisons of significant means at $p \leq 0.05$.

activity was significantly reduced in these cells, it cannot be attributed to treatment, because the cell number in the culture plates was reduced.

## CONCLUSIONS

Halogenated hydrocarbons constitute an important class of compounds used commercially and by the U.S. Air Force. These compounds possess physical properties that make them ideal candidates for hydraulic fluids, lubricants, and fire extinguishants. The principal property of these compounds that makes them attractive is their low flammability.

Previous in vivo work with PFDA indicated a 14-day 50% lethal dose ($LD_{50}$) of about 64 mg/kg in male Fischer 344 (F-344) rats [19]. The 14-day $LD_{50}$s for the tetramer and trimer CTFE acids were recently found to be approximately 88 and 396 mg/kg in male F-344 rats, respectively [20]. In the present in vitro study, PFDA was the most toxic at 0.1 mM, based on LDH leakage and cell loss from culture plates in both rat and primate hepatocytes. The tetramer CTFE acid induced significant LDH leakage only in primate hepatocytes exposed under serum-free culture conditions. However, this LDH leakage was not as great as that in PFDA-exposed rat hepatocyte cultures containing 5% serum. Exposure of either rat or primate hepatocytes to the tetramer CTFE acid also resulted in significantly reduced GSH levels. Several studies have shown a correlation between GSH depletion or GST inhibition and toxicity or carcinogenicity [17,21–24]. No significant effect on LDH leakage or GSH levels was observed in either rat or primate hepatocytes exposed to the trimer CTFE acid. Therefore, these data indicate that PFDA was the more toxic than the tetramer CTFE acid, which was more toxic than the trimer CTFE acid. These in vitro results correlate well with the previously established in vivo $LD_{50}$s for these compounds.

The greater LDH leakage observed in primate hepatocytes exposed to the tetramer CTFE acid is more than likely due to the low amount of protein in the StimuCYTE-supplemented serum-free culture medium. StimuCYTE is a low-protein culture medium supplement that can be used as an equivalent to serum to support the growth of various cell types [25]. This medium supplement was used to prevent fibroblastic overgrowth. The reduced levels of protein in these cultures may allow a higher percentage of free tetramer CTFE acid that can interact with cellular membranes or proteins, resulting in greater LDH leakage.

Interestingly, only the trimer CTFE acid significantly induced cytochrome P-450 activity in rat hepatocytes, as determined by 7-ECOD activity. The significance of this induction relative to toxicity is unknown. It is possible that the trimer CTFE carboxylic acid is effectively neutralized by the induction of a particular cytochrome P-450 isozyme that is not inducible by longer-chained halogenated carboxylic acids. Additional in vitro studies using sensitive fluorescent probes for different isozymes of cytochrome P-450 and parallel metabolic in vivo studies are required to help elucidate the involvement of cytochrome P-450 in determining halogenated fatty acid-induced hepatotoxicity.

The ability of PFDA and "parent" tetramer CTFE to induce peroxisomal proliferation and ACO activity has been previously demonstrated in the rat [4,26]. In contrast to the previous in vivo exposure of rats to the "parent" tetramer CTFE oligomer, exposure of rhesus monkeys to this CTFE oligomer did not result in any increase in peroxisomal ACO activity [27]. Recently, exposure to the tetramer CTFE carboxylic acid, but not the trimer CTFE carboxylic acid, has resulted in increased peroxisomal ACO activity in the rat [28].

In the present in vitro study, rat hepatocytes exposed to all three test agents demonstrated measurable ACO activity, whereas identically exposed primate hepatocytes did not. This species difference is in agreement with the previous in vivo studies described earlier. The increase peroxisomal β-oxidation observed in primary rat hepatocytes exposed to the trimer CTFE carboxylic acid is in contrast with that observed in vivo [28]. An in vitro system does not allow the chemical or metabolite elimination that can occur in vivo in exposed animals. Therefore, chemicals or metabolites may interact with hepatocytes in vitro, when they would not in vivo.

The increase in CAT activity in rat hepatocytes exposed to the experimental halogenated fatty acids is not surprising, since this activity has been coinduced with ACO with exposure to other classes of compounds [13]. A significant increase in CAT activity was also observed in primate hepatocytes exposed to the trimer CTFE carboxylic acids. The significance of this increased activity in primate hepatocytes is unknown. However, it does not appear that increased CAT activity in primate hepatocytes would be associated with toxicity or carcinogenicity, because other associated changes in the liver (hepatomegaly) do not occur as they do in the rat [27]. As of yet, it is unclear what the relation between peroxisomal β-oxidation induction and carcinogenicity is.

The results from this study indicate that primary hepatocytes are a useful and inexpensive research tool for screening potential peroxisome-proliferating chemicals, and for rank-order screening of chemical toxicity using in vitro biochemical and fluorescence assays. With the rapid development of novel halogenated hydrocarbon fluids for military and commercial use, it would be advantageous to develop in vitro systems that are inexpensive and reduce animal use to screen this class of compounds for toxicity.

## REFERENCES


1. Gad, S. C., Recent developments in replacing, reducing, and refining animal use in toxicological research and testing, *Fundam. Appl. Toxicol. 15*:8 (1990).
2. Gad, S. C., Rusch, G. M., Reigle, K. S., Darr, R. W., Hoffman, G. M., Peckham, J. C., and Schardein, J. L., Inhalation toxicity of chlorotrifluoroethylene (CTFE), *J. Am. Coll. Toxicol. 7*:663 (1988).
3. Plaa, G. L., Experimental evaluation of haloalkanes and liver injury, *Fundam. Appl. Toxicol. 10*:563 (1988).
4. DelRaso, N. J., Godin, C. S., Jones, H. G., Wall, H. G., Mattie, D. R., and Flemming, C. D., Comparative hepatotoxicity of two polychlorotrifluoroethylene oils (3.1 oils) and two chlorotrifluoroethylene (CTFE) oligomers in male Fischer 344 rats, *Fundam. Appl. Toxicol. 17*:550 (1991).
5. Kinkead, E. R., Kimmel, E. C., Wall, H. G., Conolly, R. B., Kutzman, R. S., Whitmire, R. E., and Mattie, D. R., Subchronic inhalation studies on polychlorotrifluoroethylene (3.1 oil), *Inhal. Toxicol. 2*:431 (1990).
6. Reddy, J. K., Azarnoff, D. L., and Hignite, C. E., Hypolipidemic peroxisome proliferators form a novel class of chemical carcinogens, *Nature 283*:397 (1980).
7. Smuckler, E. A., and James, J. L., Irreversible cell injury, *Pharmacol. Rev. 36*:77s (1984).
8. Reed, D. J., and Fariss, M. W., Glutathione depletion and susceptibility, *Pharmacol. Rev. 36*:25s (1984).
9. Grishham, J. W., and Smith, G. J., Predictive and mechanistic evaluation of toxic responses in mammalian cell culture systems, *Pharmacol. Rev. 36*:151s (1984).
10. Bradlaw, J. A., Evaluation of drug and chemical toxicity with cell culture systems, *Fundam. Appl. Toxicol.6*:598 (1986).
11. DelRaso, N. J., Mattie, D. R., and Godin, C. S., In vitro toxicity of solubilized 2,3,4-trimethylpentane I. Cytotoxicity and metabolism of TMP using primary hepatocytes, *In Vitro 25*:1031 (1989).
12. Kreamer, B. L., Staecker, J. L., Sawada, N., Sattler, G. L., Hsia, M. T. S., and Pitot, H. C., Use of a log-speed iso-density Percoll centrifugation method to increase the viability of isolated hepatocyte preparations, *In Vitro 22*:201 (1986).
13. Gray, T. J. B., Lake, B. G., Beamand, J. A., Foster, J. R., and Gangolli, S. D., Peroxisome proliferation in primary cultures of rat hepatocytes. *Toxicol. Appl. Pharmacol. 67*:15 (1983).
14. Gray, T. J. B., Beamand, J. A., Lake, B. G., Foster, J. R., and Gangolli, S. D., Peroxisome proliferation in cultured rat hepatocytes produced by clofibrate and phthalate ester metabolites, *Toxicol. Lett. 10*:273 (1982).
15. Hissin, P. J., and Hilf, R., A fluorometric method for determination of oxidized and reduced glutathione in tissues, *Anal. Biochem. 74*:77 (1976).
16. Tyson, C. A., and Green, C. E., Cytotoxicity measures: Choices and methods, *The Isolated Hepatocyte: Use in Toxicology and Xenobiotic Biotransformations* (E. J. Rauckman and G. M. Padilla, eds.), Academic Press, Orlando, 1987, p. 119.
17. Ahmed, A. E., Soliman, S. A., Loh, J.-P., and Hussein, G. I., Studies on the

mechanism of haloacetonitriles toxicity: Inhibition of rat hepatic glutathione *S*-transferases in vitro, *Toxicol. Appl. Pharmacol. 100*:271 (1989).
18. Bradford, M. M., A rapid and sensitive method for the quantitation of protein utilizing the principle of protein–dye binding, *Anal. Biochem. 72*:248 (1976).
19. Olson, C. T., and Andersen, M. E., The acute toxicity of perfluorooctonoic and perfluorodecanoic acids in male rats and effects on tissue fatty acids, *Toxicol. Appl. Pharmacol. 70*:362 (1983).
20. Kinkead, E. R., Bunger, S. K., and Wolfe, R. E., $LD_{50}$ and $LD_{10}$ oral toxicity studies of chlorotrifluoroethylene acids in F-344 rats, AD 235756, Harry G. Armstrong Laboratory, Wright-Patterson AFB, OH, 1990, p. 82.
21. Maellaro, E., Casini, A. F., Del Bello, B., and Comporti, C., Lipid peroxidation and antioxidant systems in the liver injury produced by glutathione depleting agents, *Biochem. Pharmacol. 39*:1513 (1990).
22. Potter, W. Z., Thorgeirsson, S. S., Jollow, D. J., and Mitchell, J. R., Acetaminophen-induced hepatic necrosis. V. Correlation of hepatic necrosis, covalent binding, and glutathione depletion in hamsters, *Pharmacology 12*:129 (1974).
23. Quinn, B. A., Crane, T. L., Kocal, T.E., Best, S. J., Cameron, R. G., Rushmore, T. H., Farber, E., and Hayes, M. A., Protective activity of different hepatic cytosolic glutathione *S*-transferases against DNA-binding metabolites of aflatoxin $B_1$, *Toxicol. Appl. Pharmacol. 105*:351 (1990).
24. Watanabe, P. G., Hefner, R. E., Jr., and Gehring, P. J., Vinyl chloride induced depression of hepatic non-protein sulfhydryl content and effects on bromosulphalein (BSP) clearance in rats, *Toxicology 6*:1 (1976).
25. McKeehan, W. L., Sakagami, Y., Hoshi, H., and McKeehan, K. A., Two apparent human endothelial cell growth factors from human hepatoma cells are tumor-associated proteinase inhibitors, *J. Biol. Chem. 261*:5378 (1986).
26. Harrison, E. H., Lane, J. S., Luking, S., Van Rafelghem, M. J., and Andersen, M. E., Perfluoro-*n*-decanoic acid: Induction of peroxisomal beta-oxidation by a fatty acid with dioxin-like toxicity, *Lipids 23*:115 (1988).
27. Jones, C. E., Ballinger, M. B., Mattie, D. R., DelRaso, N. J., Sekel, C., and Vinegar, A., Effects of short-term oral dosing of polychlorotrifluoroethylene (polyCTFE) on the rhesus monkey, *J. Appl. Toxicol. 11*:51 (1991).
28. Kinkead, E. R., Bunger, S. K., Wolfe, R. E., Flemming, C. D., Whitmire, R. E., and Wall, H. G., Repeated-dose gavage studies on polychlorotrifluoroethylene acids, *Toxicol. Ind. Health 7*:295 (1991).

# 9

# Tissue Slices as an In Vitro Model for Studying Heart, Liver, and Kidney Toxicity

**Paul M. Silber, Tami M. Greenwalt, and Charles E. Ruegg**

*In Vitro Technologies, Inc.*
*Baltimore, Maryland*

Tissue slices have been used as an alternative in vitro method to primary cell culture. Historically, slices were prepared with the Stadie-Riggs microtome or by hand, resulting in wide variations in thickness. These thickness variations cause thicker areas of the slice to become anoxic and necrotic following short periods (a few hours) of incubation; this problem confounded experimental results and limited the usefulness of slices as an in vitro model [1,2]. The development of the Krumdieck tissue slicer [3] allowed the production of precision-cut tissue slices of uniform dimensions. Such slices are rapidly prepared under physiological conditions, are uniform in thickness, and do not develop local regions of anoxic injury. Normal physiological processes are preserved in liver slices, as measured by protein synthesis, intracellular potassium content, and maintenance of ATP and cytochrome P-450 levels [4,5]. This slicing technique has also been used to evaluate chemical toxicity in vitro in kidney [6,7], lung [8], and heart tissues [9].

The methodology of precision-cut slicing can be easily adapted for preparing slices from the soft tissues of most species for in vitro investigations of chemical metabolism and toxicity. The specific goal of the study reported in this chapter was to determine if slices prepared from different tissues respond differently in vitro to classic target organ-specific toxicants.

## MATERIALS AND METHODS

Sprague–Dawley rats (Harlan) and New Zealand White rabbits (Hazleton) were used as a source of fresh tissue. Mercuric chloride, potassium dichromate, nickel chloride, cadmium chloride, zinc chloride, allyl alcohol, *cis*-platinum(II) diammine dichloride (displaton), *trans*-platinum(II) diammine dichloride, 3,4,5(dimethylthiazol-2-yl)-2,5-diphenyltetrazolium (MTT), and tissue culture media were obtained from Sigma Chemical Company. Methanol was obtained from Mallinckrodt, ethanol from the Warner Graham Company, magnesium chloride from Aldrich Chemical Company, and doxorubicin was purchased from Adria Laboratories. Serial dilutions of each of these test materials were made using tissue culture medium as a diluent. Precision-cut tissue slices were made using a Krumdieck tissue slicer (K & F Research).

Animals were anesthetized with pentobarbital and their tissue rapidly harvested using sterile technique. Heart, liver, and kidney were obtained from either rats or rabbits. Tissue and chemical comparisons were always made between tissues derived from the same species. Tissue cores and slices from each of these three organs were prepared by methods previously described [2,6,9]. Replicate slices were incubated in tissue culture plates in a 37°C, $CO_2$ incubator in tissue culture medium in the presence of serial dilutions of the various chemicals for 22 hr. At the end of this time, slices were harvested for MTT and histopathological analyses.

Chemically induced injury to slices was determined by measurement of MTT conversion as a marker of mitochondrial integrity according to the method of Mosmann [10]. Spectrophotometric analyses of converted MTT was performed with a UVMax microtiter plate spectrophotometer at an absorbance of 570 nm (Molecular Devices). For histopathological analyses, slices were fixed, embedded in paraffin, sectioned with a microtome and then stained with hematoxylin and eosin.

## RESULTS

The first phase of this study involved determining if the tissue slice models were suited for predicting the rank-order of toxicity of chemicals with well-documented toxicity profiles in vivo. *cis*- and *trans*-platinum were selected for study, since the nephrotoxic potency of the *cis*-isomer of is far more injurious than the *trans*-isomer to the renal proximal tubule in vivo [11]. The results of this study using rabbit kidney slices dosed with serial dilutions of these chemicals (Fig. 1) demonstrate the same rank-order of potency in vitro as observed in vivo for these stereoisomers.

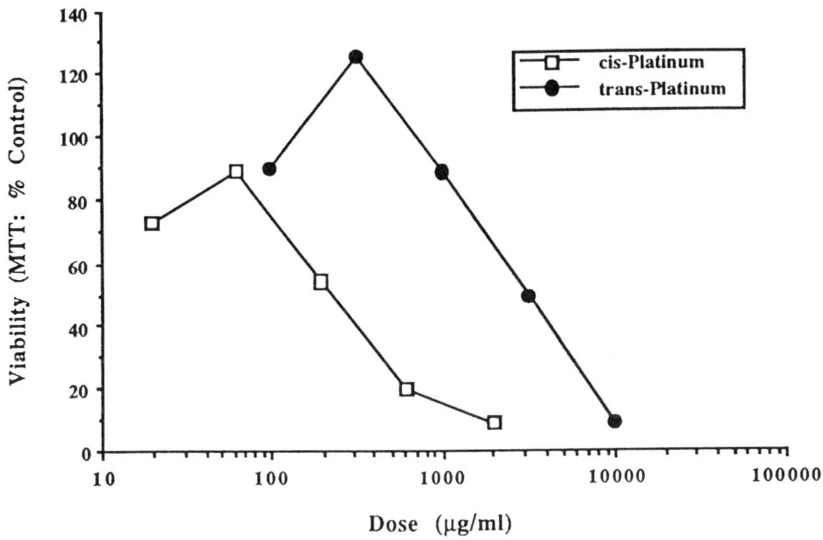


**Figure 1** Kidney slice viability: *cis*-platinum vs. *trans*-platinum.

A second experiment was conducted to determine if the tissue slice models were suited for predicting the rank-order of toxicity of another series of structurally related chemicals with well-documented toxicity profiles in vivo. In this study the potency of three alcohols with significantly different in vivo hepatotoxic potencies were selected to evaluate the susceptibility of the in vitro liver slice model to alcohol-induced toxicity. The alcohols selected for this experiment included ethanol, methanol, and allyl alcohol. Of these three compounds allyl alcohol is the most hepatotoxic in vivo by virtue of its hepatic alcohol dehydrogenase-mediated conversion to acrolein [12]. The results of this study with rabbit liver slices dosed with serial dilutions of these alcohols show that allyl alcohol is far more hepatotoxic than either methanol or ethanol in vitro, similar to the rank-order toxicity observed in vivo (Fig. 2).

The rank-order of toxicity of a series of metals was evaluated to determine if the slice model could predict the toxicity of this group of related chemicals. In this experiment the potency of six different metal salts with significantly different in vivo hepatotoxic potencies were evaluated in the in vitro liver slice model. Three of the metals selected for this experiment are considered potent toxicants, including cadmium chloride, mercuric chloride, and potassium dichromate. Zinc chloride, nickel chloride, and magnesium

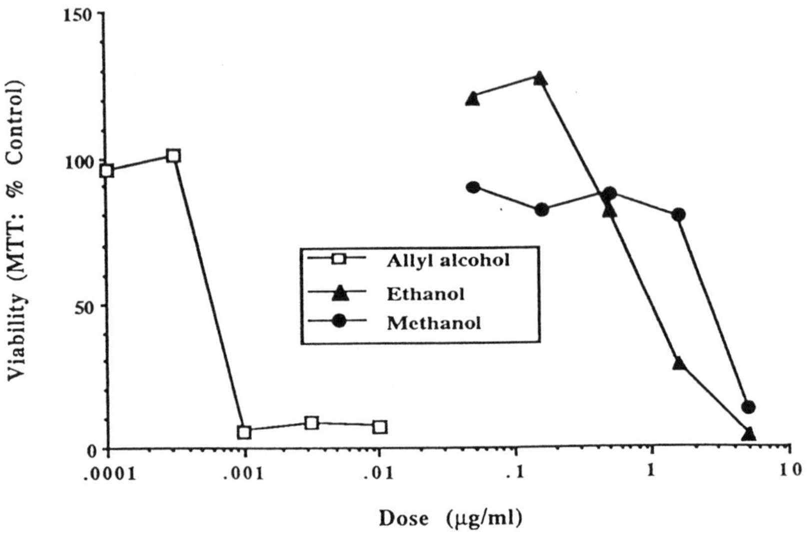


**Figure 2** Liver slice viability: alcohols.

sulfate were selected for comparison as metals with considerably less in vivo hepatotoxicity potential [13].

The results of this study in rabbit liver slices dosed with serial dilutions of these metals are presented in Figure 3. Generally, the rank-order of potency for these metals in vitro correspond qualitatively to the potency of these chemicals observed in vivo, with the possible exception of zinc chloride (see Fig. 3). This discrepancy may be due to the poor absorption of this metal salt when administered orally in vivo. Thus, the enhanced toxicity of zinc chloride in the slice system is likely due to a higher concentration of this metal in contact with the target organ (liver) in vitro than is present in vivo. This hypothesis is supported by data comparing the acute toxicity of metals administered by different routes (Table 1). Note that the toxicity of mercury is not nearly so route-dependent as is zinc, suggesting that the systemic availability of zinc is very limited when administered orally.

Although mercury and chromium follow a similar dose–response curve of toxicity in this slice system, histopathological evaluations show that these metals can target different cell types in the kidney [14]. Many chemicals have been demonstrated to target specific regions within tissue slices owing to innate differences in cellular susceptibility [4,7].

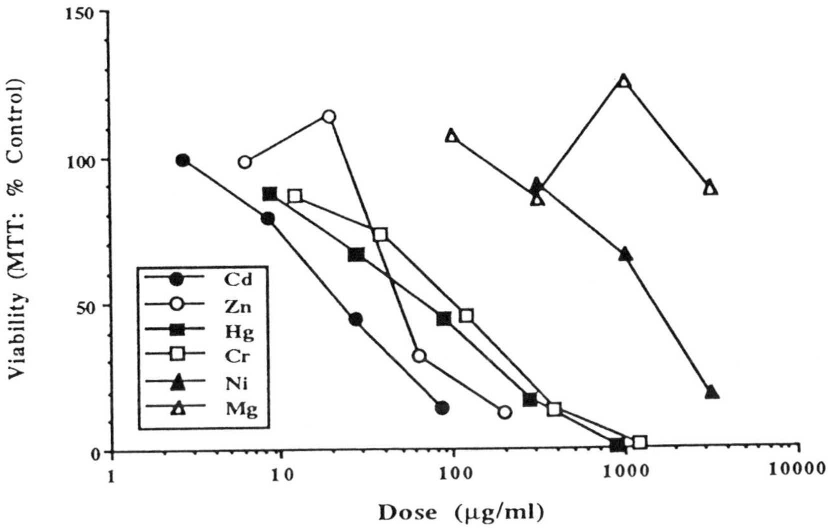


**Figure 3** Liver slice metal susceptibility.

A fundamental question in this study was to determine whether the tissue slice models were suited for evaluating and differentiating among tissue-specific toxicants. Allyl alcohol was selected as a model hepatotoxicant. This compound is thought to selectively target the liver, since it is converted to acrolein by the high activity of hepatic alcohol dehydrogenase (ADH) [12]. By way of contrast the toxicity of allyl alcohol to kidney slices was also studied; renal ADH activity is approximately one-tenth of that found in liver [15], so that the conversion of allyl alcohol to the toxic intermediate acrolein is likely to be proportionately lower in the kidney. This experiment shows that the rabbit liver is approximately one order of magnitude more sensitive to allyl alcohol

**Table 1** Mouse Acute Toxicity Data in Vivo

| Compound | Oral $LD_{50}$ (mg/kg) | Intraperitoneal $LD_{50}$ (mg/kg) |
|---|---|---|
| Zinc chloride | 350 | 31 |
| Cadmium chloride | 60 | 9 |
| Mercuric chloride | 6 | 6 |

*Source*: Data from Sigma Chemical Company.

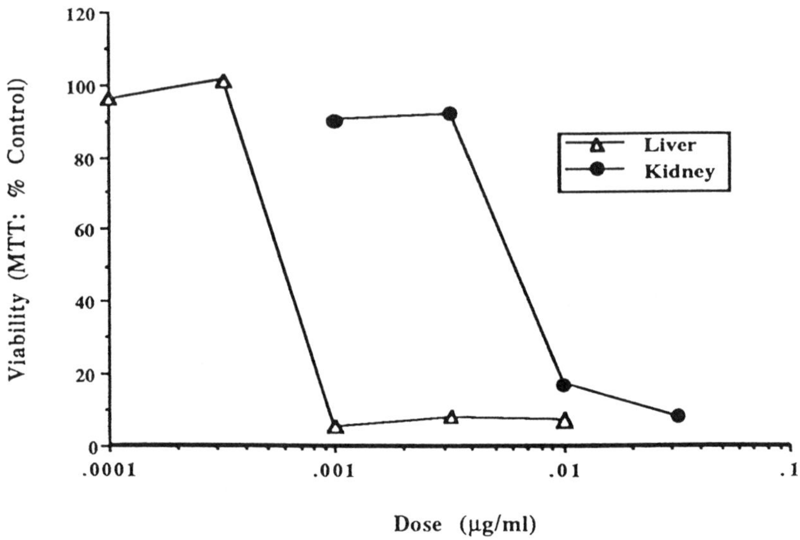


**Figure 4** Target organ susceptibility: allyl alcohol.

than is rabbit kidney (Fig. 4), modeling in vitro the same qualitative target organ susceptibility that is seen in vivo.

A separate experiment was conducted to determine if the tissue slice models were suited for evaluating and differentiating among tissue-specific toxicants, this time using doxorubicin as a model cardiotoxicant. This anthracycline chemotherapeutic agent has been demonstrated to cause a severe and irreversible cardiomyopathy, by poorly understood mechanisms [16]. By way of contrast, the toxicity of doxorubicin to kidney slices was also studied in vitro, since doxorubicin is not significantly nephrotoxic in vivo at therapeutic levels. The results of this experiment (Fig. 5) indicate that the rat heart slices in vitro are far more sensitive to doxorubicin than are rat kidney slices, modeling in vitro the same qualitative target organ susceptibility that is seen in vivo.

## DISCUSSION

The results of these studies indicate that slices from different tissues discriminate among tissue-selective toxicants. Specifically, the cardiotoxicant doxorubicin was far more potent in heart tissues than in kidney, whereas the hepatotoxicant allyl alcohol was most toxic to liver. This model also demonstrates usefulness for rank-ordering the toxicity of chemically related classes of

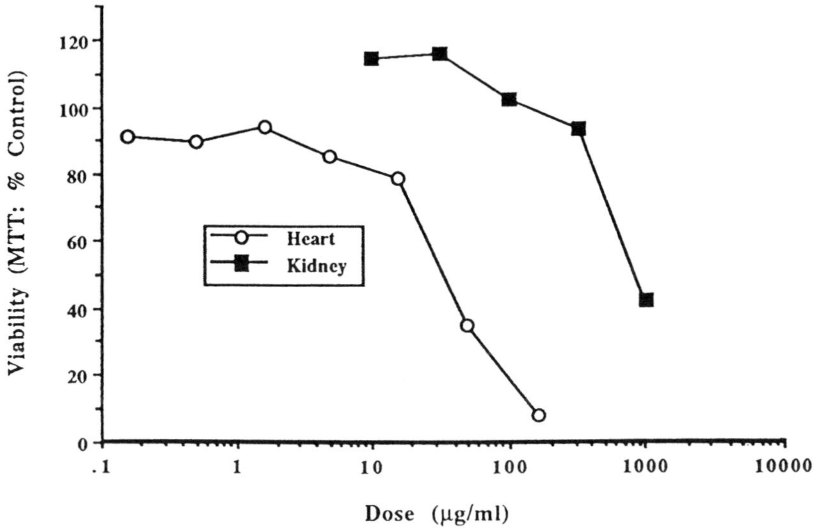


**Figure 5** Target organ susceptibility: doxorubicin.

compounds, such as metals and alcohols. For example, the order of toxicity of alcohols in liver was allyl alcohol > methanol = ethanol, whereas in kidney slices *cis*-platinum was more toxic than its isomer *trans*-platinum, and the toxicity of metals in liver rank-order as Cd > Zn > Hg ≥ Cr ≫ Ni ≫ Mg. This study demonstrates that an in vitro tissue slice battery can provide a useful tool for rapidly identifying target organ susceptibility, chemical potency, and (through the evaluation of multiple endpoints) the mechanism of action of a wide variety of chemicals.

## CONCLUSION

The usefulness of the precision tissue slice model is supported by the results of the studies described in this chapter and by previously published studies in which this model has been used to answer related questions. Sipes et al. [5] described the utility of this model for studying the toxicity of three different stereoisomers of dichlorobenzene and showed that the liver slice model accurately predicted the same rank-order of toxicity for these compounds in vitro as is observed in vivo. In a recent report, Wishnies et al. [17] described the utility of a liver–kidney slice coincubation experiment to model the interactions between these two organs in producing injurious metabolites from phenacetin in vitro. Finally, the ability to use the slice model as a tool for evaluating cell-specific injury by combining histopathological and biochemical analyses as described by Ruegg [7] serves as yet another example of some of the unique characteristics of this in vitro system.

## ACKNOWLEDGMENT

The authors wish to express their appreciation to Charles Zheng, who played a critical role in assisting with many of the studies described in this chapter.

## REFERENCES

1. Campbell, A. K., and Hales, C. N., *Exp. Cell Res. 68*:33–42 (1971).
2. Smith, P. F., Gandolfi, A. J., Krumdieck, C. L., Putnam, C. W., Zukoski, C. F., Davis, W. M., and Brendel, K., *Life Sci. 36*:1367–1375 (1985).
3. Krumdieck, C. L., Dos Santos, J. E., and Ho, K., *Anal. Biochem. 104*:118–123 (1983).
4. Smith, P. F., Fisher, R. L., Shubat, P. J., Gandolfi, A. J., Krumdieck, C. L., and Brendel, K., *Toxicol. Appl. Pharmacol. 87*:509–522 (1987).
5. Sipes, I. G., Fisher, R. L., Smith, P. F., Stine, E. R., Gandolfi, A. J., and Brendel, K., *Arch. Toxicol. 11*(Suppl.):20–33 (1987).
6. Ruegg, C. E., Gandolfi, A. J., Nagle, R. B., Krumdieck, C. L., and Brendel, K., *J. Pharmacol. Methods 17*:111–123 (1987).
7. Ruegg, C. E., Wolfgang, G. H. I., Gandolfi, A. J., Brendel, K., and Krumdieck, C. L., In *In Vitro Toxicology* (C. A. McQueen, ed.), Telford Press, NJ, pp. 197–230, 1989.
8. Stefaniak, M. S., Gandolfi, A. J., and Brendel, K., *Proc. West. Pharmacol. Soc. 31*:149–151 (1988).
9. Parrish, A. R., Shipp, N. G., Dorr, R. T., Gandolfi, A. J., and Brendel, K., *Toxicologist 11*:337 (1991).
10. Mosmann, T., *J. Immunol. Methods 65*:55–63 (1983).
11. Calabresi, P., and Parks, R. E., Jr., In *The Pharmacological Basis of Therapeutics*, 7th ed. (A. G. Gilman, L. S. Goodman, T. W. Rall, and F. Murad, eds.), Macmillan, New York, 1985, p. 1290.
12. Reid, W.D., *Experientia 28*:1058–1061 (1972).
13. Hammond, P. B., and Beliles, R.P., In *Toxicology: The Basic Science of Poisons*. 2nd ed. (J. Doull, C. D. Klaassen, and M. O. Amdur, eds.), Macmillan, New York, 1980, pp. 409–467.
14. Ruegg, C. E., Gandolfi, A. J., Nagle, R. B., and Brendel, K., *Toxicol. Appl. Pharmacol. 90*:261–273 (1987).
15. Ohno, Y., Jones, T. W., and Ormstad, K., *Chem. Biol. Interact. 52*:289–299 (1985).
16. Minow, R. A., Benjamin, R. S., Lee, E. T., and Gottlieb, J. A., *Cancer 39*:1397–1402 (1977).
17. Wishnies, S. M., Gandolfi, A. J., and Brendel, K., *Toxicologist 12*:93 (1992).

# 10

# Potential for Interspecies Extrapolation of Macrophage Chemiluminescence Data from Immunotoxicology Studies

**Robert S. Anderson and Laurie M. Mora**

*University of Maryland*
*Solomons, Maryland*

**Sandra A. Thomson**

*U.S. Army Edgewood Research, Development, and Engineering Center*
*Aberdeen Proving Ground, Maryland*

Virtually all animals yet studied contain macrophages or macrophagelike cells in one or more of their body compartments. These cells are typically phagocytic and serve to remove and sequester microbes and other nonself materials from the blood or other host tissues. In this chapter, the phagocytes that exist free in the blood, hemolymph, and body cavity fluids of mammals and invertebrates will be compared in form and function. The aim is to describe the numerous similarities between these cells from phyletically diverse species for the mechanisms by which they establish resistance to infectious disease. In keeping with the theme of current concepts and approaches on animal test alternatives, two hypotheses are proposed: Invertebrate macrophages emulate the chemically induced immunomodulatory behavior characteristic of mammalian macrophages, and invertebrate macrophages lend themselves well to in vitro immunotoxicity testing protocols because of their quantifiable activities in short-term, primary culture.

Phagocytic blood cells, macrophages, and other cells involved in antimicrobial activities generally employ intracellular hydrolytic enzymes and reactive oxygen intermediates (ROIs) to mediate cytotoxicity. Clearly, interference with these mechanisms resulting from exposure to chemical stressors could compromise an organism's resistance to infectious disease. Appropriate stimulation of cell surface receptors or ingestion of bacteria or other foreign material, can initiate increased metabolic activity (the respiratory burst) and the production of the superoxide anion ($O_2^-$) [1–4]. Superoxide is the precursor of a series of other ROIs such as hydrogen peroxide ($H_2O_2$), hydroxyl radical, and singlet oxygen. Both $O_2^-$ and $H_2O_2$ have direct cytotoxic activities [5], and $H_2O_2$ plus myeloperoxide and chloride ions can form one of the most effective antibacterial systems of leukocytes [6]. In this study, phagocytically stimulated cellular chemiluminescence, augmented by the presence of 5-amino-2,3-dihydro-1,4-phthalazinedione (luminol), was quantified as an indication of the production of ROIs. Oxidation of luminol by ROI produced during the respiratory burst results in the generation of excited aminophthalate anions that release photons during relaxation to the ground state [7]. This chemiluminescence (CL) was accurately quantified in a liquid scintillation counter adapted for single-photon monitoring.

Immunotoxicological studies with higher animals indicate that certain chemicals possess specific immunomodulatory activity toward macrophages [8–11]. Identification of these chemicals is a priority because changes in macrophage responsiveness could have profound effects on the health of the host. Macrophage CL has been suggested as a tier-one screening assay for identifying immunomodulatory chemicals, using cells from laboratory rodents [12]. Bivalve mollusks have large numbers of circulating macrophagelike cells that also actively produce luminol-augmented CL [13,14]. Investigations on chemically induced immunosuppression in hemocytes of the American oyster (*Crassostrea virginica*) are being conducted in developing animal test alternatives and in evaluating the effect of aquatic pollutants on immunocompetency of estuarine species.

## METHODS

Hemolymph samples were withdrawn from the adductor muscle of the oyster and added to plastic culture dishes. Most hemocytes adhered tightly to the plates, and 30 min later the supernatant (serum and nonadherent hemocytes) was replaced by oyster cell support medium (CSM): 5% fetal calf serum, 0.5% antibiotic–antimycotic solution, and 1 mg/ml glucose in filter-sterilized ambient estuarine water. After 2-hr incubation, the cells began to lose adherence and could be gently washed off the substrate with CSM. The cells were washed, counted, and resuspended in CSM. Scintillation pony vials were prepared; each vial contained $2 \times 10^6$ hemocytes in CSM, which were exposed to toxicant (here particulate brass dust) for 1 or 20 hr in the dark. All following steps were carried out under dim red illumination to reduce background, cell-unrelated, CL. After incubation and exposure, luminol (100 μM final concentration) was added, and the CL activity of the resting cells recorded for 5–10 min. Then heat-killed yeast cells (20 yeast cells for each hemocyte) were added to provide phagocytic stimulation, and the resultant induced CL continuously recorded for 60 min.The CL determinations were carried out in a scintillation counter modified for single-photon counting. Aliquots of all hemocyte samples were tested for viability by trypan blue exclusion immediately before CL analysis.

## RESULTS AND DISCUSSION

Representative results of dose-dependent inhibition of hemocyte CL by exposure to brass dust are represented in Figures 1 and 2. In these figures, aliquots of a single hemocyte pool were exposed to the agent for 1 or 20 hr; mean cell viability ranged from ~80 to 90% for 1-hr samples, and from ~75 to 90% for the 20-hr samples. However, the CL responses of these cells were markedly diminished by brass exposure, as shown in the figures.

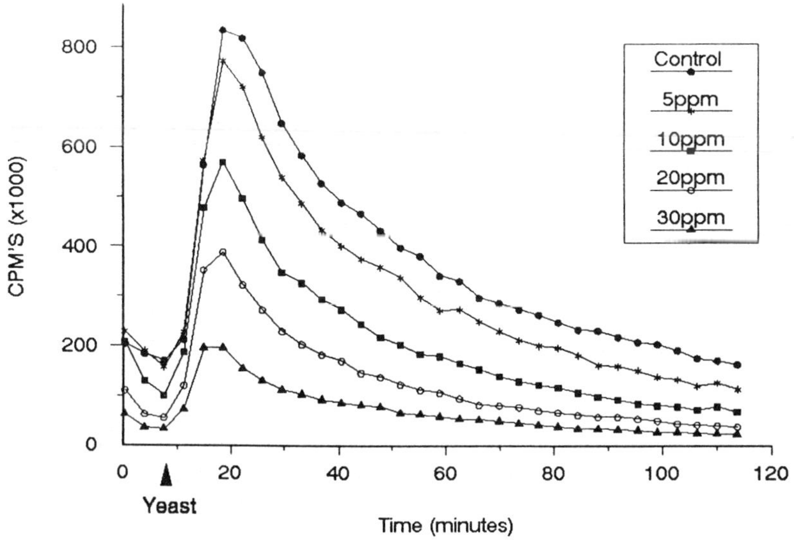


**Figure 1** Effect of 1-hr exposure to particulate brass on oyster hemocyte CL.

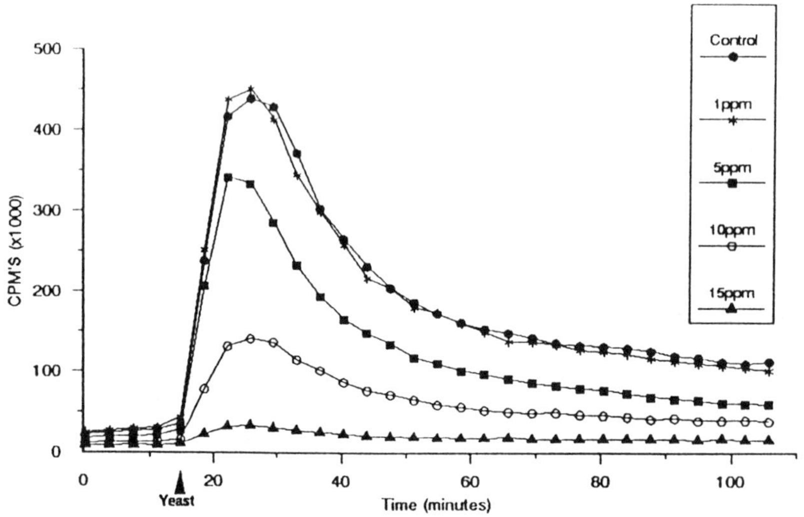


**Figure 2** Effect of 20-hr exposure to particulate brass on oyster hemocyte CL.

The figures show representative data from aliquots of a single cell pool; Table 1 presents cumulative data derived from five to six cell pools for each exposure time. The dose-dependent nature of brass dust inhibition of both resting and stimulated CL parameters is obvious, as is the increased intensity of the response when measured at 20 hr, compared with 1 hr. The data were used to estimate the brass concentration required for 50% inhibition of total CL activity ($EC_{50}$) and the concentration required to kill 50% of the cells ($LD_{50}$) were determined: 1-hr $EC_{50}$ = 27 μg/ml; 1-hr $LD_{50}$ = 273 μg/ml; 20-hr $EC_{50}$ = 9 μg/ml; 20-hr $LD_{50}$ = 67 μg/ml. Clearly, these are sublethal effects that can be measured with sensitivity.

The brass dust used in this study was composed of 73% copper, 26.5% zinc, and 0.5% aluminum. The hemocytes were exposed to the dust by particles and metal ions (or complexes) in the medium, as well as by endocytosis and subsequent breakdown of the particles. Preliminary work showed that brass leachates (400 ppm in estuarine water, all particles removed after 1-wk extraction) also could cause CL inhibition in oyster hemocytes. The leachates contained copper and zinc concentrations in excess of environmentally relevant levels. Copper ions, from copper sulfate, also produced CL inhibition after incubation with hemocytes.

**Table 1** Precentage Inhibition of Hemocyte CL by Exposure to Particulate Brass

| | 1-hr exposure | | | | 20-hr exposure | | | |
|---|---|---|---|---|---|---|---|---|
| | 5 | 10 | 20 | 30 | 1 | 5 | 10 | 15 |
| | (μg/ml) | | | | (μg/ml) | | | |
| Background | 18 ± 21 | 45 ± 6 | 72 ± 6 | 88 ± 9 | 15 ± 24 | 53 ± 24 | 73 ± 13 | 83 ± 11 |
| Peak CL | 9 ± 10 | 27 ± 11 | 56 ± 12 | 73 ± 9 | 18 ± 15 | 49 ± 30 | 76 ± 18 | 91 ± 10 |
| Total CL | 11 ± 8 | 42 ± 4 | 66 ± 5 | 76 ± 8 | 14 ± 9 | 51 ± 25 | 76 ± 17 | 91 ± 9 |

## CONCLUSIONS

Exposure to brass dust, or aqueous leachates of the dust, will mediate a dose-dependent inhibition of oyster hemocyte chemiluminescence (CL). Copper, the major constituent of this agent, was directly involved in this suppression. Since CL is correlated with cellular antimicrobial activity and hemocytes are the oyster's main line of defense against infectious diseases, metal-induced CL inhibition was interpreted as a measure of immunosuppression. Thus, it appears the CL is a useful method to identify potential environmental immunomodulators in an important aquatic organism. Similar CL findings are reported with equivalent mammalian cells exposed to sublethal concentrations of metals [12]. Therefore, it is proposed that hemocyte CL studies can provide not only information on immunotoxicants for bivalves, but also the results have a good potential for interspecies extrapolation.

## REFERENCES

1. Badwey, J. A., and Karnovsky, M. L., Active oxygen species and the function of phagocytic leukocytes, *Annu. Rev. Biochem. 49*:695–726 (1980).
2. DeChatelet, L. R., Long, G. D., Shirley, P. S., Bass, D.A., Thomas, M. J., Henderson, F. W., and Cohen, M. S., Mechanisms of the luminol-dependent chemiluminescence of human neutrophils, *J. Immunol. 129*:1589–1593 (1982).
3. Klebanoff, S. J., Oxygen-dependent cytotoxic mechanisms of phagocytes, *Advances in Host Defense Mechanisms*, Vol. 1 (J. I. Gallin and A. S. Fauci, eds.), Raven Press, New York, 1982, pp. 111–162.
4. Babior, B. M., Oxidants from phagocytes: Agents of defense and destruction, *Blood 64*:959–966 (1984).
5. Fridovich, I., The biology of oxygen radicals: General concepts. *Oxygen Radicals and Tissue Injury* (B. Halliwell, ed.), Federation of American Societies for Experimental Biology, Bethesda, MD, 1988, pp. 105–110.
6. Klebanoff, S. J., Myeloperoxidase–halide–hydrogen peroxide antibacterial system, *J. Bacteriol. 95*:2131–2138 (1968).
7. Allen, R. C., and Loose, L. D., Phagocytic activation of a luminol dependent chemiluminescence in rabbit alveolar and peritoneal macrophages, *Biochem. Biophys. Res. Commun. 69*:245 (1976).
8. Kutz, S. A., Hinsdill, R. D., and Weltman, D. J., Evaluation of chemicals for immunomodulatory effects using an in vitro antibody-producing assay, *Environ. Res. 22*:368–376 (1980).
9. Tam, P. E., and Hinsdill, R. D., Evaluation of immunomodulatory chemicals: Alteration of macrophage function in vitro, *Toxicol. Appl. Pharmacol. 76*:183–194 (1984).
10. Hilbertz, U., Kramer, U., De Ruiter, N., and Baginski, B., Effects of cadmium

and lead on oxidative metabolism and phagocytosis by mouse peritoneal macrophages, *Toxicol. 39*:47–57 (1986).
11. Snoeij, N. J., Penninks, A. H., and Seinen, W., Biological activity of organot in compounds—an overview, *Environ. Res. 44*:335–353 (1987).
12. Tam, P. E., and Hinsdill, R. D., Screening for immunomodulators: Effects of xenobiotics on macrophage chemiluminescence in vitro, *Fundam. Appl. Toxicol. 14*:542–553 (1990).
13. Bachère, E., Hervio, D., and Mialhe, E., Luminol-dependent chemiluminescence by hemocytes of two marine bivalves, *Ostrea edulis* and *Crassostrea gigas*, *Dis. Aquat. Org. 11*:173–180 (1991).
14. Anderson, R. S., Oliver, L. M., and Jacobs, D., Immunotoxicity of cadmium for the eastern oyster (*Crassostrea virginica*[Gmelin, 1791]): Effects on hemocyte chemiluminescence. *J. Shellfish Res. 11*:29–33 (1992).

# 11

# An In Vitro System for the Evaluation of Cyanide Ion Binding By Potential Candidate Antidotes

**Lemuel T. Russell and Jurgen D. von Bredow***

*Walter Reed Army Institute of Research, Washington, D.C.*

**James A. Vick**

*U. S. Food and Drug Administration*
*Washington, D.C.*

Intoxication by cyanide continues to be a military and civilian problem that requires rapid therapy with effective antidotes. The testing of potential antidotes is limited to in vivo systems, which are expensive and require the use of numerous animals to reach a decision concerning the potential effectiveness of an antidote. Since many compounds tested as possible antidotes ultimately prove to be inactive, the potential exists for the unproductive loss of experimental animals.

A sensitive and rapid in vitro system has been developed to monitor the rate and extent of binding of cyanide ions with potential antidotal compounds. In this system, known concentrations of potassium cyanide ($10^{-5}$–$10^{-3}$ M) in Tris-buffered saline, plasma, or whole blood are treated with candidate cyanide-binding agents.The rate of depletion of the cyanide ion is monitored by the specific ion electrode system, which has been described by Lebeda and Deshpande [1]. Candidate antidotes can be evaluated in whole blood, plasma, or Tris-buffered saline for the ability to interact with cyanide compared with the rate of binding to $Co^{2+}$, which has demonstrated anticyanide activity.

---

**Present affiliation*: U.S. Food and Drug Administration, Beltsville, Maryland

The in vitro technique is able to determine if the action of a potential compound is (a) through a direct interaction with cyanide, (b) through an indirect interaction with cyanide, or (c) if the compound is capable of employing both mechanisms. These in vitro evaluations make possible the preliminary estimate of potential anticyanide activity without the use of traditional animal models.

## MATERIALS AND METHODS

Potassium cyanide (Fisher Scientific) is added to Tris (100 μM)-buffered saline, horse plasma, or heparinized dog blood to develop cyanide concentrations of $10^{-5}$–$10^{-3}$ M.

Cyanide ion concentration is determined with a cyanide-specific ion electrode (cyanide combination ion Selective Electrode FK1502CN, combined with an Ion 85 analyzer; Radiometer, Copenhagen). Potential differences are displayed on a stripchart recorder (Kipp and Zonen) to monitor alterations in cyanide ion concentration.

Candidate cyanide antidotes are added to the constantly stirring cyanide solution to ensure rapid and complete interaction. Temperature is maintained at 37°C in a temperature bath monitored with a thermocouple (Physitemp) temperature probe. The pH of saline is maintained at 7.2–7.6 through the addition of 100 mM Tris buffer (tris[hydroxymethyl]aminomethane; Sigma Chemical), and the pH of the plasma and whole blood are carefully monitored throughout the addition of the unknown antidote to the in vitro system.

Formation of methemoglobin by antidotes was assessed in heparinized dog whole blood with a OSM-3 Hemoximeter (Radiometer, Copenhagen).

Test antidote reagents consist of

Cobalt chloride,
Sodium nitrite,
Hydroxylamine HCl,
Dimethylaminophenol,
*o*-Ethylquinolinium chloride,
Ethyl 4-(4-methoxyphenyl)-2,4-dioxobutyrate,
3,3′-Tetrathiobis-*N*-acetyl-L-alanine.

## RESULTS

The rate of the direct interaction of cobalt chloride with cyanide can be calculated from a plot of the disappearance of cyanide ion versus log time in seconds. The plot revealed two distinct rates of interaction:

Initial rate $= 1.344 \times 10^{-4}$ mol/sec

Second rate $= 6.413 \times 10^{-8}$ mol/sec

The rates may be expressed in a more useful manner that states that 25% of the total cyanide is bound in 1.0 sec, and 73% of the total cyanide is bound in 10.0 sec.

Of all of the compounds tested only cobalt chloride demonstrated a direct interaction with the cyanide ion. The following compounds do not interact directly with cyanide ion in solution:

Sodium nitrite
*o*-Ethylquinolinium Cl
Hydroxylamine HCl,
Dimethylaminophenol
3,3′-Tetrathiobis-*N*-acetyl-L-alanine
Ethyl 4-(4-methoxyphenyl)-2,4-dioxobutyrate

The following compounds form methemoglobin and interact indirectly with cyanide ion in solution:

Sodium nitrite
Hydroxylamine HCl
Dimethylaminophenol

In heparinized dog whole blood the rate of interaction of cyanide and methemoglobin is similar to the rate of interaction of cyanide and cobalt ion.

## DISCUSSION

The specific ion electrode system is limited in sensitivity to the range of $10^{-5}$–$10^{-3}$ M in buffered saline, plasma, and heparinized whole blood. Therefore, the range of cyanide ion concentrations and the concentration of antidotes must be maintained within an optimal range. Control of the pH with Tris buffer in saline is important,since the cyanide will volatilize in an acid medium, leading to the disappearance of cyanide at a rate similar to the cyanide-binding patterns. This phenomenon can be demonstrated with the well-known cyanide antidotes hydroxylamine hydrochloride and dimethylaminophenophenol in unbuffered saline. Both of these compounds rapidly reduce the pH of the test solution, leading to a disappearance of cyanide and an apparent interaction directly with cyanide. The same study carried out in Tris-buffered saline demonstrates no apparent interaction between these compounds and cyanide. The loss of cyanide is less apparent in plasma or whole blood samples, which moderate the reduction in pH caused by these compounds.

This is an aqueous system, with a limited capability to interact with water-insoluble compounds. This limitation may be circumvented, to some extent, by adding up to 50% methanol to the buffered saline. The cobalt chloride standard will interact with the cyanide ion in a manner similar to the complete aqueous system. Candidate antidotes (3,3′-tetrathiobis-*N*-acetyl-L-alanine and ethyl 4-(4-methoxyphenyl)-2,4-dioxobutyrate) were dissolved in 10% methanol and added to the cyanide solution.

This in vitro system responds directly to the added compound or to the indirect effect of the compound through methemoglobin formation. The potent methemoglobin formers, hydroxylamine hydrochloride, dimethylaminophenophenol and, to a lesser extent, sodium nitrite, demonstrate significant activity in this in vitro model. The system does not form metabolites and candidate antidotes, such as the *para*-aminopropriophenones, which must be metabolized in vivo to the active methemoglobin-forming metabolite, would not be expected to demonstrate activity in an in vitro whole blood assay system.

The electrode system may become less sensitive to compounds dissolved in blood or plasma [2–4], nevertheless, this electrode system responds to a change in cyanide ion concentration induced by complexation with potential antidotes, which is sufficient to determine the active nature of a potential antidote.

## CONCLUSION

A rapid and sensitive in vitro assay system has been developed that is capable of defining direct and indirect interactions of potential antidotes with cyanide ion in saline, plasma, or whole blood. This in vitro assay system may provide a useful preliminary evaluation of candidate compounds to prioritize and limit the total number of animal tests required to select successful antidotes.

## REFERENCES

1. Lebeda, F. J., and Deshpande, S. S., Potentiometric measurements of hydrogen and cyanide ions in buffered media, *Anal. Biochem. 187*:302–309 (1990).
2. McAnalley, B. H., Lowry, W. T., Oliver, R. D., and Garriott, J. C., Determination of inorganic sulfide and cyanide in blood using specific ion electrodes: Application to the investigation of hydrogen sulfide and cyanide poisoning, *J. Anal. Toxicol. 3*:111–114 (1979).
3. Egekeze, O. J., and Oehme, F. W., Direct potentiometric method for the determination of cyanide in biological materials, *J. Anal. Toxicol. 3*:119–124 (1979).
4. Kistner, J. R., Longnecker, D. E., Miller, E. D., and Lescanic, A. D., Limitations of the cyanide electrode for use in plasma and whole blood, *Anesth. Analg. 58*:457–460 (1979).

# 12

# Munitions Cytotoxicities In Vitro

**W. R. Mitchell, L. M. Dasko-Vincent, and D. R. Wellington**

*U.S. Army Biomedical Research and Development Laboratory*
*Fort Detrick, Frederick, Maryland*

Rapid, inexpensive, and reliable methods are needed for the toxicological screening of chemical substances with the potential to affect health and the environment. One such method, the neutral red (NR) cytotoxicity assay, is based on incorporation of the supravital dye neutral red into lysosomes of viable cells. The NR uptake assay can be used to detect cytotoxic or cytostatic effects of chemical substances capable of damaging cells, has been adapted to microtiter tissue culture systems, and can be analyzed by means of automated spectrophotometric microplate readers [1,2]. Neutral red cytotoxicity assays in continuous rat hepatoma H4IIE cells have been applied to a variety of environmentally important munitions and related compounds. The H4IIE cells maintain inducible oxidative microsomal enzymes and were chosen because of their application to detecting other xenobiotics in environmental and biological specimens [6].

## MATERIALS AND METHODS

The H4IIE rat hepatoma cells were obtained from the American Type Culture Collection, Rockville, Maryland, and were grown at 37°C in a 5% $CO_2$ atmosphere. Medium was Eagle's minimal essential medium (MEM) with

Earle's salts supplemented with 10% fetal bovine serum, 10% newborn calf serum, 10 mM nonessential amino acids, and 50 μg/ml gentamicin. The NR uptake assays were conducted in the same medium, except that gentamicin was raised to 250 μg/ml to accommodate nonsterile munitions. Dimethyl sulfoxide (DMSO) was the solvent for all test chemicals and was maintained at 0.25% (v/v) in NR uptake assays.

## Neutral Red Cytotoxicity Assays [2]

*Day 1.* Monolayers were trypsinized, dispersed, and 96-well microtiter plates were seeded at $9 \times 10^{-3}$ cells per well in 0.2 ml of fresh medium.

*Day 2.* Medium was aspirated, and 0.2 ml of fresh medium, containing the desired test chemical concentration, was added to each of eight replicate wells for each toxicant concentration.

*Day 3.* Medium was aspirated from all wells exposed to chemicals and solvent controls and replaced with 0.2 ml of medium containing NR (0.005%, w/v) for 4 hr at 37°C. The NR-staining solution was removed and cells were washed 2 min with a fixative (1% $CaCl_2$, 1% formaldehyde). After aspiration of the fixative solution, the dye remaining in the cells was extracted with 0.2 ml 1% acetic acid, 10% butanol, and 50% ethanol for 1.25 hr. Microtiter plates were agitated 2 min to distribute the dye, and optical absorbance measurements were made by means of an automatic microtiter plate reader.

## Measurements

Values of NR uptake in multiple wells for each toxicant dilution were plotted as mean percentages ($\pm 1\sigma$) of controls exposed only to DMSO. Midpoint cytotoxicity values ($NR_{50}$) are defined as that amount of toxicant that would reduce NR uptake to 50% of its control value, based on regression curves of the uptake data in the region of change.

## Munitions Abbreviations

2,4-Dinitrotoluene (2,4-DNT); 2,6-dinitrotoluene (2,6-DNT); 1,3-dinitrobenzene (1,3-DNB); 2-amino-4-nitrotoluene (2A-4-NT); 4-amino-2-nitrotoluene (4A-2-NT); 2-amino-6-nitrotoluene (2A-6-NT); 1-amino-3-nitrobenzene (1A-3-NB); 3,5-dinitroaniline (3,5-DiNA); 2,4,6-trinitrotoluene (2,4,6-TNT); 1,3,5-trinitrobenzene (1,3,5,-TNB); 2-amino-4,6-dinitrotoluene (2A-4,6-DNT); 4-amino-2,6-dinitrotoluene (4A-2,6-DNT); Hexahydro-1,3,5-trinitro-1,3,5-triazine (RDX); octahydro-1,3,5,7-tetranitro-1,3,5,7-tetrazocine (HMX);

1-acetylhexahydro-3,5-dinitro-1,3,5-triazine (TAX); and 1-acetyloctahydro-3,5,7-trinitro-1,3,5,7-tetrazocine (SEX).

## RESULTS AND DISCUSSION

### 2,4,6-Trinitrotoluene and Corresponding Monoamines

From plots of NR uptake versus toxicant concentrations, the midpoint cytotoxicity value for 2,4,6-TNT was determined to be 0.033 mM. The corresponding amines, 2A-4,6-DNT and 4A-2,6-DNT, demonstrated cytotoxicities similar to each other ($NR_{50}$ = 0.449 and 0.373 mM), but were at least ten times less toxic than the parent nitroaromatic compound.

### 2,4- and 2,6-Dinitrotoluene and Their Monoamines

From similar plots, 2,4-DNT ($NR_{50}$ = 0.439 mM) was at least ten times less cytotoxic than 2,4,6-TNT in the assay. Midpoint cytotoxicities for 4A-2-NT and 2A-4-NT ($NR_{50}$ = 0.269 and 0.328 mM) were only slightly lower than that of the 2,4-DNT parent compound.Likewise, 2,6-DNT ($NR_{50}$ = 0.384 mM) was also significantly less toxic than 2,4,6-TNT in the assay. Midpoint cytotoxicity for the corresponding amine, 2A-6-NT ($NR_{50}$ = 0.502 mM), occurred at a concentration only slightly higher than that of the 2,6-DNT parent.

### 1,3-Dinitrobenzene, 1,3,5-Trinitrobenzene, and Their Monoamines

Plots of NR uptake demonstrated that the most potent compound studied was 1,3,5-TNB ($NR_{50}$ = 0.0099 mM). It was more than three times as cytotoxic as 2,4,6-TNT. A microbial metabolite of the compound, 3,5-DiNA ($NR_{50}$ = 0.132 mM) showed considerably reduced midpoint cytotoxicity. A second nitrated benzene compound found in munitions wastestreams, 1,3-DNB ($NR_{50}$ = 0.298) was far less toxic than 1,3,5-TNB in the assay. Its amine 1A-3-NB ($NR_{50}$ = 1.1 mM) demonstrated considerably reduced cytotoxicity when compared with the parent compound.

### Triazines and Tetrazocines

RDX and HMX and their acetylated derivatives, SEX and TAX, showed no significant cytotoxicity at any concentration after three separate assays for each compound. The highest concentrations applied to the system (100 mg/L) far exceeded the aqueous solubilities of the compounds.

## Comparative Cytotoxicities with Two Other Screening Systems

As is shown in Table 1 for TNT and related nitroaromatic compounds, relative potencies in NR uptake assays are in the order TNB > TNT > DNB ≥ DNTs. Except for 2,6-DNT,which was somewhat more toxic than 1,3-DNB and 2,4-DNT in water fleas, toxicity orders are in general agreement. Correlation coefficients between $NR_{50}$ values and $LC_{50}$ values were 0.83 for water fleas and 0.91 for fathead minnows. For the corresponding amines in the table, with the exception of 3,5-DiNA and 1A-3-NB, $NR_{50}$ values clustered from 0.27 to 0.5

**Table 1** $NR_{50}$ Values for Munitions and Related Compounds in H4IIE Cells and Their Acute Toxicities to Water Fleas and Fathead Minnows

| | H4IIE | | | | |
|---|---|---|---|---|---|
| Compounds | $NR_{50}$ (mM) | ($-1\sigma$ to $+1\sigma$) (mM) | *r* | Water flea 48-hr $LC_{50}$ (mM) | Fathead minnow 96-hr $LC_{50}$ (mM) |
| 2,4,6-TNT-associated nitroaromatics | | | | | |
| 1,3,5-TNB | 0.0099 | (0.0083–0.012) | 0.99 | 0.013 | 0.0052 |
| | | | | 0.014[a] | 0.0024[a] |
| 2,4,6-TNT | 0.033 | (0.023–0.044) | 0.99 | 0.052 | 0.013 |
| 1,3-DNB | 0.298 | (0.247–0.354) | 0.98 | 0.295 | 0.042 |
| | | | | 0.163[a] | 0.1[a] |
| 2,6-DNT | 0.384 | (0.316–0.453) | 0.91 | 0.12 | 0.102 |
| 2,4-DNT | 0.439 | (0.408–0.472) | 0.97 | 0.261 | 0.18 |
| 2,4,6-TNT-associated amino nitroaromatics | | | | | |
| 3,5-DiNA | 0.132 | (0.101–0.164) | 0.99 | 0.084 | 0.12 |
| | | | | 0.075[a] | 0.116[a] |
| 4A-2-NT | 0.269 | (0.183–0.381) | 0.98 | 0.09 | 0.16 |
| 2A-4-NT | 0.328 | (0.204–0.480) | 0.98 | 0.15 | 0.45 |
| 4A-2,6-DNT | 0.373 | (0.298–0.458) | 0.97 | 0.027 | 0.035 |
| 2A-4,6-DNT | 0.449 | (0.354–0.547) | 0.96 | 0.023 | 0.077 |
| 2A-6-NT | 0.502 | (0.381–0.640) | 0.96 | 0.093 | 0.33 |
| 1A-3-NB | 1.1 | (1.0–1.202) | 0.96 | | |
| Triazines and tetrazocines | | | | | |
| RDX | None | | | None[b] | 4–6 mg/L[c] |
| HMX | None | | | | None[c] |
| SEX | None | | | | |
| TAX | None | | | | |

*Source*: Data from Ref. 3, unless otherwise noted. [a]Ref. 7; [b]Ref. 4; [c]Ref. 5.

mM with overlapping ranges and direct correlation with $LC_{50}$ values is not seen. Most monoamines were less toxic than the parent nitroaromatics in NR uptake and were less toxic than the parent nitroaromatics in NR uptake and fathead minnow assays, whereas the opposite was true with water fleas. None of the systems correlates with acute rat oral toxicities, for which the order is DNB $\gg$ DNTs $>$ TNB $>$ TNT, although the monoamines 2A-4-NT, 1A-3-NB, 2A-4,6-DNT are also less toxic in rats than corresponding nitro compounds. No significant cytotoxicity was noted for triazine- and tetrazocine-associated munitions in the assay. Although data are sparse, there appears to be little toxicity associated with RDX for *Ceriodaphnia dubia* and other invertebrates, and with HMX for fish, up to the limits of their solubilities [5]. In contrast, acute toxicity has been reported for RDX in fathead minnows [4].

## CONCLUSIONS

All 2,4,6-TNT-associated munitions tested and their monoamines have responded measurably in NR uptake assays in H4IIE cells. The triazine and tetrazocine munitions, RDX and HMX, did not respond, nor did their acetylated congeners SEX and TAX. Responses to 2,4,6-TNT and associated nitroaromatic compounds were essentially consistent with two other systems used to screen munition samples, whereas the monoamines they form in the environment responded differently. Coupled to compatible extraction methods, NR uptake assays are potentially valuable in detecting 2,4,6-TNT and associated compounds or their mixtures in contaminated samples, whether or not they have been reduced in the environment. The same is not true of RDX- and HMX-associated munitions.

## ACKNOWLEDGMENT

The views, opinions, and/or findings contained in this report are those of the authors and should not be construed as an official Department of the Army position, policy, or decision, unless so designated by other official documentation.

## REFERENCES

1. Babich, H., and Borenfreund, E., Structure–activity relationship (SAR) models established in vitro with the neutral red cytotoxicity assay, *Toxicol. In Vitro 1*:3–9 (1987).
2. Borenfreund, E., and Puerner, J. A., A simple quantitative procedure using

monolayer cultures for cytotoxicity assays (HDT/NR-90), *J. Tissue Cult. Methods 9*:7–9 (1985).

3. Liu, D. H. W., Spanggord, R. J., Bailey, H. C., Javitz, H. S., and Jones, D. C. I., Toxicity of TNT wastewaters to aquatic organisms. Volume II. Final report. Contract DAMD17-75-C5056. SRI International, Menlo Park, CA. AD-A142145, 1984.
4. Peters, G. T., Burton, D. T., Paulson, R. L., and Turley, S. D., The acute and chronic toxicity of hexahydro-1,3,5-trinitro-1,3,5-triazine (RDX) to three freshwater invertebrates, *Environ. Toxicol. Chem. 10*:1073–1081 (1991).
5. Rosenblatt, D. H., Burrows, E. P., Mitchell, W. R., and Parmer, D. L., Explosives and related compounds, *The Handbook of Environmental Chemistry*, Vol. 3, part G. *Anthropogenic Compounds* (O. Hutzinger, ed.), Springer-Verlag, Berlin, 1991, pp. 195–234.
6. Safe, S., Determination of 2,3,7,8-TCDD toxic equivalent factors (TEFs): Support for the use of the in vitro AHH induction assay, *Chemosphere 16*:791–802 (1987).
7. van der Schalie, W. H., The acute and chronic toxicity of 3,5-dinitroaniline, 1,3-dinitrobenzene and 1,3,5-trinitrobenzene to fresh water aquatic organisms, TR 8305. U.S. Army Medical Bioengineering Research and Development Laboratory, Fort Detrick, Frederick, MD. AD-A138408, 1983.

# 13

# Neuroblastoma–Glioma Cells as a System for Studying Drug Neurotoxicity

**Arthur D. Weissman,* Benjamin R. Crenshaw, Jr.,† and John E. Johnson, Jr.‡**

*Addiction Research Center, National Institute on Drug Abuse, National Institutes of Health Baltimore, Maryland*

In vitro cell cultures and neuronal preparations have been used with much success in the field of neurobiology to elucidate the organizing structure of complex neuronal networks that form the basis of central nervous system (CNS) function. The formation of such networks are established through chemically regulated communications between neuronal elements. These interactions result in rapid and flexible changes in neuronal structure that are vulnerable to a wide range of environmental influences. Many types of drugs can affect development by modulating chemical signaling between neurons at their synapses. In the prenatal, as well as the early postnatal period, these modifications of cellular communication can profoundly alter the final form of the adult brain and ultimately modify brain function. The use of cell culture facilitates the study of nascent neuronal structure, by allowing the experimenter to carefully regulate the chemical milieu of the neuronal elements and to simplify their chemical associations by reducing the number and types of neuronal elements. The present studies used a culture of neuroblastoma–glioma cells,

---

*Present affiliations*:
*Seton Hall University School of Graduate Medical Education, South Orange, New Jersey
†Johns Hopkins School of Medicine, Baltimore, Maryland
‡Retired

NG108-15, which possess many characteristics that enable the modeling of in vivo effects of drugs on the brain [1]. This immortal cell line incorporates some of the characteristics of a developing neuronal system, with rapid proliferation and neurite formation, as well as the establishment of a monolayer plexus of mature differentiated cells with a stable expression of many neuronal characteristics [2]. The neurochemical profile of these cells includes components of a monoamine neurotransmitter system [3], which makes them an excellent model to study neuronal–drug interactions, since many drugs have their effects through this neurochemical substrate [4]. In these experiments, a population of NG108-15 cells, were continuously exposed to various concentrations of drugs and examined for changes in morphologic appearance and function. Specifically, their viability with trypan blue exclusion, cell proliferation curves with [$^3$H]thymidine, and structural changes with light and transmission electron microscopy, were evaluated. The importance of the resulting data is that they could (a) provide information about how these drugs affect cellular structure, in general, and how a compound interacts with specific intracellular components in the brain; (b) help determine an estimate of a drug's neurotoxicity or its effects on the functional integrity of the cells in culture; and (c) give possible implications for acute or long-term consequence of a drug on the developing nervous system.

The drugs used in these studies were psychostimulants—amphetamine and cocaine. These compounds are highly addictive, with their primary sites of action in the brain [5], and they pose a significant concern for their effects on the developing human nervous system because of their abuse by some pregnant women [6]. Although the behaviorally addictive aspects are well documented, there is no clear understanding of the cellular mechanism of their actions or the consequence of their long-term effects on the central nervous system. These drugs bind to several sites in brain, including monoamine uptake transporters [5], and can alter the levels of neurotransmitters, such as norepinephrine, dopamine, and serotonin [7]. Both cocaine and amphetamine derivatives have a known toxicity associated with brief and prolonged exposure in adult [8] and neonatal humans [6], as well as in animals [7,9]. However, very few studies have examined the changes that these drugs can induce on the fine structure of neural tissue [2,3]. The first part of this study provides a detailed morphological description of psychostimulants effects on NG108-15 cells exposed at various times and doses.

## ULTRASTRUCTURAL MORPHOLOGY

The NG108-15 cell line is well characterized in terms of it ultrastructure and biochemistry [1,2,10]. Control cells contain cytoplasm rich in organelles, including mitochondria, ribosomes, endoplasmic reticulum, Golgi, dense bodies, lipid droplets, and virus particles. The dense bodies occurred in a variety of shapes and densities, with some round and of medium density; others were slightly less rounded and with much greater medium density, and still others were slightly less rounded and with much greater electron density. Elongated or rectangular dense bodies were also found. Electron-lucent regions (probably lipid droplets) were characteristic of the densest types of bodies.

After the administration of cocaine, no cytoplasmic changes were noted, at doses up to $10^{-3}$ M. Some of the cells at higher doses had numerous dense bodies (Fig. 1). Morphometric analysis was not performed because the dense bodies tended to be polarized near one edge of the nucleus and not randomly distributed throughout the cytoplasm. However, nuclear responses to cocaine were observed. After 2 days of exposure, nuclear invaginations were filled with

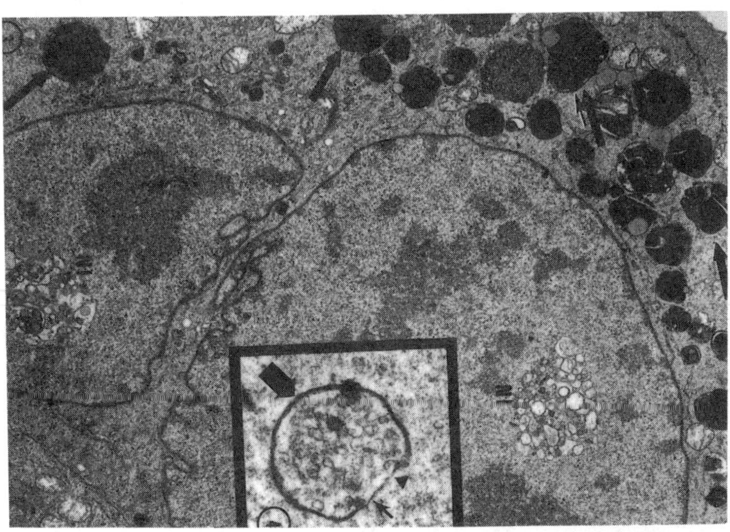

**Figure 1** Several large dense bodies (arrows) have accumulated near one end of the nucleus in the cytoplasm of this cell exposed to $10^{-3}$ M dose of cocaine for 3 days. Within the nucleus, are two inclusion bodies (NI) consisting of numerous vesicles (10,000× magnification). Insert: At 2 days of exposure to $10^{-6}$ M cocaine doses vesicles aggregated within the nucleus. A pair of membranes (large solid arrow) surrounds the vesicles, but in one area, it has been disrupted and becomes a single membrane (medium-sized solid arrow), and in one spot is broken altogether (arrowhead) (37,500× magnification).

vesicles, and the nuclear membrane surrounding the vesicles was disrupted (see Fig. 1).

Following 3 days of cocaine treatment, patches of vesicles were observed, without any surrounding nuclear membrane. The vesicles ranged in size from 0.05 to 0.8 μm. The larger ones tended to be clear-cored, whereas both clear- and dense-cored vesicles were observed in the smaller sizes (Fig. 2).

The contents of the nuclear vesicles varied from a fine granular matrix, to lipidlike, with a dense border (Fig. 3). The tubular inclusions were 0.1–0.03 μm in width and either round or ovoid. Vesicular patches were not, in general, present (see Fig. 3). Nuclear vesicles were not seen at 1 day of treatment, but at 2 days, vesicles appeared at the $10^{-3}$ M dose, and at 3 days, the vesicles were also present in cells exposed to a $10^{-6}$ M dose of cocaine. No vesiculation was seen at any time for drug exposure lower than $10^{-6}$ M.

Increases in nuclear inclusions appeared to be the primary effect of cocaine in vitro. These structures may be derived from invaginations of the inner nuclear membrane and, as in the NG108-15 cells, the nuclear membrane

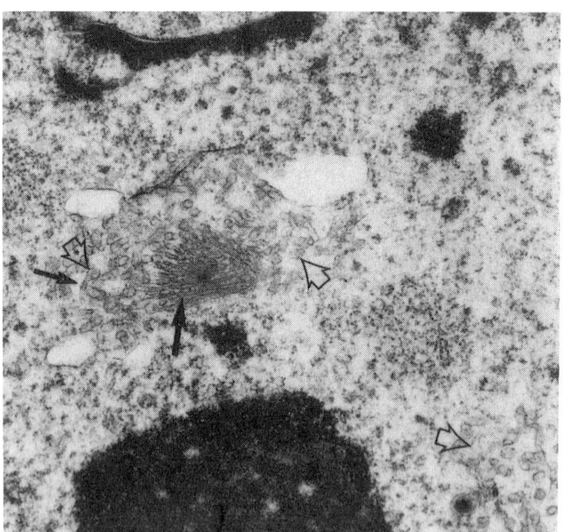

**Figure 2** This cell was exposed to a $10^{-6}$ M dose of cocaine for 3 days. Tubular structures of different sizes (solid and hollow arrows) are present within the nucleus. It is possible that the different-sized tubules are continuous with one another (arrow with asterisks) (50,000× magnification).

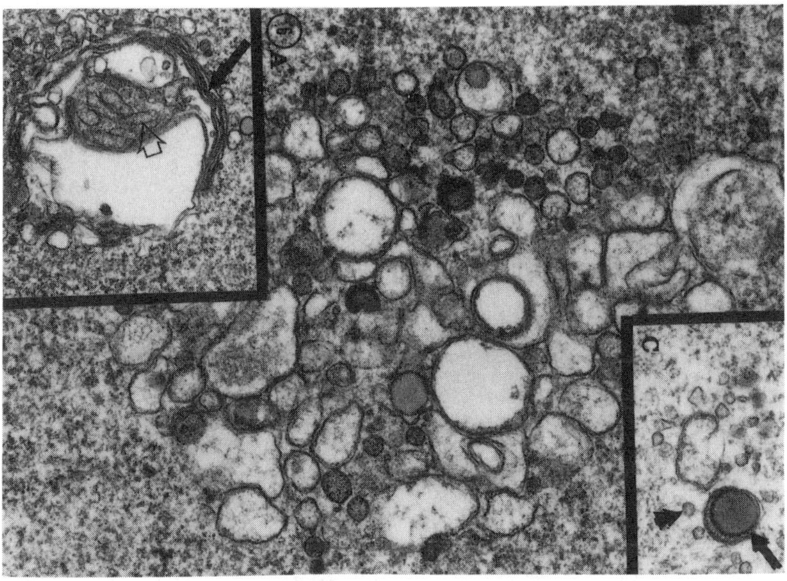

**Figure 3** (Center) Examples of nuclear inclusions in HG108-15 cells in a culture exposed to a $10^{-3}$ M cocaine dose for 3 days. Tubular structures and several membranes are shown within the nucleus. Vesicles are located nearby (25,000× magnification). *Inserts* (top left) Vesicles of varying sizes, densities, and contents, are present within the nucleus (50,000× magnification). (Bottom right) Numerous small vesicles (arrowhead) and one large dense vesicle or body (long arrow) were found in the nucleus (30,000× magnification).

could infold and detach to become a true inclusion. Nuclear inclusions have been found in various areas of the nervous system [11], and their number has been reported to increase following electrical stimulation [12]. Tubular structures also appear as inclusions in the nuclei of a variety of cell types. In hepatoma cells, the structures may form a direct tubular system between the nucleolus and perinuclear cisterna [13]. The tubular inclusions are also prominent in female reproductive cells, including mammotrophs of gerbils [14], human endometrium [15], and epithelial cells of patients with fibrotic lung disorders [16]. Although these intranuclear bodies have been found in a wide variety of cell types, their structural similarity suggest that their function may be similar in all cells. It appears that cells develop nuclear inclusions during periods of rapid growth, or in response to specific hormonal stimuli. Although

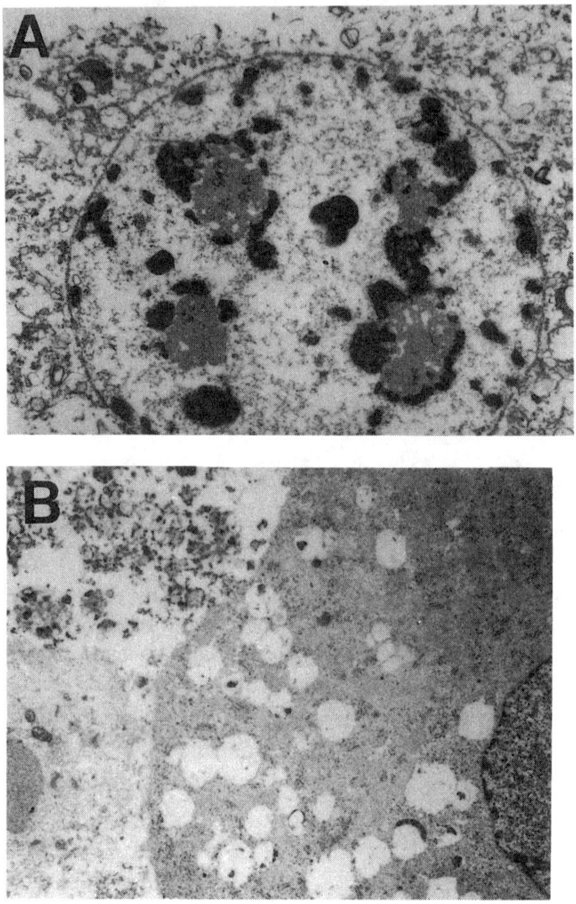


**Figure 4** Three-days treatment with (A) $10^{-3}$ M imipramine or (B) $10^{-3}$ M *d*-amphetamine NG108-15 cells showed extensive degeneration characterized by vacuolated nucleoplasm and cytoplasm (15,000× magnification).

the specific functional nature of the structures has not been determined, they have been hypothesized to be a series of common responses to stimulation of specific pathways of protein synthesis.

Other drugs that inhibit monoamine uptake (i.e., amphetaminelike compounds) showed some different effects at the same doses that were used with cocaine. These compounds appeared to be quite different in terms of the induced fine structural changes and cell toxicity. The exposure of NG108-15 cells to a $10^{-3}$ M dose of imipramine (Fig. 4A) and $10^{-3}$ M dose of amphetamine (see Fig. 4B) resulted in the disruption of cellular integrity after 3 days. The structural changes (cytotoxicity and cell death) induced by amphetamines reflected the relatively greater neurotoxic potency of these compounds seen in animal and human studies [17]. With lower doses, some of the nuclear changes observed were similar to those seen in cocaine-treated cells (Fig. 5).

## CELL PROLIFERATION

Additional experiments were performed to measure the effect of experimental drugs on cell number and mortality. The results suggest that cocaine slows the rate of cell replication while concurrently killing a portion, or perhaps select subpopulation, of cells (Fig. 6).

The amount of cellular division as measured by the total number of cells in culture is significantly reduced on the second day of $10^{-3}$ M cocaine exposure, as compared with controls. Cell mortality, as measured by the percentage of living cells in a given culture, does not appear to be significantly changed at this time and at this drug concentration. On the third day, a reduction in both cellular replication and survival was evident. The number of cells present in culture was still significantly depressed after exposure to a $10^{-3}$ M cocaine dose. In addition, the percentage of cells that were alive in the culture on the

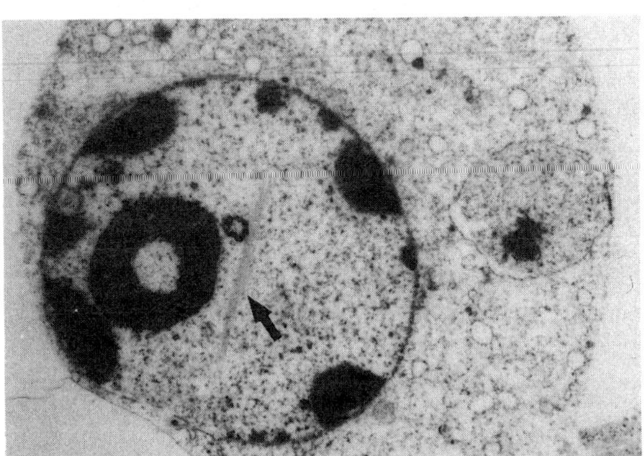

**Figure 5** At 7 days of treatment with the amphetamine analogue MDMA, $10^{-6}$ M, tubular nuclear inclusions were seen in this cell. 10,000× magnification.

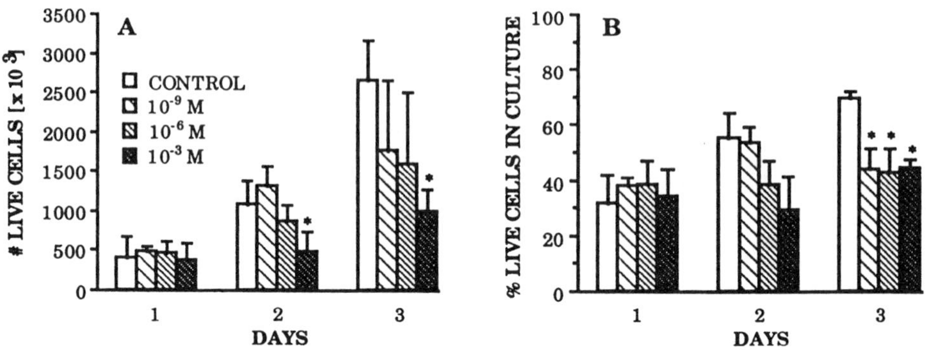


**Figure 6** (A) Number of NG108-15 cells after 1, 2, and 3 days of exposure to different amounts of cocaine; (B) the percentage of live cells in those cultures as assessed by trypan blue staining. Asterisks indicate a significant effect of cocaine, compared with controls on the same day.

third day after cocaine addition, was significantly decreased from control for all three concentrations.

To further examine drug-induced changes in cellular proliferation, DNA synthesis was measured using [$^3$H]thymidine in the cultures. This approach addressed the fine structural drug effects in the cell nucleus and the suggested differential effects on the rate of cell division and survival seen in these studies. The [$^3$H]thymidine uptake increases rapidly over the first few hours as DNA is synthesized and then levels out. A dose–response effect on cellular division was observed for a series of drugs, and their relative potency in inhibiting cell division agreed well with the data on cell counts and fine structure, as well as their potency in vivo. In general, the largest effect on cellular proliferation occurred in the first 8 hr after exposure to the drugs, with a maximum reduction to 20% of control values at the highest doses (Fig. 7). The inhibition of [$^3$H]thymidine uptake by the intermediate doses returned to control levels after 48 hr of treatment with methamphetamine, but not *d*-amphetamine. There was some stimulation of [$^3$H]thymidine uptake by both drugs at the lowest levels of exposure.

## CONCLUSIONS

These experiments have been directed at exploring the fundamental usefulness of cell cultures as a medium to examine (a) a drug's site of action at the cellular level as seen morphologically with the electron microscope; (b) the specificity

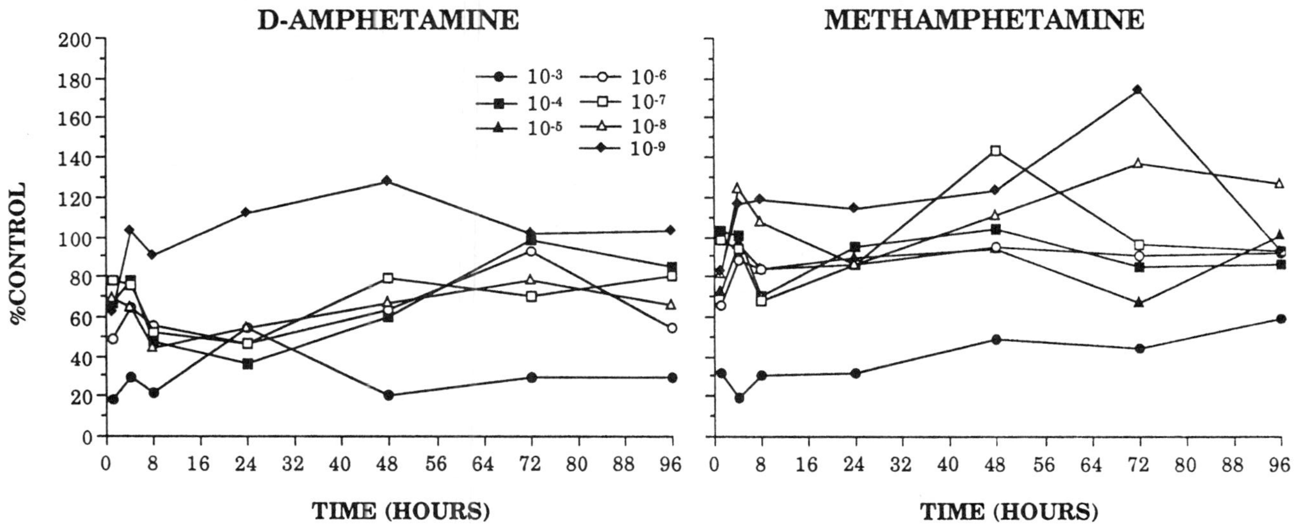


**Figure 7** NG108-15 cellular proliferation over 4 days, as measured by [$^3$H]thymidine uptake. Cells were grown in 96-well plates and allowed to proliferate for 12 hr. [$^3$H]Thymidine (5 μCi/ml) was added to the medium and incubated for 30 min. Cells were then exposed to varying doses of *d*-amphetamine or methamphetamine at time zero, and their growth was expressed as a percentage of parallel control cultures that had no drug added.

of a drug's action in neurally derived cells and its relation to known neurochemical processes, such as neurotransmitter functions. One primary question is whether a drug's effect, such as toxicity, is related to its interaction with an organ-specific system, such as the brain. Psychostimulants in our studies have major effects through the monoamine system, and their influence on morphology and cellular function may be mediated through the monoamine-reuptake process. These preliminary studies suggest this to be somewhat true in that the drug's toxic potency is reduced in cells that lack this specific monoamine system, or if that system is pharmacologically blocked. In addition, many of the psychostimulants used have shown a potency of neurotoxic effects that parallels that reported in several animal models.

The morphological data in these studies may open new avenues of research into drug effects on the developing nervous system in intact organisms. The nature of the drug-induced nuclear changes on cell growth and replication in the cell lines suggests important structural–biochemical relations. These findings that cell replication is decreased and that cell mortality is increased in a time- and dose-dependent manner with drug exposure may be related to the nuclear changes observed, which were also dose-dependent. It seems possible that some drugs could interfere with normal replication, thus implicating the production of nuclear inclusions as a marker of genetic damage. The relation of these inclusions with chromation is now unclear. On the other hand, cell mortality may reflect general neurotoxic effects of the drug. Future studies should evaluate the point at which drug-induced changes represent toxicity, compared with physiological, but nontoxic, effects.

## REFERENCES

1. Klee, W. A., and Nirenberg, M., A neuroblastoma × glioma hybrid cell line with morphine receptors, *Proc. Natl. Acad. Sci. USA 71*:3474–3477 (1974).
2. Johnson, J. E., Jr., and Weissman, A. D., Fine structural effects of cocaine on NG108-15 neuroglioblastoma cells in culture, *Brain Res. Bull. 20*:39–47 (1988).
3. Weissman, A. D., and Johnson, J. E., Jr., Assessing psychoactive drug neurotoxicity in vitro: Initial studies using NG108-15 neuroblastoma-glioma cells, *In Vitro Toxicology: Mechanisms and New Technology*, Vol. 8, *Alternate Methods in Toxicology* (A. M. Golderberg, ed.), Mary Ann Liebert, New York, 1981, pp. 387–396.
4. Taylor, D., and Ho, B. T., Neurochemical effects of cocaine following acute and repeated injections, *J. Neurosci. Res. 3*:95–101 (1977).
5. Ritz, M. C., Lamb, R. J., Goldberg, S. R., and Kuhar, M. J., Cocaine receptors on dopamine transporters appear related to drug self-administration, *Science 237*: 1219–1223 (1987).

6. Chasnoff, I. J., Burns, W. J., Schnoll, S. H., and Burns, K. A., Cocaine use in pregnancy, *N. Engl. J. Med. 313*:666–669 (1985).
7. Tonge, S. R., Permanent alterations in catecholamine concentrations in discrete areas of brain in the offspring of rats treated with methylamphetamine and chlorpromazine, *Br. J. Pharmacol. 47*:425–427 (1973).
8. Finkle, B. S., and McCloskey, K. L., The forensic toxicology of cocaine, *Cocaine: 1977* (R. C. Peterson and R. C. Stillman, eds.), National Institute on Drug Abuse, Washington, DC. 1977, pp. 153–178.
9. Bozarth, M. A., and Wise, R. A., Toxicity associated with long-term intravenous heroin and cocaine self-administration in the rat, *JAMA 254*:81–83 (1985).
10. Furuya, S., and Furuya, K., Ultrastructural changes in differentiating neuroblastoma × glioma hybrid cells, *Tissue Cell 15*:903–919 (1983).
11. Johnson, J. E., Jr., In vivo and in vitro comparisons of age-related fine structural changes in cell components, *Aging and Cell Structure*, Vol 2 (J. E. Johnson, Jr., ed.), Plenum Press, New York, 1984, pp. 37–88.
12. Seite, R., Mei, N., and Vuillet-Luciani, J., Effect of electrical stimulation on nuclear microfilaments and microtubules of sympathetic neurons submitted to cycloheximide, *Brain Res. 50*:419–425 (1973).
13. Babai, F., Tremblay, G., and Dumont, A., Intranuclear and intranucleolar tubular structures in Novikoff hepatoma cells, *J. Ultrastruct. Res. 28*:125–130 (1969).
14. Nickerson, P., Intranuclear inclusions in mammotrophs of the mongolian gerbil: Effect of low doses of estradiol benzoate and a study of females before weaning, *Tissue Cell 7*:773–776 (1975).
15. Terzakis, J., The nucleolar channel system of human endometrium, *J. Cell Biol. 27*:293–304 (1965).
16. Kawanami, O., Ferrans, V., Fulmer, J., and Crystal, R., Nuclear inclusions in alveolar epithelium of patients with fibrotic lung disorders, *Am. J. Pathol. 94*:301–322 (1979).

# 14

# In Vitro Assays for Muscle Irritation

**Sharon J. Northup**

*Baxter Healthcare Corporation*
*Round Lake, Illinois*

Medical devices have been tested for biological reactivity for more than 30 years. The initial concern was whether the new plastic materials being used for disposable syringes and catheters would adversely affect the solutions, drugs, or tissues in the practice of medical care. The designing of appropriate biological assays focused on extracting the plastics with various nontoxic solvents (saline, saline with 5% ethanol, polyethylene glycol 400, and vegetable oil) and testing the extract for biological safety, or testing the solid material directly in an animal model. The safety tests were generally adaptations of those tests commonly used for pure chemicals; namely, acute systemic toxicity and irritation potential by cutaneous, intradermal, and intramuscular routes. In the ensuing years, biological reactivity testing changed from an empirical view to a mechanistic-based approach using advances in biology, chemistry, and polymer science and technology. This chapter reviews the science of that transition and the rationale for the adaptation of cytotoxicity assays as an alternative to in vivo biological reactivity tests.

## MUSCLE IRRITATION ASSAY

The in vivo biological reactivity tests, namely the systemic injection, intracutaneous irritation, and intramuscular implant assays, are described in the

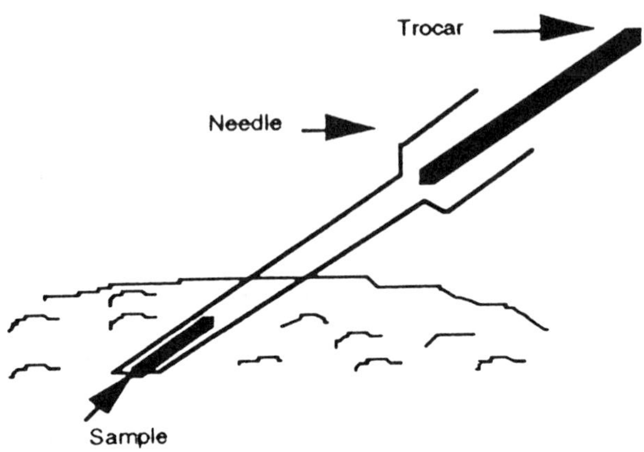


**Figure 1** Trocar method of soft-tissue implantation.

*U.S. Pharmacopeia* [1]. Experience has shown that the intramuscular irritation test was the most sensitive of the three assays. This assay involves the insertion of a small sample of solid material in the paravertebral muscles with a trocar and needle (Fig. 1). Insertion of the needle into tissue initiates a wound-healing reaction characterized by an influx of neutrophils and macrophages, which cleanup the microorganisms, dead cells, and other debris (Fig. 2). This is followed by an influx of fibroblasts that repair the damage with a mortar of collagen. The tissue reactivity (Fig. 3) is evaluated by the amount of encapsulation of the test article. The size of the wound and amount of scar tissue is dependent on the amount of tissue damage. That is, any chemicals extracted by

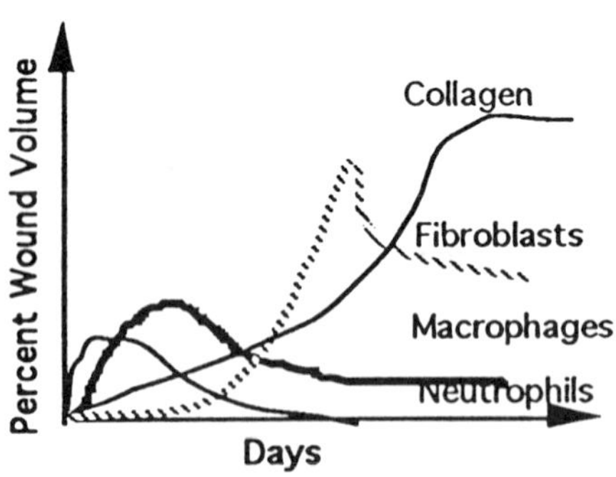


**Figure 2** Normal wound healing.

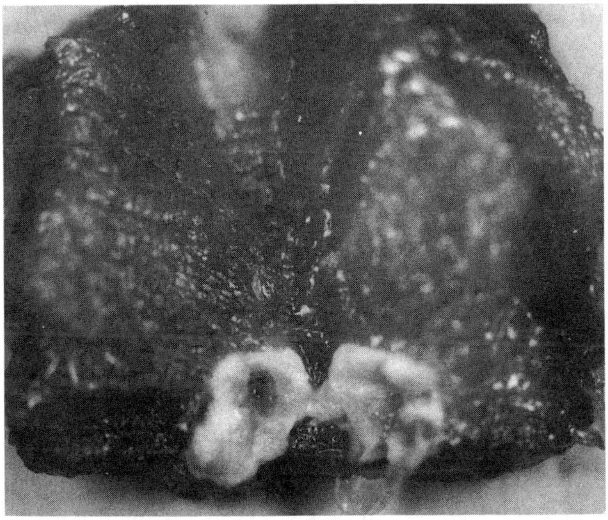

**Figure 3** Muscular irritation assay of a reactive plastic (top) illustrating fibrotic wound healing.

the tissue from the test sample can potentially kill cells if the cellular concentration attains a toxic dose.

The physicochemical mechanisms that affect intramuscular irritation are listed in Table 1. The biological reactivity between the implanted test sample and adjacent tissue is based on the target cell dose of the leachable chemicals from the implant. The quantity of chemical(s) is affected by a partitioning of the chemical in the aqueous and lipoidal solvents that constitute the tissue. The binding of lipid-soluble chemicals by the hydrophobic region of proteins enhances solubility. There is a slow, continuous dissociation of the agent into the aqueous phase, as determined by its solubility and dissociation rate from the

**Table 1** Intramuscular Irritation Assays

| |
|---|
| Target cell dose |
| Chemical partitioning affected by water and fat content of tissue |
| Limited metabolic activity |
| Limited dilution |

proteins. The limited metabolic activity and the lack of dilution by the vascular system and tissues also contribute to increase the target cell dose by affecting detoxification and elimination rates.

## BIOAVAILABILITY

The physicochemical mechanisms that affect the bioavailability [2,3] or release of the chemical(s) from the implant are listed in Table 2. The total available pool is the amount of chemical initially present in the test article. If the chemical is chemically bound to the polymer, it would not be free to migrate out of the material. Any unbound chemical(s) could migrate at a rate dependent on size, configuration, electronic charge, and such. The migration rate is closely linked with partitioning of the chemical, within the bulk phase of the material, to the surface and into the solvent phase of the tissue. The amount of chemical extracting into the tissue varies with its solubility in the tissue.

The dynamics affecting bioavailability are further illustrated in Figure 4. Any unbound chemical in the bulk phase of the material can migrate within this phase to the surface, and then out into the tissue. The rate and amount over time are affected by the partitioning among the three phases—bulk, surface, or tissue—and the solubility limit in each phase. Figuratively, one can describe these dynamics as a student moving from home (home = bulk phase) to college (college campus = surface). If the student is not happy at college, he or she moves back home (a case of no migration). Alternatively, the student graduates from college (partitioning at the surface) and gets a job in a city (city = tissue). Any additives to a plastic are theoretically expected to remain in the plastic because the plastic itself is a water-insoluble (i.e., lipophilic), solid substance. Thus, the total available pool is often much greater than the amount of chemical that is extractable in a nonlipophilic media. Additionally, the hydrophilic nature of tissue itself would limit the solubility of any lipophilic chemicals that could potentially migrate.

The migration of ethylene oxide, a sterilizing gas, from a medical device illustrates the bioavailability of extractive chemicals to tissue cells. The amount

**Table 2** Mechanisms Limiting the Bioavailability of Leachable Chemicals

| | |
|---|---|
| Total available pool | Partitioning |
| Migration rate | Solubility |

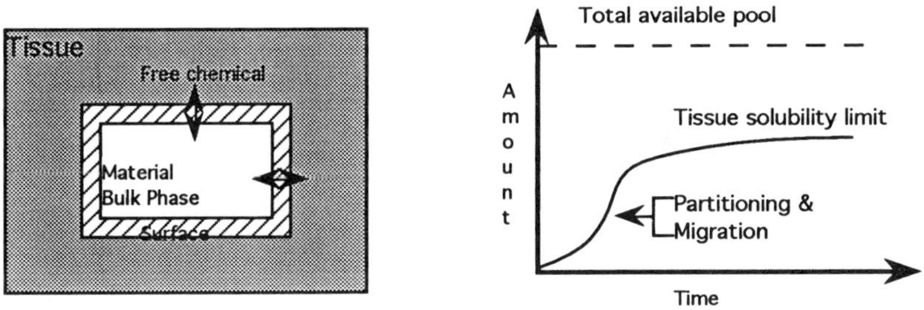


**Figure 4** Dynamics of the bioavailability of extractive chemical(s) from plastic materials implanted in tissue.

of residual ethylene oxide in the material after sterilization varies with the thickness of the material and the elapsed time since sterilization, among other factors. Thicker material will absorb a greater amount of the sterilant during the sterilization process. If the material is then placed in an open environment or tissue, the free chemical (i.e., ethylene oxide) will migrate out of the bulk phase of the material into the surrounding environment. In cells or tissue, the concentration of free chemical in each cell will determine the biological effect of the extractable ethylene oxide. The target cell dose effect of residual ethylene oxide in polystyrene, for example, resulted in substantial variance background mutagenicity rates [4].

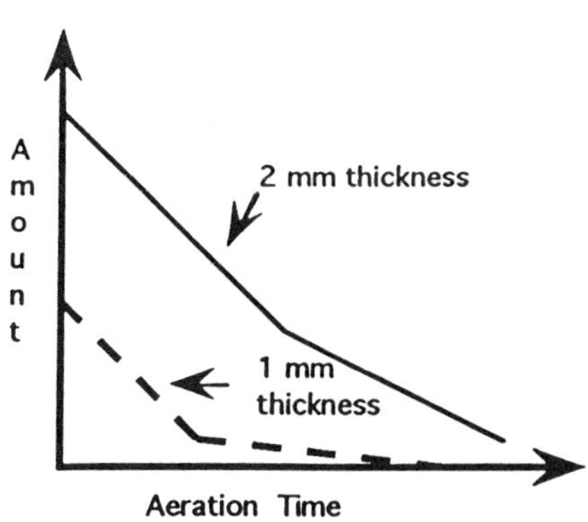


**Figure 5** Residual ethylene oxide variance with material thickness.

## IN VITRO ASSAYS

The challenge of developing an in vitro alternative to the intramuscular irritation assay initially involved selecting the appropriate parameter to model. Figure 6 outlines the pharmacological effects of a toxin or immune stimulus. The primary event is cellular injury, resulting in a change in the permeability of the cell membrane or the lysis of the cell. Either event is followed by a release of pharmacological agents, on hemodynamic effects, an so on. Thus, as a first approximation, an in vitro cell injury model gives the potential for the secondary pharmacodynamic effects. Sources of variance, such as dilution and metabolism, are eliminated in an in vitro model by a defined volume of cell medium, choice of cell line, and number of cells. This results in a controlled dose of test substance per individual cell in vitro, compared with the unknown target cell dose of in vivo assays. A fibroblast cell line was selected for the in vitro model to evaluate wound healing (see Fig. 2), an indirect measure of the amount of cell damage. These considerations led to an in vitro model using a fibroblast cell line in a near confluent culture and measurement of the cytotoxic effect of leachables from the polymeric materials.

Three different in vitro cytotoxicity models have been widely adapted by the medical device industry [5,6]. The models differ in the preparation of the test sample and the mode of exposure. In the direct-contact model, the test sample is placed on top of a monolayer of fibroblasts (Fig. 7). Chemicals are extracted from the sample by the solvent effect of the culture medium and added

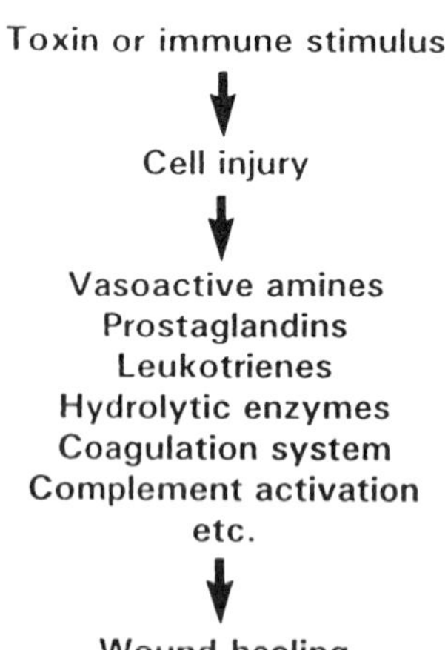


**Figure 6** Biological effects of a toxic or immune stimulus.

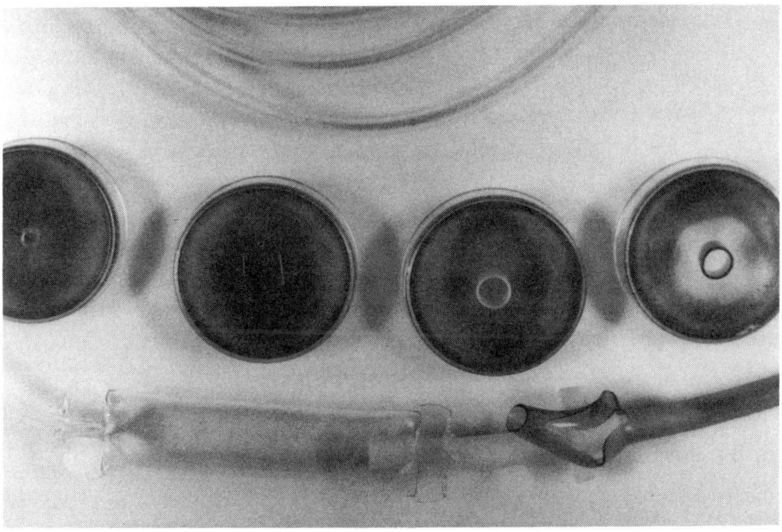

**Figure 7** In vitro direct contact assay.

serum. At the end of a 24- or 48-hr exposure period, the culture medium is removed and the remaining viable cells are treated with a histological stain. The culture dish on the right side of Figure 7 shows cytotoxicity to the cross section of tubing, as evident by the lack of stained cells surrounding the sample. Cytotoxicity was caused by extraction of a relatively low molecular weight organotin compound that was used as a stabilizer in the plastic formulation. In the agar diffusion assay (Fig. 8), a thin layer of agar or agarose mixed in the culture medium separates the test sample from the cellular monolayer. The agar functions to protect the cells from physical damage by movements of the test sample and creates a somewhat concentrated diffusion zone around the sample. Cytotoxicity is evaluated by the lack of staining of dead cells with a vital stain. In the elution test, the test sample is extracted in a nontoxic vehicle and then exposed to the cells. This method offers flexibility in the choice of extractant, and time, temperature, and duration of extraction. Cytotoxicity is evaluated by the density of cells remaining after a 24- or 48-hr exposure to the extract. The detection limit of these assays can be lowered (i.e., made more sensitive) by reducing the cell number; by using cells with a larger cell volume; by using cells in the log growth phase, rather than at near confluency; or by extending the exposure period. These variables improve sensitivity by altering the dose per

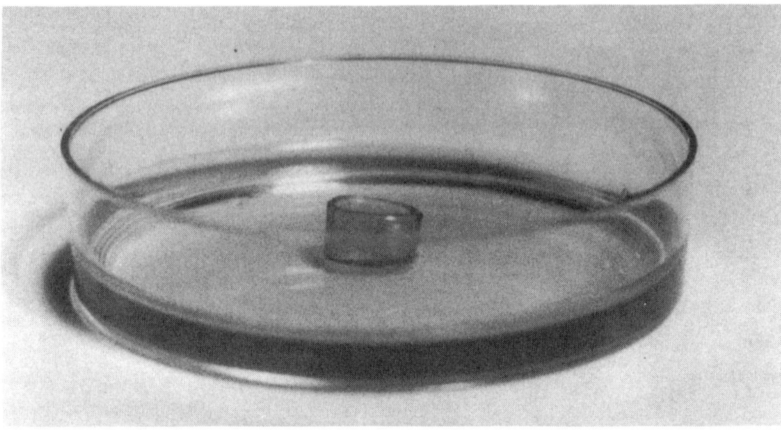

**Figure 8** In vitro agar diffusion assay.

cell or increasing the environmental stress on the cell. The target cell dose in the three assays is dependent on the total available pool of cytotoxin in the test sample, the rate of migration from the bulk and surface of the material, partitioning from the sample surface into the culture medium, and solubility in the culture medium and cells. These are the same factors affecting bioavailability in vivo following implantation.

## CONCLUSION

Two decades of success in the application of in vitro models for testing medical devices can be attributed to matching the in vivo target cell type and approximating the bioavailability of the cytotoxin. The in vitro models described herein accurately predict the acute in vivo biological reactivity of a wide variety of synthetic polymers and plastics. These models focus on reversible toxicological events free of pharmacological mediators. Adapting a mechanistic approach on the types of cytotoxins and their bioavailability during clinical use of the medical devices has ensured the validity, reproducibility, accuracy, and predictability of the in vitro assays.

## REFERENCES

1. The United States Pharmacopeial Convention, Biological reactivity tests, in vivo, *United States Pharmacopeia*, 22 ed, USP, Rockville, MD, 1989, pp. 1497–1500.

2. Sanchez, I. C., Chang, S. S., and Smith, L. E., Migration models for polymer additives, *Polymer News 6*:249–256 (1980).
3. Till, D., Schwope, A. D., Ehntholt, D. J., Sidman, K. R., Whelan, R. H., Schwartz, S., and Reid, R. C., Indirect food additive migration from polymeric food packaging materials, *CRC Crit. Rev. Toxicol. 18*:215–243 (1987).
4. Brown, A. M., Genotoxic effects of polymers on L518Y TK +/− cells [abstract], *Environ. Mutagen. Sci. 54* (1981).
5. The United States Pharmacopeial Convention, Biological reactivity tests, in-vitro, *United States Pharmacopeia*, 22nd ed., USP, Rockville, MD, 1989, pp. 1495–1497.
6. Northup, S. J., Mammalian cell culture models, *Handbook of Biomaterials Evaluation: Scientific, Technical and Clinical Testing of Implant Materials* (A. F. von Recum, ed.), Macmillan, New York, 1986, pp. 209–225.

# 15

# General Overview of In Vitro and Other Alternatives to Skin Toxicity Evaluation

**David W. Hobson***

*Battelle Memorial Institute*
*Columbus, Ohio*

The skin, the largest organ of the body, is a complex and dynamic organ composed of an outer dermis and underlying dermis (Fig. 1). Within these two primary divisions, numerous cell types and specialized adnexal structures are found. With its collection of different cellular layers, glandular components, hair follicles, and vascular elements, functioning interdependently, the role of the skin goes well beyond simply being a barrier to the environment. The skin also contributes significantly to the normal body function in other important ways: It has several physiological roles, such as controlling water loss and regulating body heat; it includes defensive roles such as protecting underlying organs, providing localized immunological defenses and photobiotic protection; it has endocrine and apocrine functions; and it performs several important metabolic functions (e.g., keratin, melanin, and collagen synthesis; carbohydrate metabolism, lipid storage and metabolism; respiration; vitamin D synthesis; and xenobiotic metabolism). The skin also contains several different types of sensory receptors that provide the nervous system with an integrated network from which to assess and respond to pain, heat, cold, pressure, irritation, or other stimuli. In a 70-kg, adult human, the surface area of the skin is estimated at 1.85 $m^2$, of which between 0.1 and 1.0%, respectively, is

**Present affiliation*: Dermatological Products of Texas, San Antonio, Texas

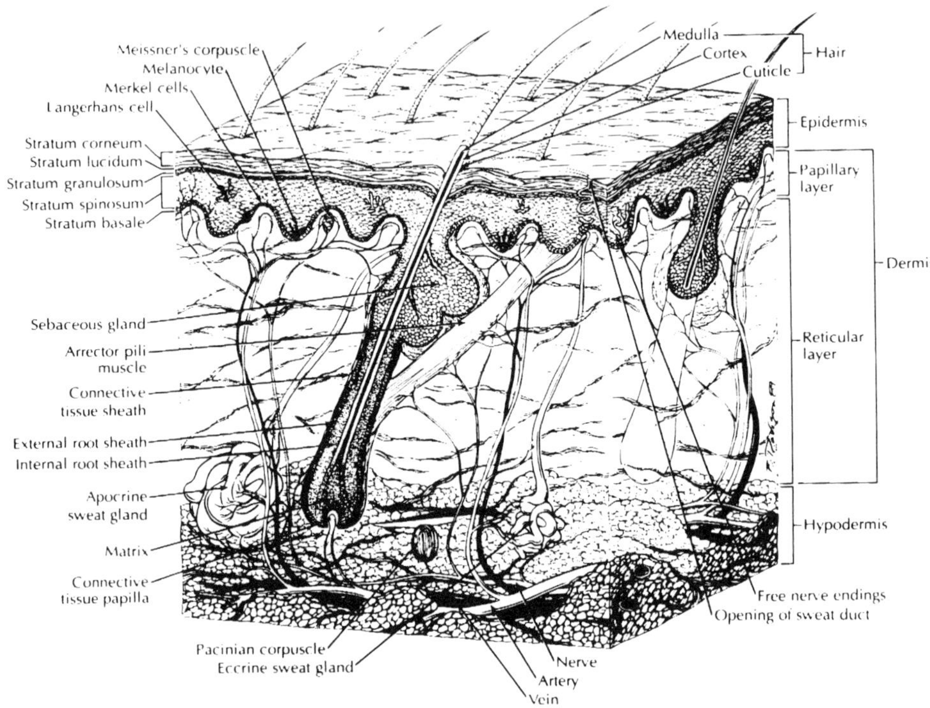


**Figure 1** A composite representation of the cross-sectional structure found in skin of various regions of the body. (From Ref. 10.)

accounted for as sweat glands and hair follicles. Skin thickness is highly variable, and ranges from 400 to 600 μm in callous areas, such as the palms of the hands and soles of the feet, to between 8 and 15 μm on the arms, back, legs, and abdomen. Under normal conditions, the stratum corneum contains a variety of lipids and waxy esters as well as being partially hydrated. The skin generally contains about 90 g of water per gram of dry tissue, but can increase its tightly bound water content three to five times following direct contact with water. Because the skin is a living integument comprised of many different cell types and connected intimately with the vascular and nervous systems, the fate of percutaneously applied chemicals can be somewhat diverse (Fig. 2), and the events occurring in dermal inflammatory, sensitization, and corrosive effects are, likewise, now known to be fairly complex (Fig. 3). Historically, animal

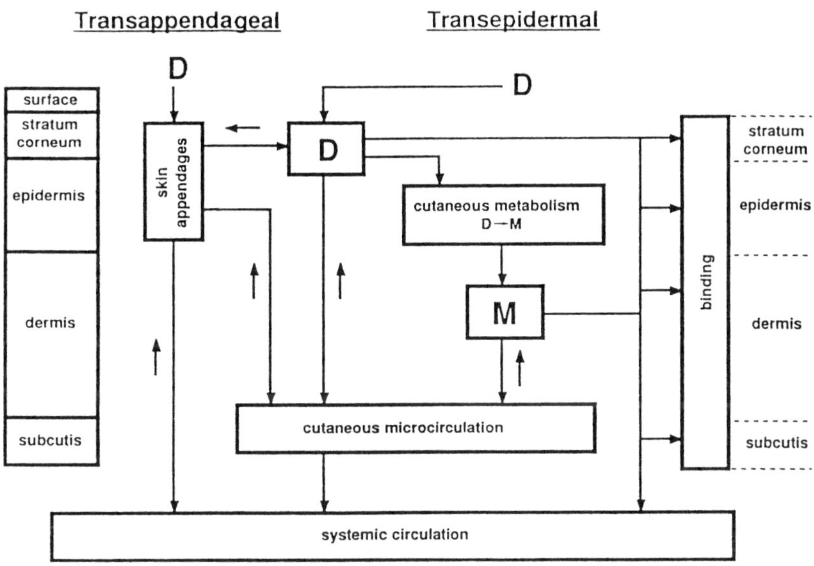


**Figure 2** Schematic showing the percutaneous fate of a compound (D) following topical exposure. M denotes cutaneous metabolites of D. (From Ref. 11.)

models, of various types, and human testing have been used to assess the skin toxicity of various chemical substances and mixtures, partially out of appreciation for the response complexity and out of convenience. In vitro procedures, on the other hand, were used to isolate and study particular aspects of the in vivo response, but have only recently started to receive serious consideration as full or partial replacements for selected in vivo models. The generally accepted types of in vitro models now include those that assist in

The selection or screening of in vivo test candidates
The prioritization of compounds for more detailed testing or for risk assessment
The investigation of activation or detoxification mechanisms
The determination of dermal penetration rates or absorptive characteristics

## TYPES OF MODELS

Because of increasing interest in the reduction, replacement, and refinement of in vivo model utilization through the introduction of alternative and in vitro procedures, the number of such models is, likewise, increasing rapidly. For-

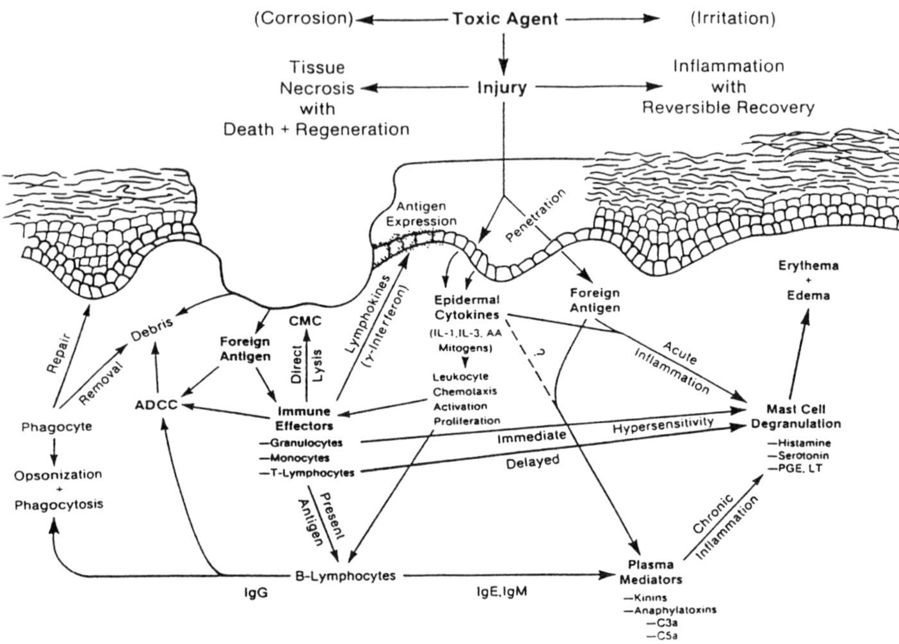


**Figure 3** The cellular and humoral mechanisms of dermal irritation and corrosion. Abbreviations: IL-1 and IL-3, interleukins 1 and 3; AA, arachidonic acid metabolites; PGE, "E" series prostaglandins; LT, leukotrienes; CMC, cell-mediated cytotoxicity; ADCC, antibody-dependent cellular cytotoxicity; Ig, immunoglobins. (From Ref. 9.)

tunately, the various in vitro or alternative model types can be classified as generally being one of five different types:

1. Quantitative structure–activity relationship (QSAR) models
2. Biochemically based models
3. Cellular models
4. Tissue models
5. Lower vertebrate models

## Quantitative Structure–Activity Relationship Models

The QSAR models are strictly computational and require, as a minimum, the following:

A verified database that contains toxicity as well as structural information

A statistical technique (usually some form of correlation analysis)
Selection of key structural parameters (e.g., ring versus no ring) for the correlation
Computer technology

In a strict sense, QSAR toxicity estimates are actually in calculo estimates rather than in vitro evaluations. The accuracy and reliability of any QSAR estimate is highly dependent on the extent and quality of the database from which the estimate is derived. For one of the better databases, which includes a total of 786 compounds, the predictive accuracy is reported to be approximately 90%, with indeterminate estimates of about 12% [1]. Because QSAR databases are dependent for validation on actual quantitative in vivo and in vitro data, it can be anticipated that the current interest in developing in vitro skin toxicity models that have as a basis some understanding of chemical structure versus biochemical target mechanics will only serve to enhance the quality of QSAR well into the future.

## Biochemically Based Models

The biochemically based models are based on the concept that chemical toxicity in the skin originates from some type of deleterious interaction between a biomolecular target molecule and a toxicant. For example, it is thought that organic arsenicals exert their initiative effects as a result of pyruvate dehydrogenase inhibition with basilar and maturing keratinocytes [2]. Surfactants are usually thought to act on membrane lipids to produce dermal inflammation and necrosis, and sulfur mustard vesication is thought to involve stimulation of DNA repair mechanisms following DNA alkylation [3,4]. The development of successful biochemically based models usually involves the identification of molecular target(s), their subsequent isolation, and the development of a controlled test procedure using the isolated system. An example of a recently developed model that employs a more collective biochemical target interaction approach is the Skintex system (Fig. 4) [5]. Because this system comprises both a biochemically definable penetration barrier as well as a complex molecular target matrix, it is probably best described as an advanced biochemically based test. Molecular interactions within the test system are characterized by increasing measurements of optical density. The scoring of test results is accomplished by plotting the optical density measurements obtained with known skin irritants against their known, in vitro, primary dermal irritation index (PDII) values and then locating unknown predictions based on their respective optical density measurements (Fig. 5). In its present form, the Skintex assay has been optimized for four different test predictive requirements:

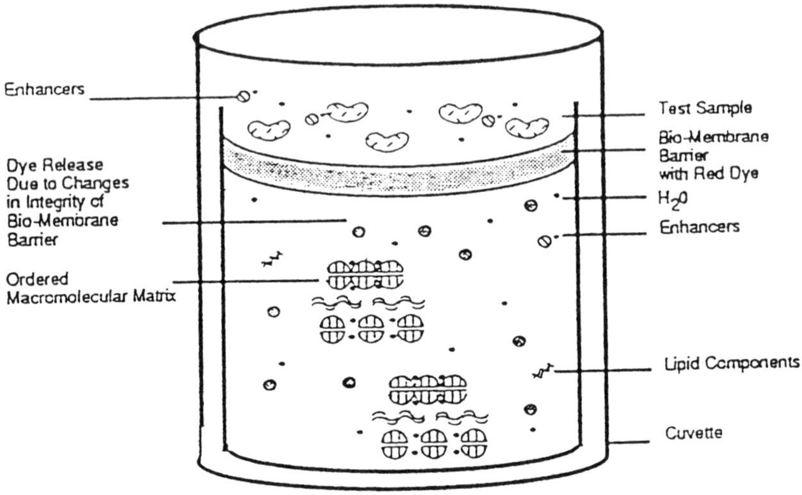


**Figure 4** A schematic representation of the Skintex assay system. (From Ref. 5.)

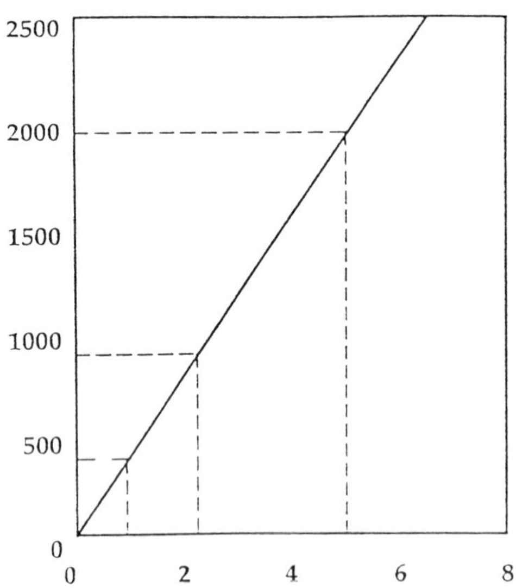


**Figure 5** The primary dermal irritation index (PDII) correlated Skintex assay scoring procedure. (From Ref. 5.)

1. Acidic or alkaline substances
2. High sensitivity
3. Human response
4. Standard chemical labeling

## Cellular Models

The cellular models currently employ human keratinocytes grown in a serum-free medium, in most instances, but models using fibroblasts, Langerhans cells, leukocytes, and mast cells are also available. Cellular models, in general, provide a greater degree of molecular integration and diversity than the biochemically based models. They are commonly used to assess the effects of chemicals on specific cell types; most commonly, the endpoint of interest is cellular viability, as determined using any of several possible assays, including neutral red dye uptake; 3-(4,5-dimethylthiazol-2-yl)-2,5-diphenyl tetrazolium bromide (MTT) dye uptake; trypan blue dye retention; propidium iodide uptake; lysosomal enzyme release; [$^3$H]thymidine incorporation; glucose utilization; and others. There are also cellular models that involve the measurement of effects on specific intracellular targets, such as DNA alkylation, membrane integrity, mitochondrial effects; and models that assess effects on specific biochemical pathways, NADH depletion, ATP synthesis, glucose metabolism, or protein synthesis. The selection of the cell model and endpoints used to assess effects should be based on the needs of the investigation and must always take into consideration the possibility of interference by direct interaction of the test chemical with some aspect of the assay (e.g., chemical interaction with dyes used to assess viability) or indirect action with some aspect of the assay environment (e.g., pH changes in the medium that affect enzymatic activity or dye color) [6]. Since human keratinocytes have recently become readily available commercially with a reasonable degree of quality control, it is reasonable to assume that relatively standardized cellular toxicity screening models using keratinocyte cultures will become popular in the near future. What remains are careful assessments of the predictive accuracy, sensitivity, and validity of these models for either generalized- or specific-screening use.

## Tissue Culture Models

Tissue culture models may consist of either natural human or animal skin explants or cultured human or animal skin "equivalent" material. Toxicity assessments using tissue culture models generally concern themselves with effects on both specific cell types as well as on structural relations and proliferative activity. Needless to say, natural skin is clearly superior over any of the cultured tissues in terms of its similarity of structure and functional relation to human in vivo skin models. Natural skin models, however,owing to individual and collection site diversity, suffer from a lack of providing a highly controlled and standardized test bed for routine toxicity assessment. This, combined with the rather opportunistic aspect of the availability of suitable quantities of natural human skin for test use, makes the development of routine screening tests based on human skin tissue culture rather impractical. Natural human tissue, when available, therefore, is used as a final comparator or standard for the development of screening tests using the more controlled and supplied-on-demand skin equivalent technologies, which have very recently become commercially available.

In selecting a human skin equivalent material for use as the basis for a screening test, several items must be considered, in addition to the obvious issue of cost per test unit. The desired endpoints must be carefully weighed against each model to determine model suitability for the given purpose. Skin equivalent models available commercially are likely to differ in potentially important structural aspects, such as the presence or absence of a basement membrane, the number of keratinocytes in epithelial "columns," degree of cornification of the keratinocyte strata, and cellular morphology of different cells used to develop the model. In a more functional sense, the models may also differ in the degree of quality control used to produce them, their structural integrity relative to natural skin, and their packaging which, in turn, may affect efficient handling of the product in screening tests.

Whether the assay uses natural human skin or a human skin equivalent, the number of possible endpoints that can be used for the toxicity assessment is much greater for tissue culture models than any other type. In addition to many of the viability, target-specific, and metabolic endpoints used for toxicity assessment in the cell culture models, tissue culture models offer the ability to study both structural changes as well as cell–cell interactive effects such as inflammatory mediator release [6].

## Lower Vertebrate Models

The use of a lower vertebrate model, more specifically, the chorioallantoic membrane (CAM) of a developing chicken embryo, has been proposed for use as an alternative model for the assessment of dermal and ocular acute irritation and corrosion testing. Even if it is not an in vitro alternative model, interest in this model has been high because of its low cost and relative simplicity. Although evaluated extensively in both the United States and Europe as a possible dermal and ocular Draize test alternative, the model seems to be currently of greater interest in Europe as a possible alternative than in the United States. The reason for this is unclear from a purely scientific perspective, but may be due to more success in validating the test against European in vivo test results and a significantly higher demand for the development, approval, and use of alternative models in Europe [7].

## ASSAY DEVELOPMENT AND VALIDATION CONSIDERATIONS

Although often there are, by design, a few technical similarities between in vitro and in vivo procedures to evaluate the adverse effects of dermal irritants and inflammatory agents, many of the in vitro test systems involve, overall, several more technically complex operations relative to their in vivo counterparts. As a result, there can be many problems and sources of error that exist for such in vitro systems that are either absent from, or less vexing for in vivo test systems. Because several in vitro test systems have been developed, evaluated, and proposed for use as possible replacements of or adjuncts to various types of in vivo tests over the past two decades, there exists a relatively large and expanding body of knowledge concerning the nature and types of problems one can expect when developing or conducting in vitro toxicity tests. Although the technical problems associated with particular tests are routinely discussed in the literature, at scientific meetings, or between individual scientists, very few publications are available that systematically present, discuss, and summarize many of the common technical problems associated with in vitro test systems. Recently, however, a technical report was issued by the Center for Alternatives to Animal Testing (CAAT) at the Johns Hopkins School of Hygiene and Public Health, as the result of findings from a technical workshop that discussed and summarized many of the technical problems encountered in the development of cell culture assays to assess chemical toxicity [8]. The report identifies several factors that may confound the results obtained from in vitro tests and notes that these factors tend to fall into two general categories: (a) those that affect biological components (i.e., cultured cells, enzymes, membranes) of the test, and (b) those that affect the dispersal or stability of the test material in the test medium. Category (a) includes factors that may unintentionally alter or influence the responsiveness of the biological component of the test to the inherent toxic effects of the test chemical, such as the following:

pH changes in the test medium, resulting from addition of the test chemical.

Alterations in test medium osmolarity at high concentrations of the test chemical.

Alterations in the composition of test medium resulting from reactions between test chemicals and medium components (e.g., protein denaturation, alteration in the availability of essential substrates or cofactors).
Confounding effects on the test system response occurring from the presence of chemicals (e.g., alcohols, dimethylsulfoxide) used as solvents for test chemicals.
Interactive effects of test chemicals and antimicrobial agents on test system response.
Complications because of evaporation of volatile test chemicals.
Interference with the endpoint measurement (e.g., the presence or formation of interfering colors in a colorimetric assay directly from test chemicals, interaction between an enzyme endpoint and the test chemical).
Influence of protein binding of test chemicals on the concentration–response curve.
Effect of test incubation time or temperature on the availability of essential nutrients.
Effect of test incubation time or temperature on the responsiveness of the test system to test chemicals.
Deterioration of test system purity and responsiveness as a result of microbial contamination.
Physical effects of insoluble particulates, formed as a result of interactions between test chemicals and components of the assay system, on the responsiveness of the test system.

The second category, category (b), of potential technical problems identified in the report includes potentially confounding factors related to the dissolution of the test chemical in the test medium. Avoidance of problems in this category involves the examination of certain specific issues before actual testing to avoid the occurrence of physical or chemical problems resulting from the incompatibility of the test chemical with the biological components of the test. The fact that many test chemicals or chemical formulations are hydrophobic and that most in vitro systems exist within an aqueous matrix in which such test chemicals are either insoluble or nondispersive is one of the principal problems of this type that can occur. Some of the most serious issues limiting the general usefulness of such in vitro systems stem from this fundamental problem of test chemical dispersal within the test matrix. For example, it is very difficult to establish the proper dose administered to a test system if the test chemical is either incompletely dispersed within, or degraded (e.g., hydrolyzed) by, the matrix. It is also difficult to determine the most appropriate means of physically introducing a test chemical into a test matrix in which it is immiscible. Aggregates of nondispersed chemicals or chemical mixtures can also have physical properties that cause mechanical damage to the test system, rather than intrinsic toxicity (e.g., detergent or crystallization effects on cellular membranes).

The incompatibility of test chemicals with components of the test medium is probably one of the most significant problems encountered with in vitro dermal toxicity assays. It also appears to be one of the most common problems. This is probably because many in vitro systems require that the test chemical be introduced into some type of liquid test matrix (e.g., buffer, cell culture medium), and even after pH, osmolarity, and solvent effects are controlled or eliminated, degradation or binding of the test chemical in the test matrix can still occur.

No matter what technical problems are encountered, identified, and then solved, during the development of an in vitro assay, the assay is only going to be of use if it is capable of producing results with an acceptable degree of validity to some standard for which its development was intended. Even though it is true that significant technical problems can and do often result in problems that affect the validity of such assays, the most difficult problem currently affecting the achievement of valid results for most assays concerns the proper means by which assays are designed and determinations of validity are made.

## CONCLUSIONS

The number of in vitro or alternative assays proposed for use in the evaluation of the potential for toxic chemicals or chemical mixtures to cause dermal irritation, inflammation, and corrosion is rapidly growing, stimulated by a continuing desire to develop valid methods for the reduction and replacement of current laboratory animal tests, as well as the introduction of new bioassay procedures and technology. Proposed alternative systems that have been shown to be responsive in some fashion to known dermal irritants, inflammatory, or corrosive agents can be categorized by their biological complexity. In ascending order of complexity, these systems include QSAR procedures, biochemical assays, cellular assays, tissue culture models, and lower vertebrate models. Unfortunately, none of these systems has been universally acceptable for use as a routine replacement for any in vivo procedure to evaluate such dermal effects. Reasons for this are numerous and varied. To some extent, this lack of acceptability appears to involve the inherent difficulty in getting any new technology intended to replace an accepted standard generally recognized and accepted by the public, scientists, businessmen, and regulators. There is, in fact, a strong and growing interest among toxicologists to utilize alternative test systems whenever possible as replacements for or adjuncts to in vivo procedures.

Unfortunately, there are several important scientific issues related to the design, operation, and validation of many of the alternatives that must be resolved before any of these systems is generally acceptable as a replacement for any in vivo test procedure. These issues vary from one proposed alternative system to another. Often a test system is identified that responds as expected to some test agents and then, after more extensive evaluation, it is unexpectedly unresponsive to others.

Thus, the first step toward the development of universally acceptable assays should be to perform studies designed to identify any significant variables that must be controlled to establish an acceptable level of response and quality control.

Despite the growing number of proposed tests, the acceptance of such tests by toxicologists as valid replacements for animal tests has been cautious. This situation is, in large part, due to the need to conduct more comprehensive and impartial validation studies involving a wide spectrum of chemicals with known in vivo dermal toxicity and for the establishment of formalized criteria to define the circumstances under which a particular alternative test is considered valid. As such, there is a need for the development of a generally acceptable validation protocol for the development and acceptance of such procedures, if they are to become widely accepted. Unfortunately, in addition to the inherent problems with the tests themselves, the funding and organization of large-scale validation studies is currently not widely available for many of the potential users of such tests, and progress toward acceptance has been slow. It is anticipated, however, that because of the increasing desire among scientists, politicians businessmen, and consumers to identify suitable alternative tests to assess dermal toxicity, validated alternative procedures will emerge, within a few years, for the assessment of specific types of cutaneous toxicity endpoints that will be widely accepted by toxicologists.

## REFERENCES

1. Enslein, K., Borgstedt, H. H., Blake, B. W., and Hart, J. B., *In Vitro Toxicol. 1*:129 (1987).
2. Hobson, D. W., Snider, T. H., Chang, M. J., and Joiner, R. L., *Toxicologist 8*:20 (1988).
3. Houk, J., Hansch, C., Hall, L. L., and Guy, R. H., Chemical structure-transport rate relationships for model skin lipid membranes, *Alternative Methods in Toxicology*, Vol. 5 (A. M. Goldberg, ed.), Mary Ann Liebert, New York, 1987.

4. Papirmeister, B., Gross, C. L., Meir, H. L., Petrali, J. P., and Johnson, J. B., *Fundam. Appl. Toxicol. 5*:s134 (1985).
5. Gordon, V. P., Kelly, C. P., and Bergman, H. C., *Toxicologist 9*:6 (1989).
6. Hobson, D. W., and Blank, J. A., In vitro alternative methods for the assessment of dermal irritation and inflammation, *Dermal and Ocular Toxicology: Fundamentals and Methods* (D. W. Hobson, ed.), CRC Press, Boca Raton, FL, 1991.
7. Leighton, J., Nassauer, J., and Tchao, R., *Food Chem. Toxicol. 23*:293 (1985).
8. Frazier, J. M., and Bradlaw, J. A., eds., *Technical Report No. 1: Technical Problems Associated with In Vitro Toxicity Testing Systems—A Report of the CAAT Technical Workshop of May 17–18, 1989*, Johns Hopkins School of Hygiene and Public Health, Baltimore, 1989.
9. Boyce, S. T., Hansbrough, J. F., and Norris, D. S., *Alternative Methods in Toxicology*, Vol. 6 (A. M. Goldberg, ed.), Mary Ann Liebert, New York, 1988.
10. Monteiro-Riviere, N. A., Comparative anatomy, physiology, and biochemistry of mammalian skin, *Dermal and Ocular Toxicology: Fundamentals and Methods* (D. W. Hobson, ed.), CRC Press, Boca Raton, FL, 1991.
11. Kao, J., In vitro assessment of dermal absorption, *Dermal and Ocular Toxicology: Fundamentals and Methods* (D. W. Hobson, ed.), CRC Press, Boca Raton, FL, 1991.

# 16

# In Vitro Dermal Toxicity Assays: Validation with Human Data

**Jeff D. Harvell and Howard Maibach**

*University of California School of Medicine*
*San Francisco, California*

In the past two decades, the scientific community has witnessed greater interest in the development of alternative in vitro assays for predicting ocular and dermal toxicity. The reasons for this relate mainly to criticisms of the most widely used test for ocular and dermal toxicity—the rabbit Draize test [1]. The criticisms directed against the in vivo rabbit test include (a) the subjective nature of the visual-scoring system employed, (b) the questionable ability to extrapolate results to humans [2–4], (c) the fact that the test is time-consuming and costly, and (d) the fact that the test results in animal distress.

There currently exist various in vitro alternatives designed to predict the dermal toxic or irritant capability of compounds in humans. These alternatives are based on a variety of methods, including cell culture, physicochemical analysis, microorganism studies, isolated tissue techniques, and computer modeling. Because the process of dermal irritation is complex and its various mechanisms have yet to be fully discerned, no single parameter has emerged as a "best predictor" of irritation. The myriad of in vitro assays, therefore, employ a wide array of endpoints designed to be predictive of irritation or toxicity. Such endpoints include cell death, decreased cellular metabolic function, decreased cellular protein production, decreased mitochondrial function, decreased fluorescence capability, decreased motility of simple protozoa, histological changes, release of intracellular enzymes, release of inflammatory mediators, damage to vascular systems, and protein denaturation. The usefulness of these endpoints are made more clear in the following paragraphs that discuss individual test methods.

Cytotoxic tests that use cell culture systems include the neutral red dye uptake assay, the 3-(4,5-dimethyl-2-thiazoyl)-2,5-diphenyl-2*H*-tetrazolium bromide (MTT) dye uptake and reduction assay, and assays that measure the release of cellular proteins or inflammatory mediators. The neutral red dye uptake assay uses a dye that is taken up by viable cells, and retained in lysosomes. Perturbations by cytotoxic agents that result in damage to either plasma membrane or lysosomal membrane uptake systems, therefore, result in decreased dye uptake. The endpoint for this assay is known as the $NR_{50}$ and represents that concentration of toxicant that reduces by 50% the uptake of dye, when compared with control cells [5]. The MTT assay correlates cell viability with both intact mitochondrial function and intact plasma membrane uptake processes [6,7]. This system uses a tetrazolium salt that, in viable cells, is reduced to a blue formazan product by the action of mitochondrial enzyme systems. Cytotoxic agents that interfere with mitochondrial function cause decreased formation of the blue formazan product. The measured endpoint for this system is an $MTT_{50}$ which, like the $NR_{50}$, represents that concentration of toxic agent that reduces the formation of formazan product by 50%.

Other cell culture-based systems utilize the release of cellular proteins, such as lactate dehydrogenase, β-glucuronidase, or alkaline phosphatase [8,9]. The release of these intracellular enzymes is representative of damage to plasma membrane retention systems by cytotoxic agents. Measurements of inflammatory mediator release, such as arachidonic acid, prostaglandins, and leukotrienes, represent the cells' response to perturbation by irritating (but not necessarily lethal) substances [10,11]. These mediators are synthesized by viable cells and released to the extracellular matrix as part of the inflammatory response to irritating substances.

Finally, other cell culture-based systems have correlated cell viability to such endpoints as total cellular protein content (Kenacid blue dye method) [12], glucose utilization (i.e., depletion of glucose from the extracellular medium) [13], and changes in cellular metabolic activity. The latter system is marketed as the silicon microphysiometer, and represents a sensitive method capable of detecting small changes in the metabolic activity of a monolayer cell system [14]. The principles of this device are based on the premise that cellular metabolic activity is reflected by the concentration of extracellular, acidic byproducts, such as lactic acid and carbon dioxide. Changes in the metabolic activity of the monolayer (as a result of toxic or irritant effects) are, therefore, represented by changes in the pH of the extracellular medium. These subtle pH changes are measured by means of a silicon-based electrode, known as a light-addressable potentiometric sensor (LAPS).

All of the aforementioned cell culture-based methods generally use monolayer cell systems; however, there now exist commercially available multilayer cell systems, collectively known as skin equivalents [15,16]. Such systems consist of multilayers of dermal fibroblasts or keratinocytes grown on a variety of substrates. In many of these commercially available models, the cells are keratinocytes or fibroblasts derived from human tissue (i.e., as a result of circumcision or breast reduction surgeries). Some of the systems contain a stratified, differentiated epithelium, which arises as a result of raising the system to the air–liquid interface. In general, the same endpoints used for the monolayer systems can also be used in these multilayer skin equivalents. In the future, the incorporation of other cell types, such as mast cells, macrophages, neutrophils, or others, will be the goal in making these systems even more predictive of the in vivo situation.

Other in vitro assay systems employ the chemical processes of bacteria as a measure of toxic effect. The Microtox system uses the luminescent bacterium *Photobacterium phosphoreum*. Toxic agents generally cause a decrease in the fluorescence capabilities of this bacterium, the degree of which can be quantitated and correlated to degree of toxicity [17]. A similarly based system uses the ciliated protozoan *Tetrahymena thermophila* and correlates toxic effect to the endpoint of decreased motility [18,19].

Since the primary events in inflammation (i.e., erythema, heat, edema) are dependent on vascular processes, it follows that investigation of the effect of irritant or toxic compounds on a vascular network would constitute a suitable in vitro alternative. Such a model is represented by the choriallantoic membrane system (CAM) [20,21]. This system uses fertilized chicken eggs, the vascular network (or CAM) of which is exposed by cutting a small "window" into the shell. Test compounds can then be applied directly to the CAM, and their effects can be graded by noting visual changes in the blood vessel network (i.e., hemorrhage, injection, coagulation).

Physicochemical models for the prediction of dermal toxicity are rare. The reason for this relates to the paucity of knowledge on the effects of toxicants at the molecular level. One model, the Skintex dermal assay system, uses protein denaturation as an endpoint [22]. The ability of toxicants or irritants to disturb the ordered array of protein components in the system is measured spectrophotometrically. This model provides a quantitative response to materials that may produce irritation by protein binding, enzyme inactivation, and a variety of other pathways in which macromolecular confirmation is altered in the initiation of dermal irritation.

All of the foregoing assays are currently in a stage of validation, the goal of which is to assess both their reproducibility (i.e., degree of interlaboratory variation) and their relevance (i.e., to what degree are these tests predictive of in vivo irritation or toxicity?) [23]. The assessment of relevance, therefore, is most often made by comparing the in vitro result with some in vivo result. Currently, most validation projects use the Draize rabbit assay as the in vivo arm of the correlate; however, the question arises: Should the "standard" for in vivo irritancy reside in a species other than humans? The reasons for the use of rabbits in toxicity testing are obvious; however, their use has been criticized because of questionable ability to extrapolate results to humans and other species [2–4]. With this in mind, it makes sense that at least some aspect of the validation process should attempt to correlate the results of in vitro dermal toxicity tests to in vivo human data.

Such has been the aim of our laboratory. One example of this approach compared the irritant capabilities of the irritants benzalkonium chloride, trichloroacetic acid, phenol, and hydrochloric acid in the human and in the Skintex dermal assay system. The Skintex system was fairly sensitive in its ability to predict the irritant potential of these compounds in humans (sensitivity 82%, specificity 71%, positive predictive value 82%) [24]. Additionally, the in vivo dose–response curves for each of the four substances was compared with the in vitro dose–response curves, and correlation coefficients calculated. The in vitro dose–response for benzalkonium chloride ($R^2 = 0.987$) and phenol ($R^2 = 0.994$) were strikingly similar to those generated in vivo, possibly indicating that the mechanisms of action in vivo and in vitro are similar for these two compounds.

Such a comparative approach with in vivo human data will, no doubt, enhance the ability to assess the true relevance of a particular in vitro system; however, this approach is not without its limitations. Probably the most significant weakness of this approach is that one is limited to using fairly innocuous compounds in the in vivo arm of the comparison. Therefore, we may be skewing the relevance assessment to less irritating or toxic compounds. On the other hand, it is generally thought that if a test is sensitive enough to detect differences in the toxic potential of compounds that are in a narrow range of irritancy, it is most likely capable of making more broad-ranged assessments. The veracity of the latter statement will no doubt be borne out by future validation projects.

Other problems encountered in the process of validation that are more general and are irrespective of the in vivo species used include the following: (a) How can an assessment be made of a particular in vitro assay's ability to predict both acute and chronic irritation? (b) Similarly, what in vivo exposure period should be used to compare results—24 hr? 72 hr? (c) Should the in vivo exposure period be an open or closed test?, and (d) How can the in vivo data be made more objective?

One solution to these questions would involve a standardized approach to making in vitro–in vivo correlations. Such an approach would test a reference set of compounds both in vitro and in vivo. The in vivo test, whether performed on animals or humans, would involve assessments of both acute irritancy (i.e., after 24 hr) and cumulative irritancy (over the course of 21 days). Also, the in vivo irritancy test should be performed under both open and closed conditions. Finally, in addition to visual assessments of irritancy (which tend to be subjective), more objective measures of irritancy (such as transepidermal water loss, blood flow, chromatography, and conductance) should also be made. By proceeding in such a manner the in vitro result can be compared with an in vivo database that encompasses several variables (time exposure, mode of application to the skin, different measures of erythema, or different measures of barrier disruption). In vitro–in vivo comparisons, therefore, are more sound, and an improved assessment of the in vitro assay's relevance is achieved.

What then is the future for in vitro assays? First, it is apparent from already completed validation steps that a single in vitro assay will not be capable of predicting the dermal irritancy for all classes of compounds. It seems unlikely that one assay could be predictive for the array of vastly different chemical structures that induce irritation by as yet undefined mechanisms. Rather, batteries will be employed that use single in vitro tests, each test differing in the molecular or physiological endpoint measured. Second, as confidence in the reproducibility and relevance of these in vitro alternatives increases, we will see a reduction in the number of animals used for toxicity study. The total elimination of animal use for toxicity study is most likely far in the future. Finally, as experience with the array of in vitro assays increases—each test with its distinct physiological or molecular endpoint–knowledge concerning mechanisms of dermal irritation and toxicity will expand concurrently.

## CONCLUSIONS

To protect humans from substances, skin irritation tests must be performed. At a minimum, the new tests must be sensitive enough to characterize the potential degree of irritation. In the future, the developing skin systems may also provide information about the mechanism by which a substance causes irritation.

It is unlikely that an in vitro system could ever be developed to mimic the complex cascade of reactions that occur in the human skin. However, an in vitro system used as an initial screening device would permit the use of a minimum number of animals for skin tests and simplify the process by which new compounds are developed.

Appropriate use of human irritancy data with standard compounds, employing occlusive and nonocclusive dosing, should permit a facile and realistic correlation between the in vitro assays and likely human experience. In vitro irritation systems are now entering a validation phase. Criteria must be established permitting identification of standards (i.e., how irritating are model compounds in humans, or animals)? Personal impressions suggest that the data must be clearly related to the method of application (occlusive or open), single-dose (primary irritant) or multiple dose (cumulative irritation); and anatomical site (face responding differently than back).

## REFERENCES

1. Draize, J. H., Woodard, G., and Calvery, H. O., Methods for the study of irritation and toxicity of substances applied topically to the skin and mucous membranes, *J. Pharmacol. Exp. Ther. 82*:377–389 (1944).
2. Kastner, G., Irritancy potential of cosmetic ingredients, *J. Soc. Cosmet. Chem. 28*:741–754 (1977).
3. Patrick, E., and Maibach, H. I., Comparison of the time course, dose response, and mediators of chemically induced skin irritation in three species, *Current Topics in Contact Dermatitis* (P. J. Frosch et al., eds.), Springer-Verlag, New York, 1989, pp. 399–403.
4. Davies, R., Harper, K. H., and Kynoch, S. R., Inter-species variation in dermal reactivity, *J. Soc. Cosmet. Chem. 23*:371–381 (1972).
5. Borenfreund, E., and Puerner, J. A., A simple quantitative procedure using monolayer cultures for cytotoxicity assays, *J. Tissue Cult. Methods 9*:7–9 (1984).
6. Mosman, T., Rapid colorimetric assay for cellular growth and survival: application to proliferation and cytotoxic assays, *J. Immunol. Methods 65*:55–63 (1983).
7. Tada, H., Shiho, K., Kuoshima, K., Koyama, M., and Tsukamoto, An improved colorimetric assay for interleukin 2, *J. Immunol. Methods 93*:157–165 (1986).
8. Korzeniewski, C., and Callewaert, D. M., An enzyme release assay for natural cytotoxicity, *J. Immunol. Methods 64*:313–320 (1983).
9. Szekers, A., Pasca, S., and Pejtsik, B., *J. Immunol. Methods 80*:1 (1981).
10. Bell, E., Gay, R., Swiderick, M., Class, T., Kemp, P., Green, G., Haimes, H., and Bilbo, P., NATO Advanced Research Workshop, Pharmaceutical Application of Cell and Tissue Culture to Drug Transport, Bandol, France, 1989.
11. DeLeo, V., Harber, L. C., Kong, B. M., and DeSalva, S. J., Surfactant induced alteration of arachidonic acid metabolism of mammalian cells in culture, *Proc. Soc. Exp. Biol. Med. 184*:477–482 (1987).
12. Knox, P., Uphill, P. F., Fry, J. R., Benford, J., and Balls, M., The FRAME multi-centre project on in vitro cytotoxicology, *Food Chem. Toxicol. 24*:457–463 (1986).

13. Mol, M. A. E., Van de Ruit, A. B. C., and Kluivers, A.W., $NAD^+$ levels and glucose uptake of cultured human epidermal cells exposed to sulfur mustard, *Toxicol. Appl. Pharmacol. 98*:159–165 (1989).
14. Parce, J. W., Owicki, J. C., Kercso, K. M., Sigal, G. B., Wada, H. G., Muir, V. C., Bousse, L. J., Ross, K. L., Sikic, B. I., and McConnell, H. M., Detection of cell-affecting agents with a silicon biosensor, *Science 246*:243–247 (1989).
15. Naughton, G., Jacob, L., and Naughton, B. A., *A Physiological Skin Model for In Vitro Toxicity Studies*—A symposium of the CAAT technical workshop of May 17–18, 1989,Vol. 7 (J. M. Frazier and J. A. Bradlaw, eds.), Johns Hopkins School of Hygiene and Public Health, Baltimore, 1989.
16. Bell, E., Parenteau, L., Haimes, H. B., Gay, R. J., Kemp, P. D., FoFonoff, T. W., Mason, V. S., Kagan, D. T., and Siderek, M., Testskin: A hybrid organism covered by a living human skin equivalent designed for toxicity and other testing, *Alternative Methods in Toxicology*, Vol. 6 (A. M. Goldberg, ed.), Mary Ann Liebert, New York, 1988.
17. Bulich, A. A., Greene, M. W., and Isenberg, D. L., Reliability of bacterial luminescence assay for determination of the toxicity of pure compounds and complex effluents, *Aquatic Toxicology and Hazard Assessment*, 4th Conference. American Society for Testing and Materials (ASTM 737) (D. R. Branson and K. L. Dickson, eds.), Philadelphia, 1981, pp. 338–347.
18. Silverman, J., Preliminary findings on the use of protozoa (*Tetrahymena thermophila*) as models for ocular irritation testing rabbits, *Lab. Anim. Sci. 33*:56–58 (1983).
19. Silverman, J., and Pennisi, S., Evaluation of *Tetrahymena thermophilia* as an in vitro alternative to ocular irritation studies in rabbits, *J. Toxicol. Cutan. Ocul. Toxicol. 6*:33–42 (1987).
20. Luepke, N. P., and Kemper, F. H., The HET–CAM test: An alternative to the Draize eye test, *Food Chem. Toxicol. 24*:495–496 (1986).
21. Bagley, D. M., Rizva, P. Y., Kong, B. M., et al., An improved CAM assay for predicting ocular irritation potential, *Alternative Methods in Toxicology*, Vol. 6 (A. M. Goldberg, ed.), Mary Ann Liebert, New York, 1988, pp. 131–138.
22. Gordon, V. C., Kelly, C. P., and Bergman, H. C., Evaluation of "Skintex," an in vitro method for determining dermal irritation, *Toxicologist 10*:78 (1990).
23. Balls, M., Botham, P., Cordier, A., Fumero, S., Kayser, D., Koeter, H., Koundakjian, P., Gunnar Lindquist, H., Meyer, O., Pioda, L., Reinhardt, C., Rozemond, H., Smyrniotis, T., Spielmann, H., Van Loy H., van der Vanne M., Walum, E., Report and recommendations of an international workshop on promotion of the regulatory acceptance of validated nonanimal toxicity test procedures. *Alternatives Lab. Anim. 18*:339–344 (1990).
24. Bason, M. M., Harvell, J., Realica, B., Gordon, V., and Maibach, H. I., In vitro correlation with human in vivo data of selected primary irritants. *Toxicol. In Vitro* (in press).

# 17

# Skin Penetration and Allergic Contact Dermatitis: Historical Perspective

**Francis N. Marzulli**

*Consultant in Toxicology*
*Bethesda, Maryland*

## SKIN PENETRATION

Systematic studies in skin penetration by topically applied chemicals were undertaken by British investigators in the 1950s. Treherne [1], followed by Ainsworth [2] and Tregear [3] conducted diffusion studies on isolated pieces of skin, using it as a complex membrane. They focused on the permeability constant as a measure of the capacity of a chemical to penetrate skin.

Originally, the skin was thought of as an inert protein–lipid structure the effectiveness of which was largely due to the outer keratinous barrier structure.

More recently, attention has been directed to the metabolic potential of the underlying living epidermis.

In vivo studies at first employed pharmacological endpoints as a measure of skin penetration. One technique that is still used involves the blanching effect of topically applied steroids as an indirect semiquantitative measure of skin penetration. This method was championed by Stoughton [4]. Over the years, work with both isolated skin and with skin in vivo has provided a better, but still incomplete picture of events involved in the skin penetration processes. The arrival on the scene of relatively inexpensive radiolabeled chemicals enabled a fuller development of both in vivo and in vitro techniques for studying skin penetration. Maibach and co-workers [5] conducted skin penetration studies on human subjects in vivo. They reported their results in terms of the percentage of applied chemical that penetrated skin. Blank [6], followed by Scheuplein [7], expanded our knowledge of skin penetration using in vitro techniques with skin as a membrane. Blank did his early work on contract with Edgewood and Scheuplein did his work at Edgewood as a Chemical Corps officer at the Medical Laboratories.

More recently, Bronaugh [8] of the U.S. Food and Drug Administration (FOA), has focused attention on metabolism as an important component of the skin penetration process. Riviere [9] uses the porcine skinflap, and Anderson and associates [10] of Wright-Patterson Air Force Base, favor pharmacokinetic modeling as the approach of the future.

A comprehensive review of this subject has been developed in a document recently released by the U.S. Environmental Protection Agency (EPA) [11]. It contains up-to-date skin penetration information needed to develop a risk assessment document. There are many others besides those mentioned, who contributed to our present understanding of percutaneous absorption.

Some of the complexities of the skin barriers and properties of the stratum corneum that relate to penetration by fat-soluble and water-soluble compounds are now known, as are the effects of occlusion, hydration, temperature, molecular size, age, and skin site. Studies in animals and in humans, in vitro and in vivo, have all contributed to our present understanding of the processes that are involved in skin penetration.

## ALLERGIC CONTACT DERMATITIS

Allergic contact dermatitis (ACD) is, as the name implies, a dermatitis or inflammatory skin response produced by skin contact with an allergenic substance. Allergic contact dermatitis, as distinguished from irritant contact dermatitis, is a delayed, immune response that results in erythema, edema, and vesiculation. The skin response resulting from contact with poison ivy is a good example of this effect.

The first requirement for the production of allergic contact dermatitis is that the allergenic substance penetrate the horny layer of skin and enter the viable epidermis where it makes contact with and binds to Langerhans' cells. These are dendritic cells in the epidermis. The Langerhans' cells direct the allergen to a regional lymph node, where interaction with T lymphocytes initiates a process that results in abundant replication of sensitized T lymphocytes. This constitutes the induction phase of the sensitization process. The next contact with the allergen results in elicitation of a hypersensitive skin response. That is, the skin responds to a lower concentration of allergen than was required before induction.

The fact that the immune system is involved and a two-step process—induction and elicitation—is required in allergic contact dermatitis, makes it more difficult to simulate this process in an alternative system that lacks the complexity of an immune system. The patch test in some form provides the main approach to assessing allergic contact dermatitis. Jadassohn [12] in 1895 is thought to be the first to formalize the patch test as a means of producing skin disease on a small scale, by applying a test chemical under an occlusive patch to identify those chemicals that are allergenic. A patient of Jadassohn with hypersensitivity to mercury was demonstrated to be allergic by means of a patch test. In this case, the patch test was diagnostic for a subject who was already sensitized. The patch test can also be used in a two-step procedure to evaluate whether a chemical is allergenic. In this type of patch testing, induction of allergic contact dermatitis is accomplished by repeatedly applying a high concentration of suspected allergen. After an appropriate time frame, during which immunogenic events take place, the elicitation or challenge phase is carried out with a nonirritant concentration to find out if the test material is indeed allergenic.

Other early workers in this field were Landsteiner and Chase [13], who suggested a role for sensitized lymphocytes in allergic contact dermatitis. Their work was followed by the development of provocative tests in both animals and humans to identify substances with a proclivity to sensitive. This work led to information on sensitization frequency and the concentrations required for these allergic contact dermatitis effects.

It was recognized early that induction of allergic contact dermatitis could be accomplished by employing frequent contacts with a test allergen. Schwartz and Peck [14] and Shelanski [15] developed tests based on this type of repeated insult. It was also recognized that high concentrations of a suspect allergen at induction could also be effective. This is the approach that Draize [16] used. Kligman [17] used sodium lauryl sulfate (SLS) to assist in the induction process. When sodium lauryl sulfate is applied to the skin, it provides subtle damage that enhances skin penetration of topically applied substances. By this means, a larger amount of the penetrant reaches the immune system, thereby facilitating the induction process. The sodium lauryl sulfate may act like a higher concentration of the test material.

Guinea pigs became the first animals of choice for test purposes. Draize [16] suggested delivery of the test allergen by intradermal injection. Buehler [18] introduced essentially a modified human patch test method using guinea pigs as surrogates. Magnusson and Kligman [19] employed Freund's adjuvant to enhance the responsiveness of the animals. More recently, mouse ear-swelling tests [20] have come into vogue. The lower cost of purchasing and maintaining mice and the greater use of mice in other immunological work has probably hastened this development.

This brief introductory history is intended only to whet the appetite of investigators who seek alternatives to present methodologies.

## REFERENCES

1. Treherne, J. E., The permeability of skin to some nonelectrolytes, *J. Physiol. (Lond.) 133*:171–180 (1956).
2. Ainsworth, M., Methods for measuring percutaneous absorption, *J. Soc. Cosmet. Chem. 11*:69–78 (1960).
3. Tregear, R. T., *Physical Functions of Skin*, Academic Press, New York, 1966, pp. 1–72.
4. Stoughton, R., Some in vivo and in vitro methods for measuring percutaneous absorption, *Prog. Biol. Sci. Relat. Dermatol. 2*:263–274 (1964).
5. Maibach, H. I., Feldman, R. J., Milby, T. H., and Serat, W. F., Regional variation in percutaneous penetration in man, *Arch. Environ. Health 23*:208–211 (1971).
6. Blank, I. H., and Scheuplein, R. J., The epidermal barrier, *Prog. Biol. Sci. Relat. Dermatol. 2*:245–261 (1964).
7. Scheuplein, R. J., Mechanism of percutaneous absorption, *J. Invest. Dermatol. 45*:334–346 (1965).
8. Bronaugh, R. L., Stewart, R. F., and Storm, J. E., Extent of cutaneous metabolism during percutaneous absorption of xenobiotics, *Toxicol. Appl. Pharmacol. 99*: 534–543 (1989).
9. Riviere, J. E., Bowman, K. F., Monteiro-Riviere, N. A., Dix, L. P., and Carver, M. P., The isolated perfused porcine skin flap (IPPSF), *Fundam. Appl. Toxicol. 7*:444–453 (1986).
10. McDougal, J. N., Physiologically-based pharmacokinetic modeling, *Toxicology 37*:60 (1991).
11. Environmental Protection Agency, *Dermal Exposure Assessment: Principles and Applications*, Office of Health and Environmental Assessment, Washington, DC, EPA/600/8-91/011B. Interim Report, January 1992.
12. Jadassohn, J., Zur Kenntnis der medikamentosen Dermatosen, *Verh. Dtsch. Derm. Ges.* V. Kongress, p. 103 (1895).
13. Landsteiner, K., and Chase, M. W., *J. Exp. Med. 73*:431 (1941).
14. Schwartz, L., and Peck, S. M., The patch test in contact dermatitis, *Public Health Rep. 59*:546–557 (1944).
15. Shelanski, H. A., Experience with and considerations of the human patch test method, *J. Soc. Cosmet. Chem. 2*:324–331 (1951).

16. Draize, J. H., Dermal toxicity, *Appraisal of the Safety of Chemicals in Foods, Drugs and Cosmetics*, Association of Food and Drug Officials of the United States, Texas State Department of Health, Austin, Texas, 1959.
17. Kligman, A. M., The SLS provocative patch test in allergic contact sensitization, *J. Invest. Dermatol. 46*:573–585 (1966).
18. Buehler, E. V., A new method for detecting potential sensitizers using the guinea pig, *Toxicol. Appl. Pharmacol.6*:341 (1964).
19. Magnusson, B., and Kligman, A. M., Allergic contact dermatitis in the guinea pig, *Identification of Contact Allergens*, C. C. Thomas, Springfield, IL (1970).
20. Gerberick, G. F., House, R. V., Fletcher, E. R., and Ryan, C. A., Examination at the local lymph node assay for use in contact sensitization risk assessment, *Fundam. Appl. Toxicol. 19*:438–445 (1992).

# 18

# Toward a Predictive Model for Allergic Contact Dermatitis

**Philip S. Magee**

*BIOSAR Research Project*
*Vallejo, California, and*
*University of California School of Medicine,*
*San Francisco, California*

**Jurij J. Hostynek* and Howard Maibach**

*University of California School of Medicine*
*San Francisco, California*

One of the most complex interactions of living skin with chemicals is expressed in the phenomenon of allergic contact dermatitis (ACD). A protein-reactive chemical (hapten), or a metabolic precursor (prohapten), first modifies a structural protein in the viable epidermis. In those cases in which an ACD response occurs, this initial challenge activates the immune system in the local lymph nodes. These nodes are responsible for draining edemous fluids when local tissues are inflamed. During an induction period, antigen recognition of the modified protein and complementary templating of the T-lymphocyte receptor site takes place. Following sensitization, the T-lymphocytes proliferate in the lymph nodes and await a second challenge of the same chemical. Unlike the first challenge, which found the immune system unprepared, the second contact can lead to a severe inflammatory reaction, accompanied by erythema and edema. Under controlled conditions, the ACD response can be reliably demonstrated in the guinea pig maximization test [1], the occluded patch test [2], the guinea pig optimization test [3,4] and, more recently, the local lymph

**Present affiliation*: Euromerican Technology Resources, Inc., Lafayette, California.

node assay of Kimber et al. [5]. There is reasonable, but not perfect, concordance among these tests for moderate to strong ACD responses.

The general complexity of the process involving in vivo metabolism and the immune system response is well discussed by Dupuis and Benezra [6]. On first inspection, the ACD problem would appear too complex and too broad in chemical scope to be approachable by structure–activity relationship (SAR) analysis. However, recent studies have shown the importance of transport and binding factors in the percutaneous absorption of drugs and ordinary chemicals [7]. If such factors are important in the delivery of ACD active compounds to the response sites in the epidermis, then SAR analysis in some form is appropriate. It was clear from the outset that regression analysis of graded ACD responses was not likely to succeed. However, if diverse compounds having known strong or weak response could be selected, a discriminatory model based on mechanistic factors could be developed. This chapter represents efforts to develop such a model.

## CANDIDATES, DATA, AND DESCRIPTORS

Candidates expressing moderate to strong ACD responses were selected from the INPRET database of Ziegler [8]. The database contains approximately 900 entries. However, many are multiple entries (formaldehyde, 20 entries) and many are trademark chemicals of uncertain identity. When sorted on strong allergens (rating = 3), there are only 74 chemicals. Elimination of inorganics, salts, dyes, corrosive, and very reactive chemicals reduces the set to 18. Twelve of these were selected for chemical diversity and by a similar process, 12 more that were moderate allergens (rating = 2). Thus, 24 diverse chemicals were clearly classified as moderate to strong ACD allergens.

To provide an equivalent class of nonallergenic compounds, entries in the *Givaudan Aroma Catalog* and the CTFA *Cosmetic Ingredient Dictionary* were explored. The protocol for selection of 24 nonallergenic compounds was similarity in molecular structure and the avoidance of perfect substructure descriptors. A perfect descriptor is one that appears exclusively in one class and, therefore, guarantees separation on a trivial bases. Success can be judged by the substructure count for the ACD/non-ACD classes: H-bond receptors, 70/57; H-bond donors, 28/34; aryl, hetero-rings, 27/21; aldehydes and esters, 10/11; unsaturation, 11/8; alkanols, 11/6; phenols and anilines, 15/3. Success was seen in most categories with the exception of the prohapten phenol and aniline structures.

Descriptors for these 48 chemicals fall into three categories: continuous, categorical, and ordinal. Continuous descriptors are used to describe mechanistic behavior based on subtle differences between molecular structures.

Under this category are the transport and binding descriptors that dictate the behavior of molecules as they pass through and partition among different bioorganic phases. Primary among these are $\log P$ (octanol/water), which frequently correlates with transport and nonspecific binding [9,10], and molar refraction (MR), which correlates the polarizability (London forces) of a molecule sliding over a molecular surface (e.g., protein) [11]. As $\log P$ (o/w) is calculable [12], it is easy to calculate the partial $\log P$s of the lipophilic (PL) and hydrophilic (PH) substructures. In recent binding studies, these descriptors have proved valuable when the binding polymer interacts preferentially with one class of the substructures. Categorical descriptors are set to (1) for the presence of a substructure and to (0) for its absence. They are also known as dichotomous descriptors, or indicator variables, in the sense that statistical significance indicates the presence of the effect being tested. Substructures selected for this study are those expected to react directly with skin structures (haptens) and those with potential for metabolic activation in the epidermis (prohaptens). Examples of the hapten structures are ICHO, ICONJ, and IX representing aldehydes, conjugated double bonds, and reactive halides. Prohapten structures are represented by IOH, IAROH, and IARNH2 for aldehyde precursors (primary alcohols) and quinones (phenols and anilines). It must be remembered that these are test descriptors with no a priori significance. The last category (ordinal) is mechanistic, as it tests for the possible significance of hydrogen-bonding of these 48 structures to protein domains in the stratum corneum and epidermis. Both descriptors are based on ordinal counts of the free electron pairs on O and N (HBA, acceptor) and the number of N-H or O-H bonds (HBD, donor). In sum total, these descriptors cover all of the anticipated effects of binding, delivery, and ACD reactivity. The categorical reactivity descriptors are crude, at best, but we are not looking for details at the quantum chemical level. We will be thoroughly satisfied if these descriptors implicate the mechanistic factors and provide a reasonable model for the a priori estimation of ACD potency.

## STATISTICAL METHODOLOGY

Considerable experimentation with statistical methods led us to select two-class, multiple regression analysis as the method of choice. The results obtained are completely consistent with those from linear discriminant analysis, and there is no need for a redundant discussion of this approach. Two-class regression is a seldom used procedure in QSAR [14] and was unfamiliar to these investigators at the outset of this problem. The statistical meaning of $r^2$ (explained variance), $s$ (equation standard deviation) and $F$ (Fisher distribution; measures equation strength) are nicely discussed by Lachenbruch [15]. The procedure is at its best when the two classes are equally populated or nearly so. Unlike discriminant analysis, it cannot be extended to three classes. Therefore, the procedures cannot be expected to separate strong, moderate, and non-allergenic; strong and moderate must be considered together as a single class. The advantages found in using the regression approach were the statistical strength measures of the descriptors (*T* values), the overall significance of the equation ($r^2$, $s$, $F$) and the ability to plot the residuals to determine exactly where each compound fell.

## RESULTS AND DISCUSSION

By a process of backward selection, the least significant variables were deleted, resulting in a highly significant regression model.

| | | |
|---|---|---|
| | ACD ($n$ = 24) Class = 1 | |
| | NON ($n$ = 24) Class = 0 | |
| | Coefficient | *T* value |
| Class = | 0.048 MR | 3.58 |
| | −0.698 PL | 5.35 |
| | −0.390 HBD | 3.00 |
| | 0.983 IHET | 3.61 |
| | 1.168 ICONJ | 5.87 |
| | 0.257 IX | 2.26 |
| | 1.367 IAROH | 5.02 |
| | 1.228 IARNH2 | 4.41 |
| | 0.928 IOH | 4.58 |
| | −0.635 (intercept) | |
| | $n$ = 45 $s$ = 0.296 | |
| | $r^2$ = 0.727 (0.853) $F$ = 10.34 | |

The dichotomous descriptors include both hapten and prohapten substructures in addition to IHET for heterocyles not known to be either. All have positive coefficients, meaning that these structures contribute to the ACD classification (class = 1). The weakest of these is IX for which halogen reactivity extends over an enormous kinetic scale. This may be a poor discriminator, as compounds such as β-bromostyrene are cosmetically inert. Of the two ordinal descriptors, only donor H-bonds (HBD) proved significant. The negative coefficient indicates that H-bonding in the stratum corneum retards the delivery of these compounds to the viable epidermis. In the continuous descriptors, PL correlated much better than log $P$ (o/w), and its strong negative coefficient clearly suggests that lipophilic groups reduce reactivity with proteins and metabolic enzymes by weakening binding and slowing partitioning into protein domains. The positive correlation with polarizable size (MR) is most interesting, as the strength of an immune response will clearly depend on the bulk of the abnormal structure targeted for antigen recognition and templating. A case in point is formaldehyde, which reacts avidly with skin proteins and, yet, is borderline in ACD response. A simple scrutiny of the two classes shows that the active set tends to be larger (MR = 30.0–93.0) than the inactive set (MR = 23.4–58.0).

Three outliers were removed from the final equation ($n = 45$). Geraniol is the *trans*-isomer of nerol, a confirmed cosmetic ingredient and has exactly the same descriptors in this study. Both are predicted to be nonallergenic, as expected. If geraniol is truly a moderate ACD chemical as indicated in the INPRET database, then the difference must be conformational and lies outside the scope of this study. Glycidic benzoate, a moderate ACD chemical, is underpredicted by this equation. It is also unique in the set of ACD compounds, and no descriptor for the reactive epoxy group was introduced.

If this missing descriptor is equated to the value for ICONJ (similar nucleophilic reactivity), then glycidic benzoate would be correctly predicted. Hydroquinone, a strong ACD chemical, has the opposite problem and was overpredicted. It is a statistical outlier because its positive value lies beyond the distributional range of the active class.

The equation shows satisfactory statistical strength as each of the nine descriptors is supported by five compounds. As the set increases in size toward $n = 100$ (50 per class), the number of required descriptors is expected to increase. This expectation comes from a very large study of the Draize test by Enslein and co-workers [16]. For $n = 200$–400, the number of descriptors required for discriminant analysis was $k = 17$–36. Because of the care with which sets are selected, the descriptors are not expected to double or triple, but simply to increase reasonably as new substructures are introduced.

Finally, it is necessary to address the indeterminate range from 0.3–0.7, as defined by Enslein [15]. Compounds of borderline strength or uncertain model prediction will fall in this range. In the present case, ethylenethiourea is the most negative ACD compound (0.477) and methyl-2-aminobenzoate is the most positive nonallergenic (0.547). Although both are incorrectly classified, neither is an outlier within the model statistics. This is nicely visualized by a residual plot of $Y$ (estimate) versus actual class, clearly showing the indeterminate range.

## MODEL VALIDATION

Ten ACD compounds not included in the model were used for validation. This included three compounds known to be borderline in response as well as compounds of both simpler molecular structure (formaldehyde) and more complex (griseofulvin, penicillin G) than those used to derive the model. The results are shown in Table 1. The notation, GPMT refers to the guinea pig maximization test [4], 2/Z and 3/Z refer to moderate and strong ACD response (INPRET DB) [8], and (+) or (±) refer to the Kimber local lymph node assay [5,17]. Seven of the ten are correctly and strongly predicted to be class 1 ACD irritants. The other three fall in the indeterminate range, 0.3–0.7 and require some discussion. Aniline is incorrectly predicted (score = 0.482) and is classed as moderate in the GPMT and borderline in the local lymph node assay. Griseofulvin is rated moderate in the GPMT, but is known to be borderline in medical practice. It can be applied topically for ringworm in most people, but may require oral administration for some. Formaldehyde has 20 entries in the INPRET database, ranging from innocuous to strong, depending on the investigator and method of challenge.

All factors considered, these ten compounds are correctly predicted by the regression model, including those expected to be borderline.

## CONCLUSIONS

As a first step toward a predictive model for ACD, this study was exceptionally fruitful. The derived relationship separating allergens from cosmetically innoc-

**Table 1** Ten ACD Compounds Not Included in the Model

| Compound | Score | Class | Rated |
|---|---|---|---|
| Aniline | 0.482 | (0) | GPMT/90X(+/−) |
| Griseofulvin | 0.566 | (1) | 2/Z |
| Formaldehyde | 0.597 | (1) | 2/Z(+) |
| Picryl chloride | 0.770 | 1 | 3/Z(+) |
| Phthalic anhydride | 0.956 | 1 | GPMT/90X(+) |
| *m*-Aminophenol | 1.094 | 1 | GPMT/100X(+) |
| Penicillin G | 1.106 | 1 | 3/Z(+) |
| Piperonal | 1.247 | 1 | 2/Z |
| *p*-Benzoquinone | 1.613 | 1 | GPMT/100X(+) |
| *n*-Propyl gallate | 1.820 | 1 | GPMT/100X(+) |

uous chemicals provides both new and confirmatory evidence for the mechanistic nature of allergic contact dermatitis. The dichotomous descriptors for hapten and prohapten substructures confirm the reactivity and metabolic activation processes so ably described by Dupuis and Benezra [6]. Unique outliers, such as glycidyl benzoate, suggest that more such descriptors will be required as the structural scope of the study is expanded. The current goal is to double the data set to 50 allergens and 50 nonallergens. The candidate selection process to accomplish this is now underway.

Of greatest importance was the quantifying of site delivery in terms of calculable mechanistic factors. Calculation from molecular structure provides a means of estimating the behavior of an untested or conceptual chemical or drug. Basing of the factors on properties related to known mechanisms gives us actual insight at the molecular level. Thus, the retarding effect of hydrogen-bonding and lipophilicity are clearly consistent with preconceived mechanisms considered to be probable. In fact, these descriptors were selected for the purpose of probing the operation of such events. Offsetting the retarding effects of stratum corneum binding and lipophilic partitioning is the unexpected discovery of a bulk effect in promoting ACD response. Although not anticipated, this effect is consistent with the vigor of the immune system response. The energetics of antigen recognition and lymphocyte templating will clearly depend on the size of the binding surface. Overall, these insights bring ACD response clearly into the same arena with drug delivery and percutaneous absorption.

The performance of the model in the validation tests is much better than expected for this intermediate stage of development. It was surprising that formaldehyde, griseofulvin, and penicillin G were so well predicted, as all are clearly outside the structural scope of the model, representing extrapolations, rather than interpolations. It was a particular surprise that the three borderline chemicals were correctly predicted as such. This provides some hope for a future model to define mechanistic differences between weak and strong allergens. For the present, however, allergens can be separated from nonallergens.

## SPECIAL DEDICATION

This chapter is dedicated to the memory of our friend and colleague, Claude Benezra, who did so much to stimulate research in this field.

## REFERENCES

1. Magnusson, B., and Kligman, A. M., *J. Invest. Dermatol.* 52:268 (1969).

2. Buehler, E. V., *Arch. Dermatol. 91*:171 (1965).
3. Maurer, T., Thomann, P., Weirich, E. G., and Hess, R., *Agents Actions 5*:174 (1975).
4. Andersen, K. E., and Maibach, H. I., *Contact Allergy Predictive Tests in Guinea Pigs*, Vol. 14. *Current Problems in Dermatology*, S. Karger, Basel, 1985.
5. Kimber, I., et al., *Toxicol. Lett. 55*:203 (1991).
6. Dupuis, G., and Benezra, C., *Allergic Contact Dermatitis to Simple Chemicals* (C. D. Calman and H. I. Maibach, eds.), Marcel Dekker, New York, 1982.
7. Magee, P. S., Critical factors in transdermal transport, *Dermatotoxicology* 4th ed. (F. N. Marzulli and H. I. Maibach, eds.), Hemisphere Publishing, New York, 1991.
8. Ziegler, V., INPRET database, Department of Dermatology, Karl Marx University, Leipzig.
9. Hansch, C., and Clayton, J. M., *J. Pharm. Sci. 62*:1 (1973).
10. Hansch, C., and Dunn, W. J. III, *J. Pharm. Sci. 61*:1 (1972).
11. Charton, M., Volume and bulk parameters, *Steric Effects in Drug Design: Topics in Current Chemistry 114* (M. Charton and I. Motoc, eds.), Springer-Verlag, Berlin, 1983, Chap. 4.
12. Hansch, C., and Leo, A., *Substituent Constants for Correlation Analysis in Chemistry and Biology*, John Wiley & Sons, New York, 1979, Chap. 4.
13. Magee, P. S., Complex factors in hydrocarbon/water, soil/water and fish/water partitioning, *The Science of the Total Environment*, Elsevier Science Publishers, Amsterdam, 1991, pp. 155–178.
14. Martin, Y. C., Holland, J. B., Jarboe, C. H., and Plotnikoff, N., *J. Med. Chem. 17*:409 (1974).
15. Lachenbruch, P. A., *Discriminant Analysis*, Hafner Press, New York, 1975, pp. 17–19.
16. Enslein, K., Borgstedt, H. H., Blake, B. W., and Hart, J. B., *In Vitro Toxicol. 1*:129 (1987).
17. Kimber, I., Hilton, J., and Botham, P. A., *J. Appl. Toxicol. 10*:173 (1990).

# 19

# A Three-Dimensional Human Skin Model for Toxicity Testing

**Dennis Triglia,* Tracy Donnelly, Inger Kidd, and Sonia Sherard Braa**

*Advanced Tissue Sciences, Inc.
La Jolla, California*

Several in vitro alternatives to animal testing are currently being studied worldwide to assess the toxicity of various chemical compounds, raw materials, and formulated products and, consequently, to reduce the number of research animals used. Few of these assay systems have been subjected to extensive validation studies to assess their intra- and interlaboratory reproducibility and relevance to in vivo toxicity data. A three-dimensional human skin model, developed by Advanced Tissue Sciences (skin$^2$ ZK1300 Barrier Function Model), consists of several layers of actively dividing, metabolically active, neonatal, foreskin-derived fibroblasts grown on nylon mesh in the presence of ascorbate, and a basal layer of epidermal keratinocytes, several layers of differentiated keratinocytes, and a stratum corneum. Fibroblasts growing within this three-dimensional framework secrete several growth factors and extracellular matrix proteins. In the current study, this substrate was employed to study the in vitro toxicity of various compounds, including petrochemicals, cosmetic and personal care products, by topical application of the test materials directly to the stratum corneum surface of the epidermis. Test agent-induced toxicity was assessed by the 3-(4,5-dimethylthiazol-2-yl)-2,5-diphenyltetrazolium bromide (MTT) (mitochondrial function) viability assay, release of the

**Present affiliation*: Houghten Pharmaceuticals, San Diego, California

inflammatory mediator prostaglandin $E_2$ ($PGE_2$), and release of the cytosolic enzyme lactate dehydrogenase (LDH) from cells of which membrane integrity has been compromised. In vivo–in vitro comparative toxicity data are very encouraging.

## MATERIALS AND METHODS

### Preparation of the Barrier Function Model

Neonatal foreskins (from routine circumcisions) were enzymatically digested with trypsin to mechanically separate the epidermis from the dermis. Further treatment yielded individual keratinocytes, which were cultured in serum-free medium and expanded in monolayer. The dermis was digested with collagenase to release fibroblasts, which were expanded in monolayer in Dulbecco's minimal Eagle's medium (DMEM) supplemented with fetal bovine serum. Fibroblasts were seeded onto medical-grade nylon mesh and grown for 26 days in DMEM (10% calf serum and 100 μg/ml ascorbate) to form a dermal model. The cells proliferate on the mesh by stretching across the interstices created by the mesh fibers. They become confluent in 2–3 weeks and synthesize a three-dimensional network of collagen and extracellular matrix proteins. Keratinocytes were then seeded ($1 \times 10^5/cm^2$) onto the dermal model. The coculture was grown submerged for 1 week, then it was lifted to the air–liquid interface for an additional 3 weeks. These cultures have the optimal morphology and barrier function. The sheets of nylon mesh are cut with a laser (Texcel; Westfield, MA) into 11 × 11-mm squares that are used as the substrate for the topical toxicity testing described in this chapter (Fig. 1). A histological cross section of the mesh depicts the multilayered fibroblasts, differentiated keratinocytes, and stratum corneum (Fig. 1). Figure 2 outlines the protocol.

### MTT Assay Procedure

The 11 × 11-mm Barrier Function Model ZK1300 mesh squares were removed from the MilliCells after test agent treatment, rinsed free of test article, and placed into six-well plates containing 2 ml of 2 mg/ml MTT diluted in serum-free DMEM-based assay medium at 37°C in 5% $CO_2$ (≥90% humidity) for 2 hr. Untreated control mesh squares were also incubated with MTT. The cultures were incubated for 2 hr, then washed twice with phosphate-buffered saline (PBS). Blue formazan precipitate was extracted from the mitochondria using 4 ml isopropanol on a shaker platform at room temperature for 1 hr. Aliquots (100 μl) of the extracted MTT solutions were diluted 1:2 with isopropanol, then transferred to 96-well plates and the optical density at 540 nm ($OD_{540}$)

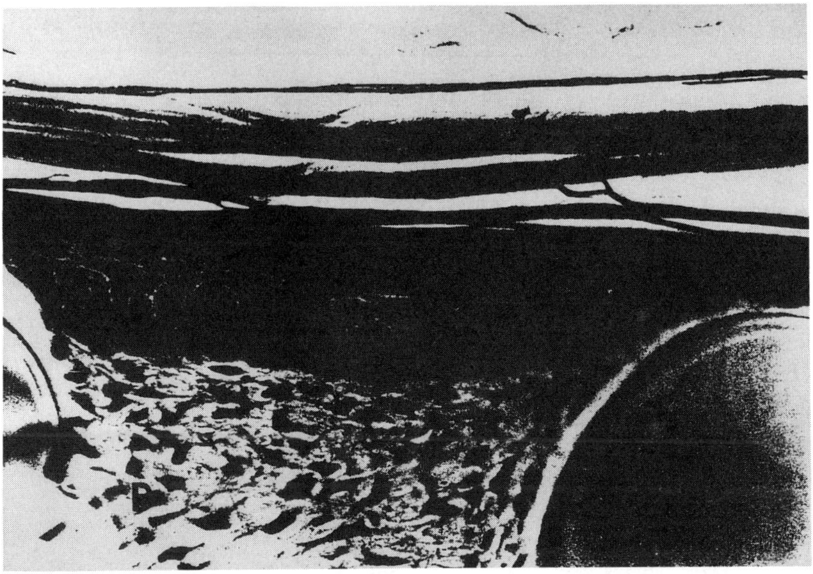

**Figure 1** Barrier function model histology. Fibroblasts surrounded by naturally secreted extracellular matrix proteins form the dermal layer (D). A proliferative basal layer, differentiated keratinocytes, and stratum corneum (SC) form the epidermis (E). A nylon mesh fiber (M) is seen at right in cross section. (H & E stain, 1000×).

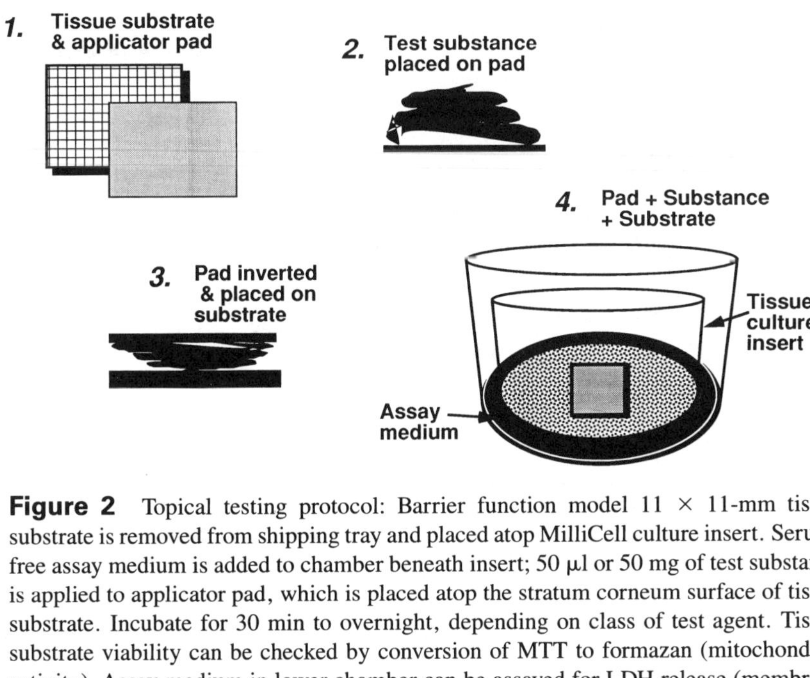


**Figure 2** Topical testing protocol: Barrier function model 11 × 11-mm tissue substrate is removed from shipping tray and placed atop MilliCell culture insert. Serum-free assay medium is added to chamber beneath insert; 50 μl or 50 mg of test substance is applied to applicator pad, which is placed atop the stratum corneum surface of tissue substrate. Incubate for 30 min to overnight, depending on class of test agent. Tissue substrate viability can be checked by conversion of MTT to formazan (mitochondrial activity). Assay medium in lower chamber can be assayed for LDH release (membrane integrity) and $PGE_2$ release (inflammatory mediator).

was determined using a microplate reader, making a blank correction to pretreated nylon mesh (without cells) that had been similarly treated with MTT. The mean $OD_{540}$ of the duplicate untreated control wells was set to represent 100% viability. Results are expressed as percentage of untreated control (Fig. 3).

## Prostaglandin $E_2$ Assay Procedure

The serum-free assay medium under the cell cultures was assayed for test agent-induced release of cellular $PGE_2$, using a commercially available $PGE_2$ enzyme-linked immunosorbent assay (ELISA) kit (Advanced Magnetics, Cambridge, MA).

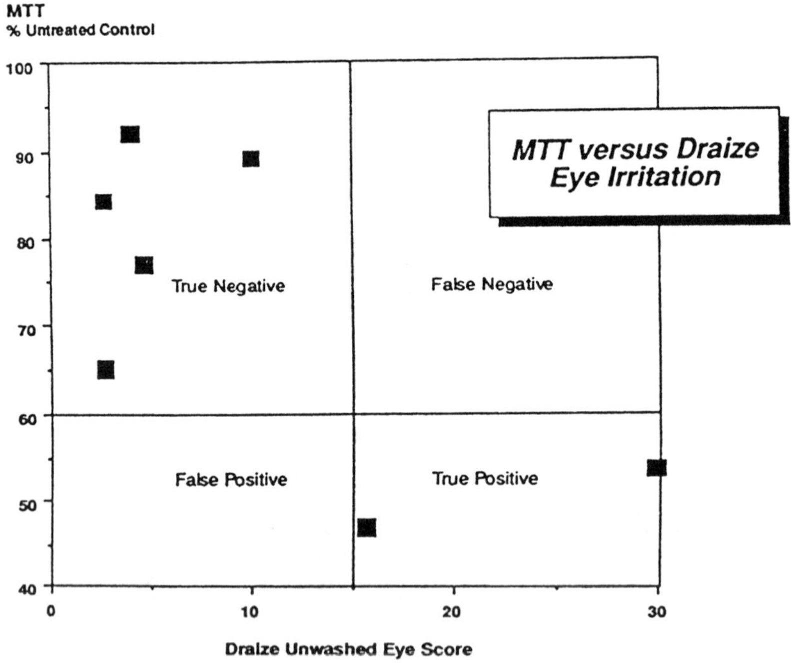


**Figure 3** Seven conditioners were applied undiluted to the stratum corneum surface of the barrier function model and incubated for 1 hr at 37°C in 5% $CO_2$ in a humidified incubator. A Draize in vivo ocular irritation cutoff score of 15 and an MTT in vitro cutoff value of 60% viability correctly classified the two positive and five negative reactions.

## Lactate Dehydrogenase Release Assay Procedure

Release of LDH from the barrier function model cells (Fig. 4) was quantified using a commercially available kit from Proteins International (Rochester Hills, MI) and a standard curve using purified human LDH purchased from Sigma (St. Louis, MO).

## Human Skin Patch Testing

A 14-day cumulative irritation test was performed using 20 subjects, who received daily patches of the test substances on their backs. Results of this testing were generously provided by Dr. Thomas Stephens of In Vitro Alternatives, Inc. (Carrollton, TX)

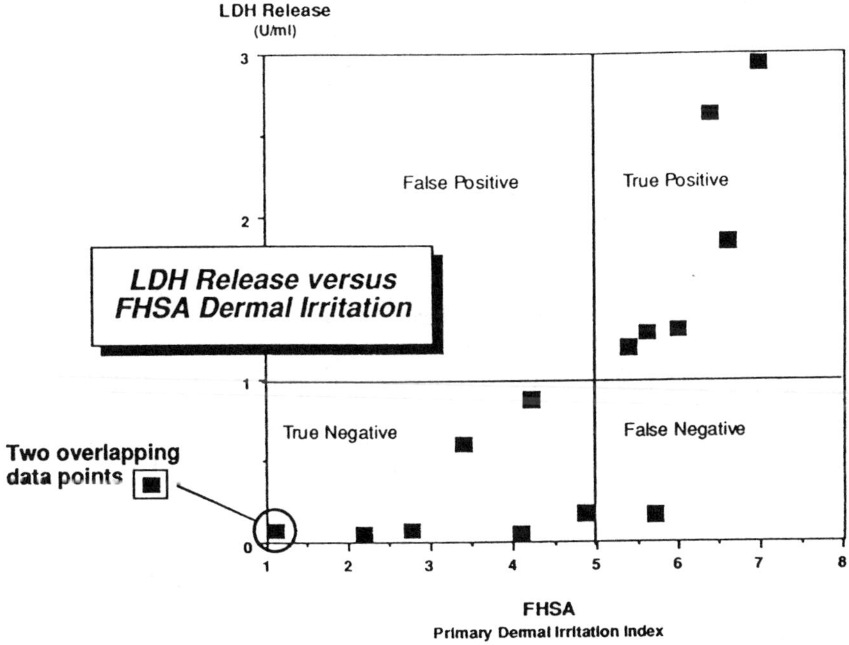


**Figure 4** Fifteen petrochemicals (8 lubricants, 2 organophosphate esters, 1 organic acid base condensate, 3 metalworking fluids, and 1 surfactant) were applied undiluted to the barrier function model and incubated for 2 hr. With cutoff values of 5.0 in the FHSA primary dermal irritation index (PDII) and 1.00 unit/ml LDH release, 6/7 positive and 8/8 negative reactions were correctly classified. The one false negative reaction (a metalworking fluid) was correctly classified as a true positive in the MTT assay (8% viable).

**Table 1** Comparison of MTT and LDH Data for Cosmetic and Personal Care Products

| Test agent (2-hr exposure) | MTT assay[a] (% untreated control) | LDH assay[a] (U/ml) |
|---|---|---|
| Eye shadow | 94 | 0.08 |
| Blush | 92 | 0.08 |
| Lipstick | 88 | 0.48 |
| Foundation | 85 | 0.16 |
| Mascara | 72 | 0.22 |
| Hair spray | 93 | 0.07 |
| Deodorant | 76 | 0.00 |
| Face powder | 91 | 0.13 |
| Facial depilatory | 63 | 2.44 |
| Perm solution | 38 | 2.10 |
| Perfume 1 | 13 | 2.39 |
| Perfume 2 | 30 | 1.41 |

[a]The MTT assay was performed directly on the BFM mesh cultures, and released LDH was quantified in the underlying serum-free assay medium.

**Table 2** Comparison of Human Skin Patch Irritation Data with In Vitro $PGE_2$ Release

| Test agent (24-hr exposure) | Human skin patch classification | $PGE_2$ release[a] (ng/0.1 ml) |
|---|---|---|
| Mask | Mild | 1 |
| Toner 1 | Less than mild | 2 |
| Sunscreen product SPF-15 | Mild | 3 |
| Facial cleanser | Mild | 5 |
| Gel product | Mild | 7 |
| Moisturizer 1 | Mild | 8 |
| Toner 2 | Mild | 9 |
| Moisturizer 2 | Mild | 10 |
| Lipstick | Nonirritating | 14 |
| Hydroalcoholic-containing botanicals | Mild | 179 |
| Facial cleanser | Moderate[b] | 324 |

[a]$PGE_2$ release was quantified as described in Materials and Methods section.
[b]Caused moderate acne in some subjects.

## CONCLUSIONS

The Advanced Tissue Sciences skin[2] Barrier Function Model ZK1300 has successfully been employed in three in vitro alternative assay systems (MTT, $PGE_2$, and LDH; Tables 1 and 2), measuring three different mechanisms of toxicity.

Toxicity data have been obtained for undiluted cosmetics, petrochemicals, and personal care products. Good correlation was demonstrated with in vivo dermal and ocular irritation data for these agents.

# 20

# The SOLATEX-PI System: An In Vitro Method to Predict Photoirritation

**Virginia C. Gordon and José Acevedo**

*In Vitro International*
*Irvine, California*

Dermal photoirritation is the production of a reversible reaction in the skin after one contact of the skin with a test substance and exposure to UV light. Photoirritation occurs when the dermal irritation response of a test material increases when the system is exposed to a dose of excitating radiation. Photoirritation is produced by all nonimmunological, light-induced toxic skin reactions with endogenous or exogenous chemicals that absorb light to an excited state and transfer energy to or react with cellular components. Ultraviolet light, including combined UVA and UVB, as well as UVA alone were used to expose samples in contact with the SOLATEX-PI system for varying times and doses. UVA has elicited most of the known in vivo photoirritant responses. In this study, UVA alone, as well as UVA and UVB combined, were used for broad-range exposure and to study possible synergistic effects of UVA and UVB.

Ultraviolet radiation consists of light with wavelengths from 200–400 nm. Radiation with short wavelengths (UVB) does not penetrate as far into the epidermis in the in vivo model. The UVA, of longer wavelengths, penetrates farther.

Photoirritation includes many mechanistic pathways [1–3]. The simplest mechanism is the alteration of the test sample to produce a toxic product, which then causes increased irritation. Other pathways require the test sample, the in vitro or in vivo system, and the UV radiation to interact. Free radicals also form (oxidation–reduction products) that then cause alterations in any components of the in vitro or in vivo system. As chemicals are excited, their excited state can interact directly with components of the in vitro or in vivo system (Fig. 1).

The basic principle of this method is the quantitation of an enhanced response in a previously described two-compartment physicochemical model of dermal irritation after exposure to UV radiation [4–6]. Appropriate controls with no UV radiation provide the background irritation response. The irritation response is quantified as the change either in a keratin–collagen barrier membrane or in a highly ordered macromolecular matrix. A red dye attached to the keratin–collagen barrier membrane (compartment 1) can be released if the integrity of this barrier is altered. Chemical irritants can perturb the matrix (compartment 2) to produce turbidity. The total response, the sum of the response in compartments 1 and 2, can be measured spectrophotometrically.

The major advantage of this model is the rapid, objective, and cost-effective methodology. The number of new chemicals introduced yearly is large. They include very diverse structural molecules, so all classes of these chemicals could produce adverse photobiological effects. Rapid screening before use would be valuable. In vitro testing should include an evaluation of photoreactivity of a new chemical in the presence of biomacromolecules, such as SOLATEX-PI, as well as complementary testing in cellular phototest systems.

The SOLATEX-PI system permits an evaluation of chemicals and formulations with essential biomembranes and biomacromolecules. The role of biomacromolecular change in adverse photobiological effects is very important. Formation and reaction of intermediates and damage, and reversible and irreversible binding to biomacromolecules provide an indication that adverse light-induced effects can be expected.

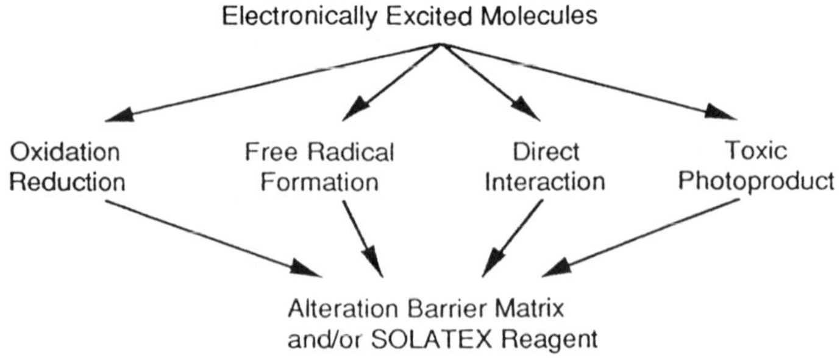


**Figure 1** Photoirritation mechanism.

In the SOLATEX-PI model, when UV radiation alters the test sample or formulation in contact with the two-compartment model, the electronic excitation can produce changes in the barrier membrane or matrix greater than those produced by the test sample in the absence of UV radiation. This enhanced response is the photoirritant response (Fig. 2). The changes can occur by free radicals, toxic photoproducts, oxidation–reduction, or direct interaction of the excited chemical with the barrier membrane or reagent matrix (Table 1).

## MATERIALS AND METHODS

The SOLATEX-PI system uses the principles of the SKINTEX test method as a model for in vitro dermal irritation. SKINTEX is based on a two-compartment physicochemical model of dermal irritation that incorporates a membrane barrier and a reagent matrix sensitive to perturbation by chemical irritants [1–3]. This is called the barrier membrane–reagent matrix system.

The barrier membrane–reagent matrix has been standardized into a rapid, accurate, and reproducible assay. The possible sources of interference in this system were evaluated. The procedure to detect interference was incorporated into the protocol to minimize false-negative results.

### SOLATEX-PI Barrier Membrane

A buffered salt solution at pH 8.0 of 10% keratin and 1% collagen was bound to cellulose acetate with 0.1% glutaraldehyde as 25°C for 1 hr. After washing the support in distilled water, basic red 2 was attached to the keratin–collagen matrix with 0.1% glutaraldehyde for 10 min at 25°C. Indicator barrier matrices were washed exhaustively in distilled water and dried. The matrices were stored at 4°C and have a 180-day shelf-life. Circles are cut and formed into a well within a plastic disk.

### SOLATEX Reagent Matrix

A lyophilized powder, containing globulins, collagen, glycosaminoglycans, free fatty acids, amino acids, phospholipids, and buffer salts, was rehydrated. The rehydrated reagent was stable for 15 days at 4°C.

### SOLATEX Photoirritant Method

A prototype method was used to establish the relation between the in vitro response for known photoirritants in the presence and absence of UV radiation. Test materials were exposed to UVA or UVA in combination with UVB. This procedure permitted testing of pure liquids and solids at multiple concentra-

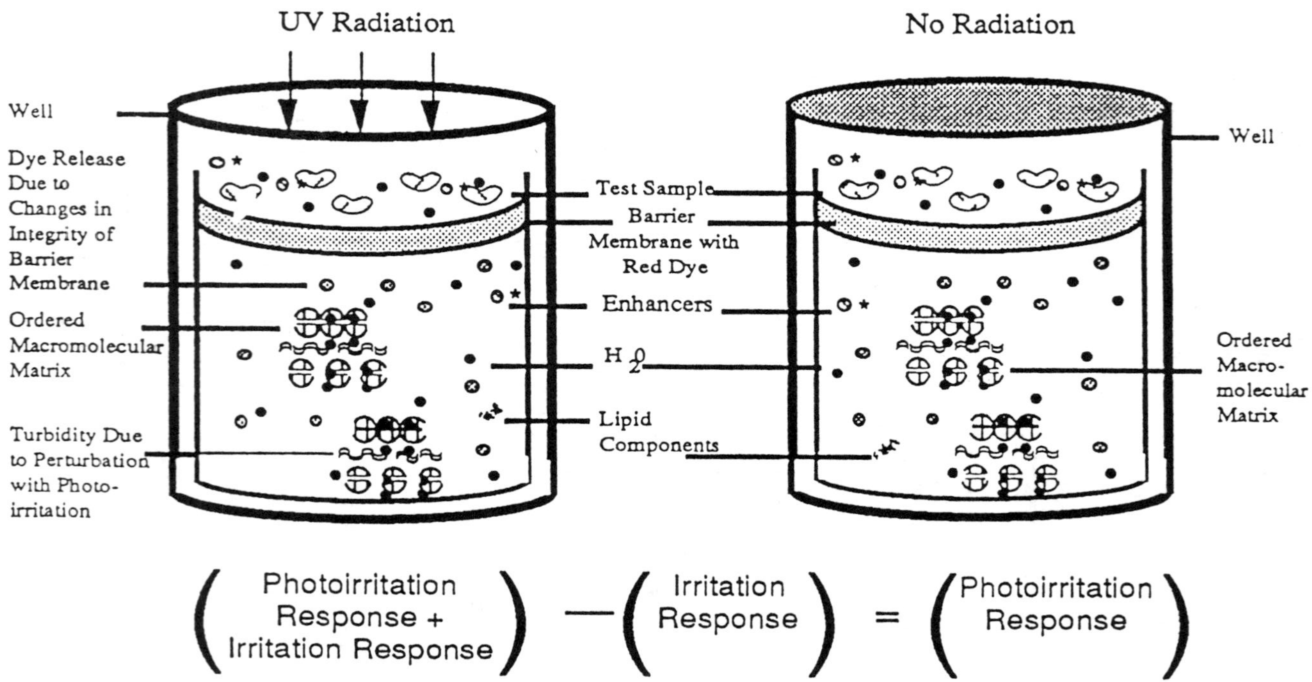


**Figure 2** SOLATEX-PI model.

**Table 1** Mechanism and Targets

| Compound | Mechanism | Target |
|---|---|---|
| Psoralen | Direct addition photodynamic | DNA<br>Proteins<br>Membranes<br>Ribosomes |
| Phenothiazine | Toxic photoproduct | DNA<br>Membranes |
| Porphyrins | Photodynamic | Membranes<br>DNA<br>Proteins |
| Acridine orange | Photodynamic | Membranes |
| Anthracene | Photodynamic | Membranes<br>DNA |

*Sources*: Refs. 4–8.

tions. Test samples were applied directly to the barrier membranes, which were then inserted into reagent or diluent filled wells. Test samples, calibrators, and controls were incubated at 25°C for 24 hr in the presence and absence of UV light. After incubation the net optical density (OD) at 400 nm was determined. Materials that exhibited a greater than 50% increase of their response with UV over their baseline without UV were considered potential photoirritants.

## Ultraviolet Light Sources

The UV light source used for UVA was a 20-W (General Electric F20F12BL) and for UVB a 20-W sunlamp (Westinghouse FS20). These light sources were housed in a temperature-controlled solarium developed and designed by In Vitro International. To maintain constant UV intensity, the irradiance of the UVA and UVB lights was determined before each screen with a radiometer Model ILI400 from International Light, with either a UVA or UVB sensor. In most experiments, the system received a UVA dose of 0.6–1.2 W/cm$^2$. The distance of the system from the lights was adjusted to assure these intensities. In experiments using UVA/UVB light exposure, the UVB dose of 0.1–0.5 W/cm$^2$ was used.

## Replicate Analysis

Replicate analysis of the response of background and a known irritant with no UV light demonstrated a baseline of −30 to +30% across a range of baseline irritation of nonirritant to mild irritant. A borderline range from 30 to 50% requires reanalysis of test sample at lower concentrations to qualify the prediction of photoirritation for a test sample.

## Range Qualification of Assays

A minimal irritant or solvent must produce a reproducible net change in the SOLATEX-PI system with UV in relation to the nonirradiated control. A mild irritant must produce within a 30% net increase of the Solatex-PI system with UV in relation to the nonirradiated control. For the entire range of irritancy from nonirritant to mild irritant, uniform changes with exposure to UV must be observed. A known photoirritant sulfanilamide at concentrations from 1 to 10% must produce an increase of 80–260% over background in a qualified assay after exposure to UV radiation in relation to the nonirradiated control.

## Ultraviolet Absorption by High Concentrations of Photoirritants

Very high concentrations of potential photoirritants can block UV exposure into the chemical in contact with the skin in vivo or in contact with the barrier membrane–reagent matrix system in vitro.

## Test Compounds and Formulations

The 65 materials used for evaluation of this assay were obtained from Sigma Chemicals, Aldrich Chemical Company, or from test materials submitted to the laboratory by outside laboratories. Pure chemicals, vehicles, mixed vehicles, dilutions, as well as complete mixtures were analyzed.

## In Vitro Photoirritant Classes

The in vitro photoirritant classes are nonphotoirritant, borderline, photoirritant, and strong photoirritant (Table 2). These classes correspond to different enhancements (net percentage increase) of the dermal irritation response.

The compounds chosen to evaluate the in vitro photoirritation test were categorized on reported adverse effects in humans [7–13] (see Table 2). Guinea pigs, minipigs, hairless mice, rats, or rabbits which can generally be used in tests for phototoxics [7–13]. Often in vivo responses in these other species are quantitatively and qualitatively different from the comparable responses in humans (see Table 2). Many photosensitizers may be phototoxic if an endpoint different than erythema is used as an indicator of phototoxicity [13].

Test samples were prescreened to determine a level of the test material or formulation that did not cause moderate or severe irritation. At least two different concentrations were tested. Photoirritant (PI) studies in vivo also must be performed at concentrations that produce a minimal level of dermal irritation. This permits the best evaluation of an enhanced dermal irritation response with exposure to UV radiation (Table 3).

## RESULTS

The use of UVA, or UVB with UVA, was evaluated in these studies. Recent studies investigating photosensitization have used UVA and UVB for broad-range screening. The presence of UVA and UVB can be made more sensitive, and UVB may cause changes that deactivate or alter a molecule electronically excited by UVA.

Studies were performed with UVA, or a combination of UVA and UVB. When the known photoirritant sulfanilamide at 1.5% was analyzed, UVB alone demonstrated the largest net percentage increase at 16 hr. At 24 hr, UVA and UVB demonstrated a similar increase. At 24 hr of UVA alone, about 66% of this increase was observed (Fig. 3). Similar results were obtained in a study of sulfanilamide at 5%. At 8 hr with UVB alone, the largest net percentage increase was observed for 5% sulfanilamide (Fig. 4).

Results were summarized at two to ten concentrations. Only results from qualified assays for which the background control and irritant control produced minimal changes with exposure were evaluated. The known in vivo photoirritant controls also had to produce a PI response for the assay to qualify. The qualification of an assay is summarized in Figure 5, with the dose–response curve for 3-nitrobenzoic acid.

A complete presentation of results is summarized in Table 4, in which the relevant in vivo results from Table 2 are included for comparison.

## CONCLUSION

In this chapter, the use of a quantifiable in vitro dermal irritation model for photoirritancy testing is presented. The in vitro method described provides a simple and rapid method to screen for photoirritants. Photoirritants, which act on different cellular sites of action, produce significant increases in dermal irritation in vitro on exposure to UV light in SOLATEX-PI. The results show the group of strong photoirritants produced increases in the dermal irritation response of 100–300%. Their potential photoirritancy was correctly indicated in the in vitro test. Test samples in the idiosyncratic and negative group of in

**Table 2** In Vivo Database

| Chemical | Concentration (%) | Light source | Photoirritant[a] |
|---|---|---|---|
| Acetone | | UVA | Non-PI |
| Acetophenone | | UVA | PI |
| Acid, 3-Nitrobenzoic | | UVA | PI |
| Acid, acetylsalicylic | | UVA | Non-PI |
| Acid, anisic | | UVA | PI |
| Acid, kynurenic | | UVA | PI |
| Acridine orange | | UVA | PI |
| Amiodarone | | UVA | PI |
| Anthracene | | UVA | PI |
| Benzaldehyde | 1 | UVA | PI, PS |
| Benzoic acid | 5 | UVA | I |
| Benzophenone | 1 | UVA | PS |
| Bergamot oil | | UVA | PI |
| Bithionol | 1 | UVA | PI, PS |
| Chloride, benzalkonium | 0.10 | UVA | Non-PI |
| Chloroquine | | UVA | PI |
| Chlorpromazine | 1 | UVA | PI, PS |
| Chlorpromazine | 0.1 | UVA | PS |
| Chlorpromazine | | UVA | PS |
| Coke, heavy | 10 | UVA | PI |
| Cold cream | 100 | UVA | Non-PI |
| Coumarin | | UVA | Non-PI |
| Coumarin, 6-methyl | 1 | UVA | PS, PI |
| Coumarin, 7-methyl | | UVA | PI, PS |
| 5-Chlorosalicylanilide | 1 | UVA | PI |
| Dimethylchlortetracycline | | UVA | PI |
| Dinitrochlorobenzene | | UVA | Non-PI |
| Doxycycline | | UVA | PI |
| Eosin Y | | UVA | PI |
| Ethylene glycol | 100 | UVA | Non-PI |
| ETOH/PPG | 100 | UVA | Non-PI |
| Fluoranthene | | UVA | PI |
| Fluorene | 1 | UVA | PI |
| Glycerol | 100 | UVA | Non-PI |
| Halogenated salicylanilides | | UVA | PS |
| Household black liquid | 10 | UVA | Non-PI |
| Hydrochlorothiazide | 1 | UVA | PI |
| Hydrocortisone | | UVA | Non-PI |
| Isopropyl palmitate | 1 | UVA | Non-PI |

**Table 2** Continued

| Chemical | Concentration (%) | Light source | Photoirritant[a] |
|---|---|---|---|
| Isopropylamine | 1 | UVA | Non-PI |
| Methylene blue | 1 | UVA | Non-PI |
| Mineral oil | 100 | UVA | Non-PI |
| Moisturizer | 100 | UVA | Non-PI |
| NaCl | 0.90 | UVA | Non-PI |
| Nalidixic acid | | UVA | PI |
| Oil, crude | 10 | UVA | Non-PI |
| Oil, slurry | 10 | UVA | Non-PI |
| PABA | | UVA | PS |
| Phenanthrene | | UVA | PI |
| Phenergan | | UVA | PI |
| Phenothiazine | | UVA | PI |
| Piroxicam | | UVA | PI |
| PPG | 100 | UVA | Non-PI |
| Promethiazine | | UVA | PI |
| Psoralen, 5-methoxy | 1 | UVA | PI |
| Psoralen, 8-methoxy | 1 | UVA | PI |
| Pyrene | | UVA | PI |
| Quinidine | | UVA | PI, PS |
| Salicylamine | | UVA | Non-PI |
| Salicylic acid | | UVA | Non-PI |
| Skin lotion | 100 | UVA | Non-PI |
| Sulfanilamide | 1 | UVA | PI |
| TCSA | | UVA | PI, PS |
| Tetracycline | | UVA | PI |
| Thiourea | | UVA | PI |
| Triton-X | 10 | UVA | Non-PI |
| Tween 80 | 10 | UVA | Non-PI |
| Water | 100 | UVA | Non-PI |

[a]PI, photoirritant; PS, photosensitizer.

vivo produced little effects. A low number of false-positive results are beneficial for photoirritation screening.

Many test materials that are known photoirritants are not detected in the yeast phototoxicity assay. These materials manifest a persistent erythema, rather than the hyperpigmentation produced by tars and furocoumarin, readily detected in yeast phototoxicity assay. Photoirritant test materials observed in

**Table 3** In Vitro Photoirritant Classes

| Net increase (%) | In vitro class |
|---|---|
| <30 | Non-PI |
| 30–50 | Border |
| 51–100 | PI |
| >101 | Strong PI |

vivo that produce response in SOLATEX-PI, but not in the photo yeast assay, are sulfanilamide, acetylsalicylic acid, and dimethylchlortetracycline [14,15]. The microbial test system may correlate with one of several types of photoirritation. Phototoxicity studies with cells do not demonstrate sensitivity to chlorpromazine, tetracycline, or sulfanilamide [14]. The psoralens are readily

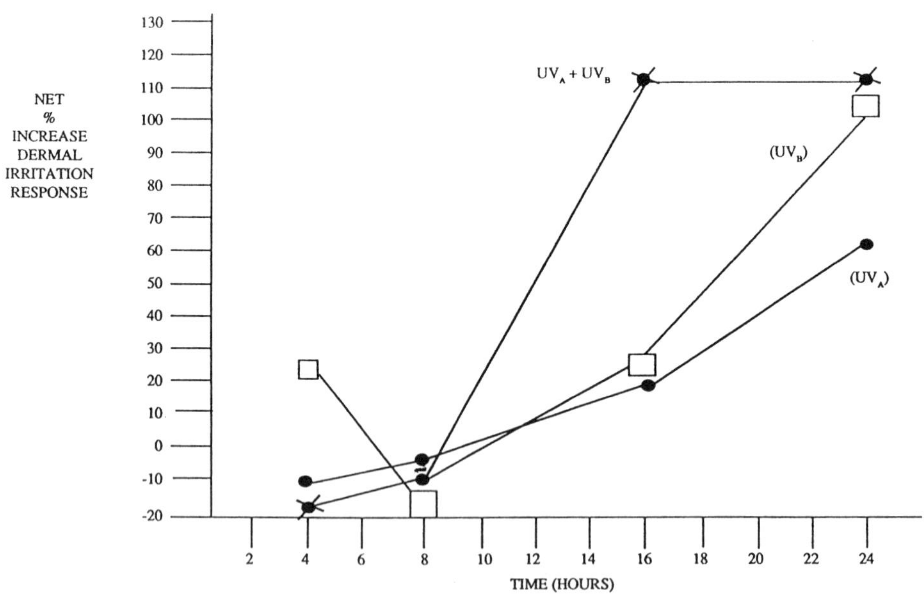


**Figure 3** Net percentage increase produced by UVA (●), UVB (□), and UVA + UVB (✱) with 1.5% sulfanilamide in SOLATEX-PI.

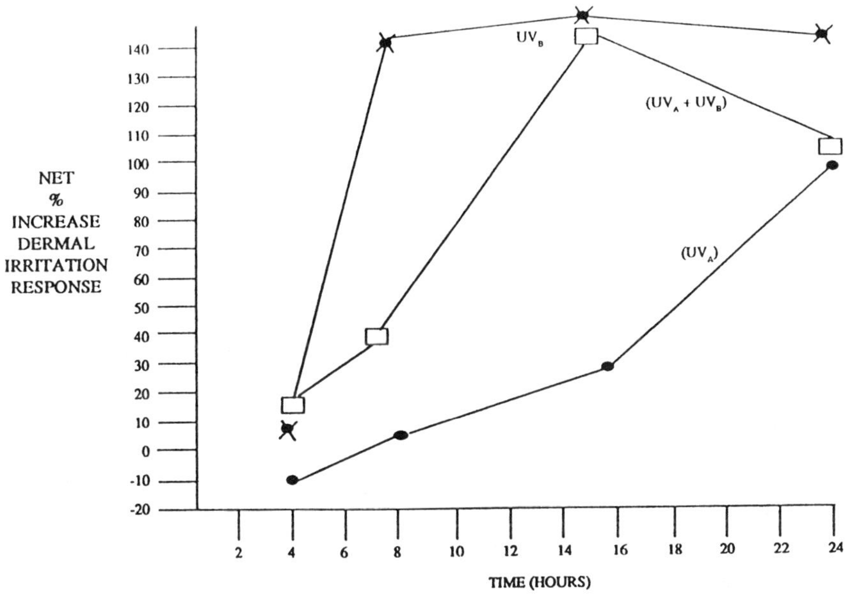


**Figure 4** Net percentage increase produced by UVA (●), UVB (□), and UVA + UVB (✱) with 5% sulfanilamide in SOLATEX-PI.

demonstrated in the yeast phototoxicity assay, which has the capability to predict mechanisms that involve DNA changes, but are not demonstrated in SOLATEX-PI. The coumarins are readily discerned in both systems [15].

To look at the complete UV spectrum, it is important to perform studies with UVB as well as UVA light [16]. Because of the lethal effect of UVB, yeast phototoxicity tests can be performed only with radiation greater than 320 nm. SOLATEX-PI can incorporate evaluation throughout the UV spectrum.

No one test system has the complete ability to predict the photoirritant potential of chemicals, fragrance oils, natural products, formulations, and sunscreens [15,16]. Some of these discrepancies are due to variations of absorption, enzyme-induced transformation, as well as cutaneous or metabolic inactivation. The SOLATEX-PI system demonstrated the capability to predict many photoirritants other than the psoralens. This model is a useful screen for predictive photoirritant testing. The major advantage of this model is the rapid, objective, and cost-effective methodology. The number of new chemicals that are introduced yearly is large. They must include an evaluation of photoreac-

| | | |
|---|---|---|
| Sample | : | Acenapthene |
| Class | : | Coal Tar |
| Analysis date | : | 08/09/91 |
| NTC # | : | DFU32 |
| Protocol | : | SOLATEX PI UVA-UVB |
| User name | : | Kathryn |
| Sample classification | : | PHOTOIRRITANT (3.0%) |

| Name | UV OD | No UV OD | Samp % | Net % | Qualification |
|---|---|---|---|---|---|
| SX0 | 480 | 258 | .86 | | Qualified |
| SXI | 932 | 520 | 79 | -7 | Qualified |
| PI 1 | 780 | 271 | 188 | 102 | Qualified |
| PI 5 | 880 | 256 | 244 | 158 | Qualified |
| PI 8 | 795 | 259 | 207 | 121 | Qualified |
| PI 10 | 941 | 308 | 206 | 120 | Qualified |
| Solvent | 549 | 298 | 84 | | Qualified |
| Sample | 651 | 253 | 157 | 73 | Qualified |
| Sample | 1044 | 394 | 165 | 81 | Qualified |

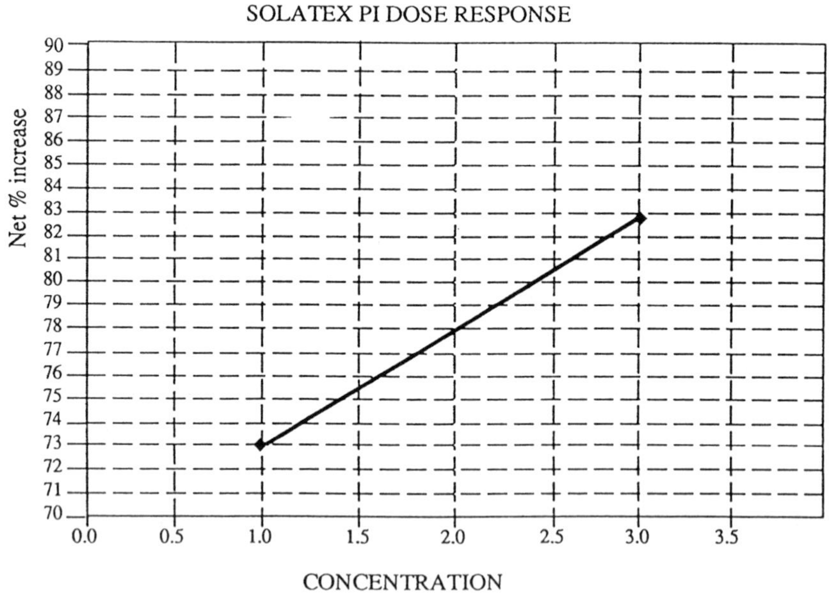


**Figure 5** Photoirritant response of acenapthene.

**Table 4** Summary of In Vitro and In Vivo Results

| | UVA ONLY | |
|---|---|---|
| Test sample | Class in vitro | In vivo response |
| Acetone | Non-PI | Non-PI |
| Acetophenone | PI | PI, PS |
| Acid, 3-nitrobenzoic | Strong PI | PI |
| Acid, acetylsalicylic | Nonstrong PI | PI |
| Acid, anisic | Strong PI | PI |
| Acid, benzoic | Non-PI | Non-PI |
| Acid, salicylic | Strong PI | Non-PI |
| Acridine | Strong PI | PI |
| Anthracene | Strong PI | PI |
| Benzalkonuim chloride | Non-PI | Non-PI |
| Benzaldehyde | Non-PI | PS |
| Benzophenone | Non-PI | PS |
| Bergamot oil | Non-PI | PI |
| Bithional | PI | PI, PS |
| Chloroquine | Non-PI | PI |
| Chlorosalicylanilide | PI | PI |
| Chlorpromazine | Strong PI | PI, PS |
| Cinnamyl alcohol | Non-PI | Non-PI |
| Coke, heavy | Strong PI | PI |
| Conditioner 2 | Non-PI | Non-PI |
| Coumarin, 6-methyl | Strong PI | PI, PS |
| Coumarin, 7-methyl | PI | PI, PS |
| Cream deodorant | Non-PI | Non-PI |
| Crude oil | Non-PI | Non-PI |
| Doxycycline | Strong PI | PI |
| Eosin Y | PI | PI |
| ETOH/PPG | Non-PI | Non-PI |
| Eythlene glycol | Non-PI | Non-PI |
| Ethylhexyl methoxy cinnamate | Non-PI | Non-PI |
| Fluorene | Strong PI | PI |
| Glycerol | Non-PI | Non-PI |
| Heavy coker gas oil | PI | PI |
| Hexachlorophene | Non-PI | Non-PS |
| Household black liquid | Non-PI | Non-PI |
| Hydrachlorothiazide | PI | PI |
| Insecticide | Non-PI | Non-PI |
| Isopropyl palmitate | Non-PI | Non-PI |
| Isopropylamine | Non-PI | Non-PI |

**Table 4** Continued

| | UVA ONLY | |
|---|---|---|
| Test sample | Class in vitro | In vivo response |
| Methylene blue | Non-PI | Non-PI |
| Mineral oil | Non-PI | Non-PI |
| Moisture balm | Non-PI | Non-PI |
| Moisture base | Non-PI | Non-PI |
| Nalidixic acid | PI | PI |
| Octyl salicylate | Non-PI | Non-PI |
| Oil slurry | Non-PI | Non-PI |
| PABA | Non-PI | Non-PI |
| Phenanthrene | PI | PI |
| Phenothiazene | PI | PI |
| Phenylbutazone | PI | PI |
| Piroxicam | PI | PI |
| PPG | Non-PI | Non-PI |
| PG | Non-PI | Non-PI |
| Psoralen, 5-methoxy | Non-PI | PI |
| Psoralen, 8-methoxy | Non-PI | PI |
| Quinidine | PI | PI |
| Retinoic acid | Non-PI | Non-PI |
| Salicylamide | Non-PI | Non-PI |
| Salicylanilide | Non-PI | Non-PI |
| Sulfanilamide | Strong PI | PI, PS |
| TCSA | PI | PI, PS |
| Tetracycline | Strong PI | PI |
| Triton X-100 | NQ | Non-PI |
| Tween-80 | Non-PI | Non-PI |
| Water | Non-PI | Non-PI |

tivity of a new chemical in the presence of biomacromolecules, such as SOLATEX-PI, as well as complementary testing in cellular phototest systems.

The SOLATEX-PI system permits an evaluation of photoreactivity of chemicals and formulations with essential biomembranes and biomacromolecules. The role of biomacromolecular change in adverse photobiological effects is very important. Formation and reaction of intermediates and damage, and reversible and irreversible binding to biomacromolecules provide an indication that adverse light-induced effects can be expected. With this evidence of photobiological effects, an evaluation can be continued with bacterial or cell systems. These methods can be integrated as part of a protocol for in vitro testing of the photoirritant potential of chemicals.

## REFERENCES

1. Harber, L. C., and Friend, J. V., *J. Invest. Dermatol.* *58*:327–342 (1972).
2. Lock, S. O., and Friend, J. V., *Food Chem. Toxicol.* *24*:289–793 (1986).
3. Machino, H., Shiraishi, S., and Miki, Y., *Arch. Dermatol. Res.* *279*:125–129 (1986).
4. Gordon, V. C., et al., Presented at SOT, Miami Beach, FL (1990).
5. Gordon, V. C., and Acevedo, J., (1991).
6. Soto, R. J., Servi, M. J., and Gordon, V. C., Presented at SOT, Miami Beach, FL (1990).
7. Epstein, J. H., et al., *Arch. Derm.* *97*:236 (1963).
8. Freeman, R. G., and Knoro, J. M., *Arch. Derm.* *97*:130 (1968).
9. Kochevar, I., et al., *Photochem. Photobiol.* *36*:65–69 (1982).
10. Stern, R. S., *New Engl. J. Med.* *309*:186–187 (1983).
11. Kaidbey, K. H., and Kligman, A. M., *Arch. Derm.* *117*:258–263 (1980).
12. Wirestrand, L. E., and Ljunggren, B., *Photodermatology* *5*:201–205 (1988).
13. Ramsay, C. A., *J. Invest. Derm.* *72*:99–102 (1979).
14. Marzulli, F., and Miabach, H., Results from independent industrial laboratories, *J. Soc. Cosmet. Chem.* *21*:655–715 (1970).
15. Daniels, F., A simple microbiological method for demonstrating phototoxic compounds, *J. Invest. Dermatol.* *44*:259 (1965).
16. Duffy, P. A., Bennett, A., Roberts, M., Flint, O. P., *Mol. Toxicol.* *1*:579–587 (1987).

# 21

# Feasibility of Fluorescence Assays in Human Skin Equivalents with the CytoFluor 2300

**Millard M. Mershon and Charles B. Millard**

*U.S. Army Medical Research Institute of Chemical Defense*
*Aberdeen Proving Ground, Maryland*

**Jeffrey R. Cook,* Laura M. Patrone,† Laura S. Rhoads,‡ and Robert G. Van Buskirk**

*State University of New York at Binghamton*
*Binghamton, New York*

Antivesicant drugs are being sought for the protection of soldiers from attacks with blistering chemical agents. Therefore, vesicant damage in humans must be modeled to permit measurement of damage and damage reduction during selection of safe and effective antivesicant compounds. A suitable model should be useful for screening such candidate compounds. Suitability for the study of the underlying mechanisms of sulfur mustard (HD) is also desirable.

The hairless guinea pig model displays gross and microscopic effects of HD that are uniform, graded, and measurable [1]. However, use of animal models is to be avoided if alternative models can be used. A search for alternative methods was based on the concept that artificial human skin might substitute for animals and humans. Methods for HD vapor exposure [1] were used with LSE (Living Skin Equivalent, Testskin, provided by Organogenesis,

*Present affiliations*:
*Anheuser-Busch Eye Institute, St. Louis Medical Center, St. Louis, Missouri
†Mayo Clinic, Scottsdale, Arizona
‡U.S. Army Medical Research Institute of Chemical Defense, Aberdeen Proving Ground, Maryland

Inc., Cambridge, MA). Gross, biochemical, and histological [2] signs of HD damage were observed. The development of an analytical spectrofluorometric method for observing some signs of other insults to related cells [3] was the basis for initiation of the present effort.

## OBJECTIVE

The long-term objective of this project is to determine the feasibility of utilizing human skin models, fluorescent probes, and the CytoFluor 2300 instrument (Millipore Corp., Bedford, MA) to measure toxic effects induced by HD, antivesicant compounds, or both. The present effort explores concurrent use of these methods to measure damage in cells. Questions that guided this effort were

1. Do probe or CytoFluor methods measure damage in skin substitutes?
2. Can dermal substrates bind enough of applied fluorescent probe dyes to obscure the fluorescent signals from epidermal cells?
3. Does conventional transportation damage skin substitutes?
4. Which site-specific fluorescent dyes might reveal HD damage?

## APPROACHES

To address these questions at a time when Testskin was unavailable, specimens of immature human epidermal model (HEM, three to four cells thick on gel–membrane) were transported to Aberdeen Proving Ground, Maryland (APG) from Binghamton, New York, under three different sets of conditions. Effects of transportation were checked with two dyes. When available, Testskin was used intact, or without epidermis, to address the first four questions. Eight fluorescent probes were tested as candidates for use with human epidermal models.

## MATERIALS AND METHODS

### Human Skin Substitutes Grown In Vitro

LSE (Testskin) and LDE (Living Dermal Equivalent) were furnished by Organogenesis, Inc., Cambridge, MA. Another human epidermal model was grown by using methods described below:

Millipore inserts with Millicell CM microporous membranes (Millipore Corp., Bedford, MA) were overlaid with a gel coating (hereafter referred to as "gel" or "gel–membrane"). The details of this technique have been patented by Millipore Corp.

Normal human epidermal keratinocytes (NHEK), as second-passage cell strains and 0.025% w/v typsin/0.01% w/v EDTA solution (CC5013) were obtained from Clonetics Corporation (San Diego, CA). The NHEK were maintained in serum-free keratinocyte growth medium (KGM) during growth on plastic of cell culture dishes. The NHEK, dissociated with CC5013, were suspended in KGM with 10% fetal calf serum/1.5 mM calcium chloride, and seeded onto gel–membranes of inserts at a cell density of 3–6 $\times$ $10^5$/cm$^2$. Seeded inserts were held in medium-filled wells of culture plates while cells were allowed to stratify for 5–7 days. The medium was changed daily while the stratifying NHEK were maintained at 37°C, in 5% carbon dioxide/95% air. To achieve mature HEM, the apical medium is removed from inside the insert, exposing the apical cell surface to air for 5–7 days of stratum corneum formation [4]. In this effort, HEM preparations were used before they could be matured in $CO_2$/air.

## Fluorescent Probe Assays of Human Epidermal Model and Testskin

Eight different assays were accomplished with Testskin and HEM tissues. Seven assays had protocols nearly identical with those described in detail by Cook et al. [4]. Briefly, Testskin was washed three times in Dulbecco's modified Eagle's medium (DMEM) without phenol red (GIBCO-BRL, Grand Island, NY) at 37°C. Similar washing was accomplished using HEM or gel–membrane (dermal model).

Each skin model was incubated for 1 hr with one of seven fluorescent dye probes or exposed to sodium fluorescein (details follow). All probes are identified by names or acronyms used by the vendor (Molecular Probes, Eugene, OR). The acetoxymethyl ester (AM) confers membrane permeability to probes identified with the −AM suffix [5]. Once inside cells, the AM is cleaved by cellular esterases to make the dye less membrane permeable [5]. The dyes and concentrations used were: calcein-AM (10 μg/ml), CMFDA (10μg/ml), neutral red (100 μg/ml), BCECF-AM (50 μg/ml), Fluo3-AM (25 μM), rhodamine-123 (10 μg/ml) or 2′,7′-dichlorofluorescin diacetate (100 μM). After incubation, each preparation was washed three more times with DMEM and read in the CytoFluor 2300, using the filter sets and gain levels listed in Table 2.

To determine dye uptake of epidermal substrates, fluorescence readings from full-thickness models were compared with readings from substrates without epidermis. Signals from LSE were compared with those from LSE after epidermis was stripped from underlying LDE with a fine forceps (subsequent to dye loading of the LSE). Signals from HEM were compared with those from gel–membrane.

Sodium fluorescein was used to assess epidermal permeability by using a protocol different from the one used with the seven fluorescent probe assays listed earlier. Briefly, after washing the LSE or HEM three times in DMEM, sodium fluorescein (10 μg/ml; 200-μl; Sigma Chemical Co., St. Louis, MO) was placed on the Testskin in an apical chamber. The apical chamber was made by using the silicone sealant, assay rings, and directions provided with Testskin, thereby creating a well to hold dye away from the medium under the Testskin. After 120 min (30 min for the HEM) the medium in the basal compartment was assayed within the CytoFluor 2300 using the filter sets and gain levels listed in Table 2. The sodium fluorescein present in the media (basal compartment), relative to the controls, was considered to be a measure of epidermal permeability. Gel–membrane was used as the dermal control for HEM. The LDE was the dermal control for LSE (LSE was not available to be stripped).

## RESULTS AND DISCUSSION

### Effects of Transportation on Immature Human Epidermal Models

The HEMs were assayed (for damage to epidermal cells) in New York before transportation, after 5 hr in transit (to APG, MD) followed by 18 hr of incubation, or after 5 hr in transit during a total time of 24 hr without incubation. Data given in Table 1 suggest that the nonviable control (gel–membrane) may *retain* slightly more (100 unit changes) calcein-AM as time

**Table 1** Transportation Effects on Plasma Membrane and Intercellular Space

| Site specificity | Dye | Control[a] | 5-hr trip[a] | 24-hr trip[a] |
|---|---|---|---|---|
| Gel–membrane fluorescence | Calcein-AM | 100 | 200 | 300 |
| HEM plasma membrane | Calcein-AM | 4200 | 2000 | 1600 |
| Dye leakage through gel–membrane | Sodium fluorescein | 4300 | 4300 | 6300 |
| Dye leakage between HEM cells | Sodium fluorescein | 500 | 2200 | 6450 |

[a]CytoFluor 2300 fluorescence units (values rounded off); $n = 3$.

enroute increases. However, the viable HEM appear to *leak* calcein-AM at increasing rates (2000+ unit changes) with transit. The sodium fluorescein assay suggests that gel–membrane offers a slight barrier to dye leakage, but even this barrier is *lost* after 24-hr enroute. Normal resistance of HEM tissue to intercellular dye leakage appears to be *lost rapidly* with more transportation stress. One interpretation of these observations is that travel may induce HEM cells to pull apart, one from another, while plasma membrane damage remains minimal.

## Observations on Fluorescent Dye Distributions

Testskin samples were incubated with eight dyes of differing site specificities to see if fluorescent dye binding to dermal tissue would hinder attempts to read changes in epidermis (question 4). Table 2 shows most of the fluorescent signals from the first six dyes as coming from LSE epidermis. Similar comparisons with HEM and gel–membranes also showed most of the signal in HEM epidermis.

Table 2 shows results with the free radical dye and with sodium fluorescein different from those of the other dyes. Differences between test conditions appear to account for these results. Dermal cells are likely to be damaged during use of forceps to strip and discard the epidermal portion of LSE. The current hypothesis is that extra dye is bound by free radicals generated in damaged dermal cells, accounting for greater fluorescence in the dermal substrates than in the whole LSE (5000 vs. 1693). It remains to be seen if HD exposures act similarly. The values obtained with sodium fluorescein diverge, because the signal is from dye that passes between the cells not through their plasma membranes. Table 2 shows LSE as an excellent barrier to sodium fluorescein (*A,C*) until epidermis is removed (*B*).

This effort was designed to explore the feasibility of using human skin models, fluorescent probes, and the CytoFluor 2300, in concert, to measure damage such as HD may be expected to induce. Current answers to the questions that guided experimental design are given in Table 3.

# DISCUSSION OF OBSERVATIONS AND OPTIONS

It has been reported that HD vapor changes the histological profile of Testskin [2]; recent unreported observations by the same pathologists suggest that immature HEM is similarly affected by HD. Our preliminary data indicate that calcein-AM is lost from LSE and HEM after exposure to HD vapor. The HEM [4] and individual cells [3] have shown cell-associated and site-specific changes

**Table 2** Fluroescent Probe/CytoFluor 2300 Assays of LSE[a] and LDE[b]

| Parameter | Dye | *A* LSE + dye | *B* LDE + dye | *C* LSE only | *D* Net signal | *E* S/N ratio | *F* Filter set | *G* Gain |
|---|---|---|---|---|---|---|---|---|
| Plasma membrane integrity | Calcein-AM | 3004 ± 294[c] | 1303 ± 196 | 521 ± 66 | 1701 | 3.18 | 485/530 | 4 |
| Glutathione | CMFDA | 8804 ± 293 | 1475 ± 60 | 103 ± 13 | 7329 | 6.34 | 485/530 | 3 |
| Lysosomes | Neutral red | 6540 ± 186 | 648 ± 55 | 116 ± 2 | 5892 | 12.07 | 508/645 | 4 |
| Intracellular pH | BCECF-AM | 3575 ± 270 | 1346 ± 156 | 521 ± 66 | 2229 | 3.7 | 485/530 | 4 |
| Intracellular $Ca^{2+}$ | Fluo3-AM | 4685 ± 327 | 2091 ± 161 | 1875 ± 15 | 2594 | 13.01 | 500/550 | 5 |
| Activity of mitochondria | Rhodamine-123 | 5732 ± 252 | 1494 ± 228 | 16 ± 2 | 2594 | 3.87 | 500/550 | 2 |
| Free radicals | 2,7-Dichlorofluorescein | 1693 ± 200 | 5000 ± 534 | 103 ± 13 | −3307 | 0.32 (see text) | 485/530 | 3 |
| Epidermal permeability | Sodium fluorescein | 15 ± 0 | 8383 ± 517 | 14 ± 0 | 8368 | 8369 (see text) | 485/530 | 3 |

[a]LSE (Living Skin Equivalent) and LDE (Living Dermal Equivalent) are registered trademarks of Organogenesis, Inc, Cambridge, MA.
[b]In this case, LSE with epidermis removed to leave LDE.
[c]± Standard error of the mean.
*A* Includes autofluorescence from epidermis and LDE (or leakage).
*B* Omits epidermal autofluorescence (or measures LDE leakage).
*C* Omits dye fluorescence (signals autofluorescence of media).
*D* Net epidermal autofluorescence, *A-B* (*A-B*, $Na^+$ fluorescein).
*E* Signal/noise ratio, *A-C*/*B-C*.
*F* Filter sets used in the CytoFluor 2300 with given dye.
*G* CytoFluor 2300 sensitivity level (gain) used for given dye.

**Table 3** Preliminary Observations on Feasibility

| No. | Question asked | Current hypothesis | Basis |
|---|---|---|---|
| 1 | Probe/CytoFluor method with LSE? | Signal is readable. | Table 2 |
| 2 | Excessive binding of dyes to LDE, gel? | Dyes bind epidermis 3–12× stripped LSE. | Table 2 |
| 3 | HEM damaged in transit? | Adhesion loss, plasma membrane damage. | Table 1 |
| 4 | Which dyes best show damage? | None excluded, tests with HD required. | Table 1<br>Table 2 |

in fluorescence after damage and incubation with various probes. Data from Table 1 suggest that immature HEM responds in the same way.

Sulfur mustard vapor exposures of HEM were made to obtain assurance that the HD effects will be measurable by CytoFluor and fluorescent dye methods. The HD exposures of HEM were made by the vapor cup method [1], as adapted for Testskin [2], with a 24-hr time interval between the HD challenge and incubation with calcein-AM. Transportation damage to the cultures compromised plans to compare fluorescence units with HD and controls. However, the unreported results appeared to be consistent with HD damage to plasma membranes and fluorescent dye leakage. Additional exposures are planned with dilute HD.

There are many virtues of fluorochrome probes that make them suitable choices for multiple endpoint assays [6]. First, some can be used to double-label cells [3]. Second, they are nonradioactive. Third, fluorochrome diversity exceeds that of radioactive probes [5]. Fourth, unlike radioactive probes, fluorochromes can be visualized with a fluorescence microscope, a spectrofluorometer, or both. Fifth, activity/site-specific fluorochrome signals can be analyzed in situ in living cells—a feature not currently possible with radioactive tracers. Sixth, since probes are being created for new purposes [5], it may soon be possible to label cells with several different probes to reveal concurrent pathobiological responses during a cytotoxic episode. Finally, such testing has promise for saving both time and money by facilitating objective analysis of a variety of responses with a single instrument.

Simultaneous uses of two probes have been reported with both cellular monolayers [3] and HEM [4]. Our data are not yet sufficient to determine that current human skin models will be suitable candidates for a CytoFluor-based, automated, multiple endpoint, fluorescent probe assay of antivesicant compounds. However, existing evidence encourages further efforts to develop such an approach.

## CONCLUSION

Incubation of damaged and control human skin models with eight site-specific fluorescent probes resulted in differences of induced or reduced fluorescence, as shown by CytoFluor 2300 measurements; the results suggest that these methods may be usable to measure damage from HD.

## REFERENCES

1. Mershon, M. M., Mitcheltree, L. W., Petrali, J. P., Braue, E. H., and Wade, J. V., Hairless guinea pig bioassay model for vesicant vapor exposures, *Fundam. Appl. Toxicol. 15*:622–630 (1990).
2. Petrali, J. P., Oglesby, S. B., and Justis, T. A., Morphological effects of sulfur mustard on a human skin equivalent. *J. Toxicol. Cutan. Ocul. Toxicol. 10*:315–324 (1991).
3. Essig-Marcello, J., and Van Buskirk, R. G., A double-label in situ cytoxicity assay using the fluorescent probes neutral red and BCECF-AM, *In Vitro Toxicol. 3*:219–228 (1990).
4. Cook, J. R., Gabriels, J., Patrone, L. M., Rhoads, L. S., and Van Buskirk, R.G., A human epidermal model that can be used in an automated multiple endpoint assay, *J. Alternative Lab. Anim. 9*:313–323 (1992).
5. Dr. Richard Haugland, Molecular Probes, Eugene, OR, personal communication (1991).
6. Van Buskirk, R. G., The evolution of multiple endpoint assays, *In Vitro Toxicology—The Versatility of Fluorochromes*, Pharmaceutical Manufacturing International, Sterling Publishing Group, London, 1991, pp. 81–85.

# 22

# Ocular Testing: Historical Perspectives

**Van M. Seabaugh**

*U. S. Environmental Protection Agency*
*Washington, D.C.*

John Henry Draize (1900–1992) was a pioneer for eye irritation testing procedures. He was employed by the Department of Defense at Edgewood Arsenal in Maryland from 1935 to 1939. Among other things, his job was to study military chemicals for potential ocular effects. The Draize protocol [1] is similar to Friedenwald's [2], but John's main contribution to ocular testing was the scoring system. The Draize scoring system [1,3] gave numerical equivalents for brief subjective statements for cornea, iris, and conjunctivae. In 1939, Draize moved to the Food and Drug Administration (FDA). When the FDA needed an eye test in the 1940s to test drugs and cosmetics, John had the answers. John Draize retired from the FDA in 1963, and Frank Marzulli left Edgewood Arsenal to take over the vacated Branch Chief job at the FDA. They worked together for 1 month before John's retirement.

What are some of the historical perspectives for toxicology? In 1986 at the Society of Toxicology's (SOT) 25th annual meeting, copies of past scientists' photographs from the National Library of Medicine were displayed. The

---

The views presented are those of the author, and do not necessarily reflect the opinions or policy of EPA or any other federal agency.

Presented in part at the Symposium on Current Concepts and Approaches on Animal Test Alternatives, Feb. 4–6, 1992, Aberdeen Proving Ground, MD.

oldest photograph on display was that of Paracelsus. Paracelsus (1493–1541) was a Swiss physician and a professor of medicine. In 1990, John Doull (Professor of Toxicology and Pharmacology, University of Kansas Medical School) gave his *Historical Perspectives in Toxicology* lecture at a symposium entitled *New Horizons in Molecular Toxicology*. John said that perhaps some of us would say that primitive man conducted the first toxicological study when he fed some of his food to his dog to see if the food was safe to eat. To add to this dialogue, a quick look in the library can give a glimpse of the historical perspectives for animal use in research and testing.

According to recorded history, the pig was the first animal to be used for medical research (Garland's work, 1500). As early as the 1600s, medical research was being conducted on humans. Galvandis (1600s) conducted medical research on humans. Rembrandt's paintings, as early as 1632, depicted medical research being conducted on people. Some of the kings of European countries had professional tasters to see before hand if the food to be consumed by the king and his family would kill or make the tasters ill. Claude Bernard (1800–1860) wrote a book entitled *An Introduction to the Study of Experimental Medicine*. From 1861 to 1889, history tells us about (a) the studies of Louis Pasteur, who studied chemistry, bacteriology, diseases of the silkworms, rabies, and anthrax; (b) Robert Koch and F. K. Clein, who studied infectious diseases. They wanted to know what killed the small animals, and did it apply to humans? (c) The Russian physiologist, Pavlov, who studied associated and conditioned responses; (d) Cooper Curtis, who studied Texas fever in cattle.

In 1892, newspapers advertised that they would pay 25 cents each for dogs or cats to be used in medical research. In 1875, J. C. Dalton published a book entitled, *Experimenting on Animals*. In 1982, James Turner published a book entitled, *Reckoning With the Beast*, and the question is asked if we are the beasts? About 1876, Henry Byrd founded the first American Society for Anticruelty to Animals (ASPCA, New York, NY), and George T. Angell founded the Massachusetts Antivivisectionist Society. In 1890, an attempt was made to obtain federal legislation for the treatment of animals. The legislation did not pass.

At the Food and Drug Administration Museum, there are pictures of its poison squad. In 1904, FDA was still using human employees to test food preservatives (e.g., boric acid, salicylic acid and derivatives, and formaldehyde) for toxicity.

What has happened in recent years? In 1959–1960, the drug thalidomide changed the course for animal testing. The Society of Toxicology (SOT) was founded in 1961. In 1963, the Public Health Services granted a policy statement entitled, *Guide for the Care and Use of Laboratory Animals*. In 1966, the Animal Welfare Act was created. In the 1970s, meetings were held by the National Institutes of Health and others on the use of animals.

In 1973, the Society of Comparative Ophthalmology (SOCO) was founded, and from that time on has been one of the leaders in educating scientists and the public on in vitro and in vivo ocular testing. In 1977, the American College of Toxicology was founded, and in 1980, Peter Singer published a book entitled *Stop the Slaughter of the Innocent*.

The Cosmetic, Toiletry, and Fragrance Association (CTFA) held workshops on the use of animals. One of them, held in 1980, was entitled *Proceedings of the CTFA Ocular Safety Testing Workshop: In Vitro and In Vivo Approaches*. We know about the work done at the Rockefeller University to help find ocular alternatives for animal testing. The Office of Technology Assessment of the United States Congress, published a book entitled *Alternatives for Animal Use in Research, Testing, and Education* [4]. The book discussed a range of options for congressional action for public policy concerned with the use of animals. The Catholic University continues to conduct workshops on tissue and cell culture techniques for alternative testing, and the Johns Hopkins University Center for Alternatives to Animal Testing (CAAT) program is also an on-going program. Other information has been presented at workshops, meetings, and publications on the issues of in vitro and in vivo ocular testing.

The present status of in vitro and in vivo ocular and dermal irritation testing is perhaps best summarized in the position paper issued by the Society of Toxicology (SOT) published in *Fundamentals and Applied Toxicology* [5] which was prepared by the Animals in Research Committee of the Society of Toxicology and approved by the SOT Council. Some opinions from the SOT position paper are given here and in the next three paragraphs. Although many in vitro alternatives to eye and skin irritation tests are under development, none have been adequately validated. However, many of these alternatives being developed show promise and may be useful as initial screening procedures. When these procedures become completely tested and validated, the need for whole-animal testing may be reduced. Until that time, however, conventional tests will continue to be used. Special attention should be given to the design and conduct of these studies in an attempt to reduce the number of animals used, and to minimize any discomfort to the animals.

This paper [5] also suggests that, as part of the ongoing test programs, investigators have refined and modified their existing in vivo techniques. These include prescreens to identify corrosive or severely irritating materials. Thus, materials with very acidic or alkaline properties, having a high potential for irritation, would be excluded from testing in animals. In addition, since it is reasonable to conclude that severe skin irritants would also be severe eye irritants, knowledge of results of a skin irritation test could preclude the need for ocular testing.

The more than 60 proposed nonintact animal tests available for predicting eye irritation have been divided into six different categories by Frazier et al. [6]. These categories are cell toxicity, inflammation or immune system response, effects on recovery or repair processes, effects on cellular or tissue physiology, cell morphology, and other, including biochemical endpoints or structure–activity analysis.

According to the Society of Toxicology [5], many of these in vitro tests hold promise, but none are yet adequately validated or evaluated for a broad range of chemical materials. In addition, none can yet provide the scientific reliability or predictive accuracy that would be required of a new test for regulatory or legal acceptability [5].

## REFERENCES


1. Draize, J. H., Dermal toxicity, *Food Drug Cosmet. Law J. 10*:722–725 (1955).
2. Friedenwald, J. S., Hughes, W. F., and Herrman, H., *Arch. Ophthamol. 31*:379–383 (1944).
3. Draize, J. H., Dermal toxicity, *Food Drug Cosmet. Law J. 10*:722–725 (1955).
4. U. S. Congress, Office of Technology Assessment, Alternatives to Animal Use in Research, Testing, and Education, U.S. Government Printing Office, OTA-BA-273, Washington, DC, February 1986.
5. Society of Toxicology, Position paper. Comments on the $LD_{50}$ and acute eye and skin irritation tests. *Fundam. Appl. Toxicol. 13*:621–623 (1989).
6. Frazier, J. M., Gad, S. C., Goldberg, A. M., and McCalley, J. P., *A Critical Evaluation of Alternatives to Acute Irritation Testing*, Mary Ann Liebert, New York, 1987.

# 23

# Overview of In Vitro Ocular Irritation Test Systems and an Evaluation of Their Status

**Shayne C. Gad**

*Gad Consulting Services*
*Raleigh, North Carolina*

The ocular irritancy test has been the most controversial animal test in toxicology from the standpoint of both the public perception and the humane use of animals. There has been some grounds for this concern, but testing practices have changed significantly since 1980. These changes have reduced (but not eliminated) grounds for public concern.

There have also been scientific grounds for reconsidering the appropriateness of the test, as currently performed in rabbits, and these have further stimulated attempts to develop in vitro alternatives to the rabbit eye-based test. The technical rationales for seeking alternatives can be summarized as follows:

1. Strict Draize scale testing in the rabbit assesses only three eye structures (conjunctiva, cornea, iris) and traditional rabbit eye irritancy tests do not assess cataracts, pain, discomfort, or clouding of the lens.
2. In vivo tests assess only inflammation and immediate structural alterations produced by irritants (not sensitizers, photoirritants, or photoallergens, or route-specific lethality).
3. Technician training and monitoring are critical (particularly owing to the subjective nature of evaluation).
4. Rabbit eye tests do not perfectly predict results in humans, if the objective is either the total exclusion of irritants on an absolute basis (that is, without

false-positive or false-negative results) or the identification mild irritants. Some (such as Reinhardt et al. [1]) have claimed that these tests are too sensitive for such uses.

5. There are structural and biochemical differences between rabbit and human eyes that make extrapolation from one to the other difficult. For example, Bowman's membrane is present and well developed in humans (8- to 12-mm thick), but not in the rabbit, possibly giving the cornea greater protection.
6. Lack of standardization of protocols and test procedures.
7. Large biological variability between experimental units (i.e., individual animals).
8. The available animal data are available as large, diverse, and fragmented databases that are neither readily available nor comparable.

These factors have made ocular irritation the most fertile area of alternative test development. There are now more than 70 proposed in vitro tests systems that can be considered in six different classes: morphology, cell toxicity, cell and tissue physiology, inflammation and immunity, recovery and repair, and miscellaneous. Some examples of tests in each of these categories are presented in the following [2].

## IN VITRO ALTERNATIVES FOR EYE IRRITATION TESTS

### Morphology

1. Enucleated superfused rabbit eye system [3].
2. Balb/c 3T3 cells:morphological assays (HTD) [4].

### Cell Toxicity

Adhesion or Cell Proliferation

1. BHK cells: growth inhibition [1]
2. BHK cells: colony formation efficiency [1]
3. BHK cells: cell colony-forming assay [5]
4. SIRC cells: colony-forming assay [5]
5. Balb/c 3T3 cells: total protein [6]
6. BCL-D1 cells: total protein [7]
7. Primary rabbit corneal cells: colony-forming assay [8]

Membrane Integrity

1. LS cells: dual dye staining [9]
2. Thymocytes: dual fluorescent dye staining [10]

3. LS cells: dual dye staining [11]
4. RCE-SIRC-P815-YAC-1/Cr release [12]
5. L929 cells: cell viability [13]
6. Bovine red blood cell: hemolysis [14]
7. Mouse L929 fibroblasts: erythrocin C staining [15]
8. Rabbit corneal epithelial and endothelial cells: membrane leakage [16]
9. Agarose diffusion [17]
10. Living dermal equivalent (LDE) [18]

Cell Metabolism

1. Rabbit corneal cell cultures: plasminogen activator [19]
2. LS cells: ATP assay [20]
3. Balb/c 3T3 cells: neutral red uptake [4]
4. Balb/c 3T3 cells: uridine uptake inhibition assay [21]
5. HeLa cells: metabolic inhibition test (MIT-24) [22]
6. MDCK cells: dye diffusion [23]

## Cell and Tissue Physiology

1. Epidermal slice: electrical conductivity [24]
2. Rabbit ileum: contraction inhibition [25]
3. Bovine cornea: corneal opacity [26]
4. Proposed mouse eye: permeability test [27]

## Inflammation and Immunity

1. Chorioallantoic membrane (CAM)
   a. CAM [28]
   b. HET–CAM [29]
   c. CAM VA
2. Bovine corneal cup model: leukocyte chemotactic factors [30]
3. Rat peritoneal cells: histamine release [31]
4. Rat peritoneal mast cells: serotonin release [32]
5. Rat vaginal explant: prostaglandin release [33]
6. Bovine eye cup: histamine (Hm) and leukotriene $C_4$ ($LTC_4$) release [34]

## Recovery and Repair

1. Rabbit corneal epithelial cells: wound healing [35]

## Other

1. Eyetex assay [36,37]

2. TOPKAT computer-based: structure–activity (SAR) [38,39]
3. Tetrahymena: motility [40]
4. Microtox

The potential advantages and disadvantages of in vitro test systems are summarized as follows:

## Advantages

1. Avoid complications and potential confounding or "masking" of findings of in vivo studies.
2. Exposure levels and conditions at target sites can be better controlled.
3. Test condition standardization can be better than for in vivo studies.
4. Reduction in animal usage or in pain to experimental animals.
5. Ability to directly study some target tissue effects on a real-time basis.
6. Reduced requirements for test agents.

## Disadvantages

1. Lack of ability to evaluate longer-term effects.
2. Limited ability to simulate and evaluate integrated organismic level effects.
3. May not reflect influence or agent absorption, distribution, metabolism, and excretion effects.
4. May not detect unique systemic effects resulting from exposure by this route.

In number 4, are those cases in which animals die very quickly after being dosed in an eye irritation test. It has been observed that a dosed rabbit died before the second dose was administered [S. C. Gad, personal communication; 41]. If ocular exposure is likely, and a significant volume of the material is to enter use by commerce, at least one animal should receive exposure by this route.

All of the in vitro methods that are available show promise of at least some uses. They have surplanted animal testing to some degree, but not nearly as extensively as they might. The first hurdle has been to have them "validated." Although individual investigators have "calibrated" the performance of their own methods using standards of known irritancy (generally of known irritancy in rabbits whereas humans remain the species of concern), this has not filled the need for scientific test validation. Rather, large intralaboratory tests of multiple methods in multiple laboratories are required. At least seven such large trials, as summarized in Table 1 have been completed.

**Table 1** Multilaboratory Evaluations

| Organization | No. methods | No. chemicals | Comparator in vivo data |
|---|---|---|---|
| CFTA (US) | 23 | 20 | Rabbit |
| EEC (European) | 7 | 21 | Rabbit |
| FRAME (UK) | 6 | 56 | Rat |
| MEIC (SW/EUROPEAN) | 87 | 10–50 | Human |
| PMA (US) | 7 | 35 | Rabbit |
| SDA: phases I and II | 14 | 23 | Rabbit |
| ZEBET (GE) | 2 | 35 | Rabbit |

The results of the multilaboratory evaluations to date can be summarized as follows:

1. Ellipsoid effect: predictive ability for negative and severe irritants or corrosives is generally good, but moderates show poor correlation, with in vivo data.
2. To date, batteries do not seem to show improved performance over "best" single tests.
3. Single-test performance is significantly influenced by the structural–functional class of the chemical.
4. Success of validation programs will depend on expectations or acceptance criteria.

One must also remember that the conduct of in vivo eye irritation tests has also been significantly modified by alternatives, such as [42]:

1. Prescreening for extreme irritants and corrosives by pH determinations and dermal irritancy tests
2. Use of the less overly sensitive and less stressful low volume test
3. Use of fewer animals

Finally, one must remember that despite their limitations, the rabbit eye tests have so successfully identified severe ocular irritants that there is no evidence of such a product entering the market. The often overlooked fourth *R* (after reduction, replacement, and refinement) is operative here. It is the *r*esponsibility of the industrial toxicologist to prevent unsafe products from entering the marketplace.

In summary, significant progress has been made in developing and evaluating in vitro tests for eye irritation. Several of these tests could probably serve successfully in place of whole-animal tests in screening out those industrial compounds that are not intended to have any eye irritation, yet still may present an unacceptable hazard because of their severe irritation potential. However, no single in vitro test (or combination of tests) exists that can currently or in the near future replace all whole-animal testing.

## REFERENCES

1. Reinhardt, C. A., Pelli, D. A., and Zbinden, G., Interpretation of cell toxicity data for the estimation of potential irritation, *Food Chem. Toxicol. 23*:247–252 (1985).
2. Fraizier, J. M., Gad, S. C., Goldberg, A. M., and McCulley, J. P., *A Critical Evaluation of Alternatives to Acute Ocular Irritation Testing*, Mary Ann Liebert, New York, 1987.
3. Burton, A. B. G., York, M., and Lawrence, R. S., The in vitro assessment of severe eye irritants, *Food Cosmet. Toxicol. 19*:471–480 (1981).
4. Borenfreund, E., and Peurner, J. A., A simple quantitative procedure using monolayer cultures for cytotoxicity assays (HTD/NR-NE), *J. Tissue Cult. Methods 9*:7–10 (1984).
5. North-Root, H. Yackovich, Demetrulias, F. J., Gucula, N., and Heinze, J. E., Evaluation of an in vitro cell toxicity test using rabbit corneal cells to predict the eye irritation potential of surfactants, *Toxicol. Lett. 14*:207–212 (1982).
6. Shopsis, C., and Eng, B., Uridine uptake and cell growth cytotoxicity tests: Comparison, applications and mechanistic studies, *J. Cell Biol. 101*:87a (1985).
7. Balls, M., and Horner, S. A., The FRAME interlaboratory program on in vitro cytotoxicology, *Food Chem. Toxicol. 23*:205–213 (1985).
8. Watanabe, M., Watanabe, K., Suzuki, K., Nikaido, Sugarhara, T., Ishii, I., and Konishi, H., In vitro cytotoxicity test using primary cells derived from rabbit eye is useful as an alternative for Draize testing, in *Progress in In Vitro Toxicology* (A. M. Goldberg, ed.), Mary Ann Liebert, New York, 1988, pp. 285–290.
9. Schaife, M. C., An investigation of detergent action on in vitro and possible correlations with in vivo data, *Int. J. Cosmet Sci. 4*:179–193 (1982).
10. Aesechbacher, M., Reinhardt, C. A., and Zbinden, G., A rapid cell membrane permeability test using fluorescent dyes and flow cytometry, *Cell Biol. Toxicol.* (in press) (1986).
11. Kemp, R. V., Meredith, R. W. J., Gamble, S., and Frost, M., A rapid cell culture technique for assaying to toxicity of detergent based products in vitro as a possible screen for high irritants in vivo, *Cytobios 36*:153–159 (1983).
12. Shadduck, J. A., Everitt, J., and Bay, P., Use of in vitro cytotoxicity to rank ocular irritation of six surfactants, *In Vitro Toxicology*, Vol. 3, *Alternative Methods in Toxicology* (A. M. Goldberg, ed.), Mary Ann Liebert, New York, 1985, pp. 641–649.
13. Simons, P. J., An alternative to the Draize test, *The Use of Alternatives in Drug Research* (A. M. Rowan and C. J. Stratmann, eds.), Macmillan Press, London, 1981.
14. Shadduck et al., 1987.

15. Frazier, A. M., Update: A critical evaluation of alternatives to acute ocular irritancy testing, *Progress in In Vitro Toxicology* (A. M. Goldberg, ed.), Mary Ann Liebert, New York, 1988,pp. 67–75.
16. Meyer, D. R., and McCulley, J. P., Acute and protracted injury to cornea epithelium as an indication of the biocompatibility of various pharmaceutical vehicles, *Progress in In Vitro Toxicology* (A. M. Goldberg, ed.), Mary Ann Liebert, New York, 1988, pp. 215–235.
17. Barnard, N. D., A Draize alternative, *Anim. Agenda 6*:45 (1989).
18. Bell, E., Parenteau, N. L., Haimes, H. B., Gay, R. J., Kemp, P. D., Fofonoff, T. W., Mason, V. S., Kagan, D. T., and Swiderek, M., Testskin: A hybrid organism covered by a living human skin equivalent designed for toxicity and other testing, *Progress In Vitro Toxicology* (A. M. Goldberg, ed.), Mary Ann Liebert, New York, 1988,pp. 15–25.
19. Chan, K. Y., An in vitro alternative to the Draize test, *In Vitro Toxicology* Vol. 3, *Alternative Methods in Toxicology* (A. M. Goldberg, ed.), Mary Ann Liebert, New York, 1985, pp. 405–422.
20. Kemp, R. V., Meredith, R. W. J., and Gamble, S., Toxicity of commercial products on cells in suspension: A possible screen for the Draize eye irritation test, *Food Chem. Toxicol. 23*:267–270 (1985).
21. Shopsis, C., and Sathe, S., Uridine uptake inhibition as a cytotoxicity test: Correlation with the Draize test, *Toxicology 29*:195–206 (1984).
22. Selling, J., and Ekwall, B., Screening for eye irritancy using cultured Hela cells, *Xenobiotica 15*:713–717 (1985).
23. Tchao, R., Trans-epithelial dermiability or fluorescein in vitro as an assay to determine eye irritants, *Progress in In Vitro Toxicology* (A. M. Goldberg, ed.), Mary Ann Liebert, New York, 1988,pp. 271–284.
24. Oliver, G. J. A., and Pemberton, N. A., An in vitro epidermal slice technique for identifying chemicals with potential for severe cutaneous effects, *Food Chem. Toxicol. 23*:229–232 (1985).
25. Muir, C. K., A simple method to assess surfactant-induced bovine corneal opacity in vitro: Preliminary findings, *Toxicol. Lett. 23*:199–203 (1984).
26. Muir, C. K., Flower, C., and VanAbbe, N. J., A novel approach to the search for in vitro alternatives to in vivo eye irritancy testing, *Toxicol. Lett. 18*:1–5 (1983).
27. Maurice, D., and Singh, T., A permeability test for acute corneal toxicity, *Toxicol. Lett. 31*:125–130 (1986).
28. Leighton, J., Nassauer, J., Tchao, R., and Verdone, J., Development of a procedure using the chick egg as an alternative to the Draize rabbit test. *Product Safety Evaluation*, Vol. 1, *Alternative Methods in Toxicology* (A. M. Goldberg, ed.), Mary Ann Liebert, New York, 1983,pp. 165–177.
29. Luepke, N. P., Hen's egg chorioallantoic membrane test for irritation potential, *Food Chem. Toxicol. 23*:287–291 (1985).
30. Elgebaly, S. A., Nabawi, K., Herkbert, N., O'Rourke, J., and Kruetzer, D. L., Characterization of neutrophil and monocyte specific chemotactic factors derived from the cornea in response to injury, *Invest. Ophthalmol. Vis. Sci. 26*:320 (1985).

31. Jacaruso, R. B., Barlett, M. A., Carson, S., and Trombetta, L. D., Release of histamine from rat peritoneal cells in vitro as an index of irritational potential, *J. Toxicol. Cutan. Ocul. Toxicol. 4*:39–48, 1985.
32. Chasin, M., Scott, C., Shaw, C., and Persico, F., A new assay for the measurement of mediator release from rat peritoneal mast cells, *Int. Arch. Allergy Appl. Immunol. 58*:1–10 (1979).
33. Dublin, N. H., De Blasi, M. C., et al., Development of an in vitro test for cytotoxicity in vaginal tissue: effect of ethanol on prostanoid release, *Acute Toxicity Testing: Alternative Approaches*, Vol. 2, *Alternative Methods in Toxicology* (A. M. Goldberg, ed.), Mary Ann Liebert, New York, 1984, pp. 127–138.
34. Benassi, C. A., Angi, M. R., Salvalaio, L., and Bettero, A., Histamine and leukotriene $C_4$ release from isolated bovine scheracharoid complex: A new in vitro ocular irritation test, *Chim. Acta* (accepted) (1986).
35. Jumblatt, M. M., and Neufeld, A. H., A tissue culture model of the human corneal epitheliym, *In Vitro Toxicology*, Vol. 3, *Alternative Methods in Toxicology* (A. M. Goldberg, ed.), Mary Ann Liebert, New York, 1985, pp. 391–404.
36. Gordon, V. C., and Bergmen, H. C., Eyetex, an in vitro method for evaluation of optical irritancy, National Testing Corporation Report, 26, 1986.
37. Soto et al., 1988.
38. Enslein, K., Estimation of toxicology end points by structure–activity relationships *Pharmacol. Rev. 36*:131–134 (1984).
39. Enslein, K., Blake, V. W., Tuzzeo, T. M., Borgstedt, H. H., Hart, J. B., and Salem, H., Estimation of rabbit eye irritation scores by structure–activity equations, *In Vitro Toxicol. 2*:1–14 (1988).
40. Silverman, J., Preliminary findings on the use of protozoa (*Tetrahymena thermophila*) as models for ocular irritation testing in rabbits, *Lab. Anim. Sci. 33*:56–59 (1983).
41. Moss, J. N., Hsu, A., Lange, B., and Scribner, H. E., The need for animal testing—report of a case, *Food Chem. Toxicol. 29*:475 (1991).
42. Gad, S. C., and Chengelis, C. P., *Acute Toxicology: Principles and Methods*, Telford Press, Caldwell, NJ, 1988.

# 24

# The Role of In Vitro Tests in Assessing the Safety of Cosmetics and Consumer Products

**Thomas J. Stephens and E. Tiffany Spence**

*Thomas J. Stephens & Associates, Inc.*
*Carrollton, Texas*

Over the last 12 years, the cosmetics industry in the United States has become the primary target of the animal rights movement. Beauty care companies are particularly susceptible to these attacks because these products are often viewed as nonessential [1].

The success of the animal rights movement is attributable to its persistence in changing public opinion about the need for safety testing in animals. Its strategy has consisted of public awareness programs, company boycotts, and seeking sponsorship at the state legislative level for bills to limit the use of Draize eye and skin tests. Several multinational cosmetics firms have, as a result, declared moratoriums or bans on animal testing.

The size of the beauty care market is currently estimated at 45 billion dollars [2]. The development of an alternative strategy to eliminate the need for animal tests is essential for cosmetic companies to remain competitive. The purpose of this chapter is to discuss alternative approaches for assessing the safety of cosmetics and how in vitro tests can help improve product quality.

## WHAT IS IN VITRO TESTING? WHAT CAN IT DO?

In vitro testing is testing "in glass," or rather, in a test tube, as opposed to in vivo, in a living animal. Although in vitro testing incorporates the use of living systems, it does not use whole, living, sentient animals. Cell cultures lines may be used, as well as single-cell organisms, such as bacteria and protozoa, to aid in the assessment of a product's safety.

Many in vitro tests using cell cultures still depend on animals for a source of cells and tissue. It is important to keep in mind that the sacrifice of one animal (if any sacrifice is needed at all) provides an enormous quantity of cells for in vitro testing, thereby greatly reducing the number of animals that would have been used for in vivo testing.

In today's product safety testing programs, in vitro data are often compared with in vivo data to assess the in vitro test's performance. The ideal in vitro test would correlate perfectly with in vivo data; by the same token, the ideal in vivo test would perfectly predict a product's effect in humans after accidental or intentional exposure to eye and skin. This is not always true. No existing in vitro or in vivo test can predict a product's safety 100% of the time. Still, in vitro tests have been reliable and sensitive testing tools. Although there is no complete substitute for human or animal data, in vitro technology can be effectively implemented by many companies for screening and evaluating products. In addition to reducing and replacing animal testing, there are the issues of time and money that strongly support the notion of in vitro testing.

## SAFETY-TESTING PROGRAMS FOR CONSUMER PRODUCTS

Most cosmetic companies have product development schedules in which safety testing programs are conducted in an accelerated rate. These aggressive schedules are maintained to support the seasonal sale periods. It is not uncommon for safety programs for cosmetics to be completed in a 4- to 6-month period on multiple versions of the same product to protect the marketing launch date. The total development time is usually less than a year.

Some cosmetic companies execute product safety testing programs using a tiered approach consisting of preclinical testing (phase 1), clinical testing (phase 2), and usage studies (phase 3). This tiered approach has the advantage of providing early safety information on products before large capital investments in equipment and packaging are made. It also provides valuable information about the safety of a product before conducting product usage studies. Marketing studies range from controlled usage tests within the company, to national test markets. As a rule, the larger the market, the greater the need for completed safety tests and exposure data before clearing a product. It would be unethical to give experimental cosmetic formulations to consumers in a home-marketing study without having knowledge of whether the formulation could be an oral toxin or an eye or skin irritant.

Most preclinical-testing programs are designed to evaluate short-term effects, such as oral toxicity, ocular and skin irritation, and allergic contact dermatitis potential, of prototype formulations and product ingredients before human panel testing. In a few instances, a new ingredient may require subchronic and chronic testing.

## Preclinical Testing (Phase 1)

Preclinical testing involves conducting animal tests in compliance with Good Laboratory Practice regulations. Usually, a minimum of two prototype formulations of each product to reach the marketplace are examined in preclinical testing.

Safety departments usually allocate 30–45 working days of the product development schedule to complete the in-life phase of these tests before recommending which prototype formulations should proceed to clinical testing. Approximately 10–20% of the total safety budget for the product may be spent on preclinical tests. The characteristics of preclinical testing can be found in Table 1.

## Clinical Testing (Phase 2)

Clinical testing involves conducting safety tests on humans under the direction of doctoral level scientists and board-certified physicians. Good Clinical Practice regulations are prescribed for these studies to protect the rights and safety of subjects involved in such investigations and to help assure the quality and integrity of the data generated from these studies. The data provided from preclinical testing serve as a building block for justification of human testing.

**Table 1** Preclinical Testing: Phase 1

| |
|---|
| 30–45 working days |
| Multiple formulations |
| Good Laboratory Practice regulations |
| Cost about 10–20% of total budget |
| Oral toxicity |
| Ocular irritation |
| Comedogenicity |
| Skin irritation |
| Allergic contact dermatitis |
| Phototoxicity and photoallergenicity |

**Table 2** Clinical Testing: Phase 2

| |
|---|
| 60–80 working days |
| Multiple formulations |
| Good Clinical Practice regulations |
| Cost about 30–40% of total budget |
| Cumulative skin irritation |
| Human allergenicity |
| Comedogenicity |
| Phototoxicity and photoallergenicity |

Occasionally, preclinical and clinical programs may be conducted in overlapping time periods. These situations are justified when the composition of similar formulations has already been assessed in preclinical-testing programs.

Clinical-testing programs usually last 60–80 working days and assess multiple prototypes of similar formulations. These tests evaluate skin irritation, allergenicity, comedogenicity, and acnegenicity. Typically 30–40% of the safety budget for a product is spent in clinical testing (Table 2). Approval for national test marketing often occurs after completion of the clinical program.

## Claims Substantiation (Phase 3)

Successful marketing of a cosmetic product often depends on the claims that can be made about a product. Popular claims often associated with cosmetics and skin care formulations include "safe for contact lens wearers," "safe for

**Table 3** Claims Substantiation: Phase 3

| |
|---|
| 30–60 working days |
| Single formulation |
| Physician as investigator |
| Good Clinical Practice regulations |
| Cost about 60–70% of total budget |
| Usage studies |
| Periocular application tests |
| Acnegenicity studies |
| Special user group tests |

sensitive skin," and "moisturizes dry skin." Many of these claims are often substantiated with clinical studies conducted under board-certified physicians (Table 3). Successful completion of the three testing phases provides justification for national release of a product.

## CHANGES IN SAFETY TESTING

Although some consumers are delighted with the availability of cruelty-free products, others have raised the question of whether or not these products are actually safe. The answers depend on the philosophy of the manufacturer. Although consumers (including those supporting the animal rights movement) want safe products, there is no consensus among interested parties on how to achieve this goal.

Some manufacturers produce and sell cosmetics with little or no safety testing. They assume that the product is safe because the ingredients are natural, as opposed to synthetically manufactured, chemicals. This could not be further from the truth. Natural ingredients contain complex molecules that may induce allergic, phototoxic, or photoallergic responses. The size of a company has little relation to whether or not it conducts safety testing.

Some companies that have declared bans or moratoriums on animal testing have developed alternative strategies that allow the replacement of animal testing using other sources of information. This may entail analysis of structure–activity relationships (SARs) of the ingredients, historical safety data, suppliers' information, and clinical data. Many products can be safely released into the market with this approach (Table 4).

**Table 4** Modified Preclinical Testing: Phase 1

| |
|---|
| 15–30 working days |
| Multiple formulations |
| Good Laboratory Practice regulations |
| Cost about 5–10% of total budget |
| Structure–activity relationship |
| Supplier's safety data |
| Historical safety records |
| Small panel usage tests (10–50 people) |
| Some in vitro testing |

Not all companies can follow this approach. Pharmaceuticals, medical devices, and insecticides are examples of product types that must comply with specific test requirements for approval from a regulatory agency. Animal testing is usually the key component of the testing requirements.

## IN VITRO ALTERNATIVES FOR THE DRAIZE EYE TEST

In vitro tests have been classified into six groups based on endpoint measurement [3]. These endpoints include morphology, cellular toxicity, cell and tissue physiology, inflammation and immunity, recovery and repair, and other. The classification of other has been replaced with the term phenomenological. This refers to in vitro tests that are based on endpoints that are not biologically relevant to ocular irritation, yet have shown promise (through statistically based correlative-testing programs) of being useful. Examples of these tests include the *Tetrahymena thermophila* motility assay and the luminescent bacteria test. Both tests, which are based on changes in primitive organisms, have shown promise as ocular-screening assays for surfactant-containing products and raw materials.

The selection of an in vitro test battery for predicting ocular irritation should be based on mechanistically relevant changes to the anterior chamber of the eye. Examples or important endpoints include alteration in protein structure (irreversible corneal damage), cytotoxicity, inflammation (e.g., prostaglandins and vasoactive amines), and recovery. A test battery should stress multiple endpoints, not necessarily multiple tests. A single test that can measure multiple endpoints will be time-efficient and more cost-effective than the use of several tests that measure a single endpoint.

A popular test design for assessing the ocular irritation potential of test materials in tissue culture and reconstituted human skin models is varying the concentration, while keeping the exposure time fixed. Results are expressed in terms of the effective concentration affecting 50% of the cells ($EC_{50}$ value).

Another test design that more closely mimics the Draize eye test is measuring cellular change at a fixed exposure time and concentration of a test material. The concentration and volume selected for this protocol should be similar to anticipated accidental exposure conditions. This type of design often reduces or eliminates false-positive reactions.

The corneal wound closure model offers a mechanistically based test to predict recovery [4]. This test is based on the ability of cultured cells to migrate and re-cover the standard wound that was inflicted on the model.

The adaptation of light addressable potentiometric sensor (LAPS) technology to a flow chamber offers an exciting advancement in studying intercellular recovery after pulse exposure to a test material. Good correlations have been reported between in vivo ocular irritation tests and in vitro tests using the microphysiometer [5]. It is premature, however, to assume that changes in acidification rate can accurately predict ocular recovery. This skepticism is based on the fact that reversible recovery of the cornea involves the movement of water over hours to days into and out of the stromal layer. The presence of water is the result of an inflammatory reaction mediated by immune cells and vasoactive cells. A relation must first be established between the acidification rate and inflammation before this can be viewed as a relevant, mechanistically based endpoint.

## MANAGING IN VITRO TESTING

For some toxicologists, the decision to use in vitro tests during the development cycle of a product would be unthinkable. The reasons most frequently quoted include "lack of regulatory acceptance," "the technology is too expensive," and " in vitro tests cannot replace animal testing." For these scientists, the use of in vitro tests is viewed as experimental and probably presents a risk, rather than benefit. Historically, this response is to be expected with the application of any new idea or technology.

The first step in managing in vitro testing is to set realistic goals for their application and use. In vitro tests do not yet possess the level of sophistication and biological integration to be used as stand-alone, replacement technology. In vitro tests, however, when applied in a cost-effective screening program, can provide valuable information about prototype formulations and strengthen the decision to approve the release of a product.

The three logical places to use in vitro testing are at the preclinical testing phase (phase 1), at the claims substantiation phase (phase 2), and for quality control checks during product manufacturing (phase 3). During the preclinical testing phase, in vitro testing results as well as a formulation review may serve as justification for further evolution and controlled inhouse panel testing of prototype formulations.

The application of in vitro tests to claims substantiation programs can be useful in screening formulations for acceptance before clinical testing on sensitive-skin individuals. Standard patch design techniques alone are inadequate for judging the mildness of cosmetics, especially for sensitive-skin individuals and panelists with preexisting dermatoses, such as seborrheic dermatitis, acne, rosacea, and atopic diathesis.

A research project was recently undertaken to examine some of the mediators released from injured reconstituted human skin models (Organogenesis, Inc.) [6]. These skin models contain various cell types that could initiate an inflammatory response to cosmetics. The use of in vitro skin models eliminates the rapid leakage of these mediators that occurs in vivo [7]. The mediators examined include histamine, serotonin, prostaglandin $E_2$ ($PGE_2$), and interleukin-1α (IL-1α). The properties of these mediators are listed in Table 5.

Two different skin cleanser formulations labeled experimental cleansers A and B were evaluated in this research. The cleansers varied in raw materials and surfactant types and levels (Table 6). Sodium lauryl sulfate and tissue culture medium were included in the in vitro tests as positive and negative controls, respectively. The results of clinical assessment of these two cleansers can be found in Table 7.

Results from this research showed that the irritation properties of these cleansers were reflected in differences in cytotoxicity and the quantity of histamine (mast cells), serotonin (mast cells), and $PGE_2$ (fibroblasts and keratinocytes). Interleukin-1α did not appear to be a discriminating marker for skin irritation. Studies from other laboratories have shown that IL-1α is more frequently associated with allergic contact dermatitis, rather than irritant contact dermatitis [8]. This research shows that the use of selected biological markers can be used to screen surfactant-containing formulations before conducting more costly clinical testing on panelists with sensitive skin.

The application of an in vitro test for biological reactivity is a new concept. The criteria for the release of a cosmetic product are usually based on

**Table 5** Inflammatory Mediators

| |
|---|
| Histamine is vasoactive amine released from mast cells as a result of physical- or chemical-induced injury. The in vivo effects of histamine are erythema and edema as well as increased capillary or permeability and a partial interruption of vascular endothelium. |
| Serotonin (5-hydroxytryptamine) exists in a performed state in mast cells. This vasoactive amine causes capillary dilation, increased permeability, and smooth-muscle contraction. |
| Prostaglandin $E_2$, an inflammatory eicosanoid, is released when the plasma membranes of cells are perturbed. $PGE_2$ causes increased permeability, dilation of capillaries, and alterations in pain threshold. |
| Interleukin-1α is a ctyokine produced by keratinocytes in response to injury. IL-1α is a 15 kDa glycoprotein that plays a central role in the initiation and development of an immune response to antigen through its direct effects on the activation, growth, and differentiation of T and B lymphocytes. |

**Table 6** Cleanser Compositions

| Cleanser A | Cleanser B |
|---|---|
| Water | Mineral oil |
| Anionic surfactants | Amphoteric surfactants |
| Amphoteric surfactants | Emollients |
| Emollients | Botanical extracts |
| Preservatives | Preservatives |
| Botanical extracts | Color |
| Color | |

visual inspection, pH, viscosity, odor, and other physical measurements. An in vitro test, when included into a QA release criteria, can detect unexpected changes in the safety of the product, thereby strengthening the quality assurance process. Further in vivo testing may still be necessary, however.

## CONCLUSIONS

The application of in vitro testing to the product development process offers companies the opportunity to improve product quality in a timely, cost-efficient manner. Too often we think of in vitro tests simply as replacements for animal testing. This has limited the acceptance of the technology. Most scientists agree that in vitro tests alone cannot replace the Draize eye test.

Many cosmetics companies have eliminated animal testing through the use of historical safety data, raw materials suppliers' information, and in vitro

**Table 7** Clinical Trials

| Clinical Trials | Cleanser A | Cleanser B | SLS |
|---|---|---|---|
| Cumulative irritation ($n = 20$) | Moderate irritation | Less than mild irritation | Moderate irritation |
| Allergic contact dermatitis (ACD) ($n = 150$) | No ACD | No ACD | No ACD |
| Comedogenicity ($n = 25$) | None | None | None |
| Acnegenicity ($n = 150$) | Positive | Negative | Not tested |
| Usage study subjective symptoms ($n = 150$) | Stinging, burning, itching | Mild itching | Not tested |

SLS, sodium lauryl sulfate

testing. Acceptance of in vitro testing will not only require validation, but also educational programs designed to show corporate toxicologists how to use the data for assessing the safety of products.

## REFERENCES

1. McGill, D. C., Cosmetics companies quietly ending animal tests, *New York Times*, Aug. 2 (1989).
2. Artzt, E. L., Whither, Procter & Gamble in the cosmetic business, *Drug Cosmet. Ind. 48*. Feb (1992).
3. Frazier, J. M., Gad, S. C., Goldberg, A. M., and McCulley, J. P., *A Critical Evaluation of Alternatives to Acute Ocular Irritation Testing*, Vol. 4, *Alternative Methods in Toxicology* (A. M. Goldberg, ed.), Mary Ann Liebert, New York, 1987.
4. Jumblatt, M. M., Simmons, S. J., and Neufeld, A. H., Corneal epithelial wound closure: A tissue culture model of ocular irritancy. *In Vitro Toxicology: Approaches to Validation*, Vol. 5, *Alternative Methods in Toxicology* (A. M. Goldberg, ed.), Mary Ann Liebert, New York, 1987.
5. Parce, J. W., Owicki, J. C., Wada, H. G., and Kercso, K. M., Cells on silicon: The microphysiometer, *In Vitro Toxicology: Mechanisms and New Technology*, Vol. 8, *Alternative Methods in Toxicology* (A. M. Goldberg, ed.), Mary Ann Liebert, New York, 1991.
6. Stephens, T. J., Spence, E. T., and Sanders, C. R., The use of reconstituted human skin as a model to study irritant contact dermatitis, American Academy of Dermatology annual meeting, Dec. 1991.
7. Rikimuru, T., Nakamura, M., Yano, T., Beck, G., Habicht, G. S., Rennie, L. L., Widra, M., Hirshman, C. A., Boulay, M. G., Spannhake, E. W., Lazarus, G., Pula, P. J., and Dannenberg, A. M., Mediators initiating the inflammatory response released in organ culture by full thickness human skin exposed to the irritant, sulfur mustard, *J. Invest. Dermatol. 96*:888–897 (1991).
8. Thestrup, K., Larson, C. G., and Ronnevig, J., The immunology of contact dermatitis, *Contact Dermatitis 20*:81–92 (1989).

# 25

# In Vitro and Other Alternatives in Inhalation Toxicology: Monitoring Biological Markers of Cellular and Biochemical Response

**Richard D. Thomas**

*National Research Council, National Academy of Sciences*
*Washington, D.C.*

In recent years, much has been learned concerning the cellular and biochemical mechanisms of lung response to both chemical insult and disease. This chapter examines the rapidly developing field of cellular interactions and biochemical mechanisms of respiratory response. Its focus is the work we have completed at the National Research Council (NRC) of the National Academy of Sciences: *Biologic Markers in Pulmonary Toxicology* [1]. Other markers reports in this series may also be of interest [2–4].

There are several possible sources of biological markers that may be monitored in in vitro systems to understand lung and respiratory tract toxicity. These techniques are relatively new and still entail some problems in their application for studying biological models. Many of these new techniques and model systems will be described in this book (see Chaps. 26–30).

The following is an overview of the cellular and the acellular components of the respiratory system that may be monitored for biological markers of toxic response.

## MACROPHAGES

The preponderant cell in bronchoalveolar lavage (BAL) fluid from human and animal subjects is the macrophage. In some species [5], lymphocytes can be present in small numbers. Neutrophils, eosinophils, and mast cells might also be present as a result of an inflammatory response.

Human alveolar macrophages (AMs) are composed of several populations that can be distinguished by density. In general, denser AMs are less mature and resemble blood monocytes more than do the less dense AMs. The denser cells are more potent producers of a soluble factor that inhibits fibroblast proliferation [6] and of interleukin-1 (Il-1) [7], and they are more efficient accessory cells for antigen-induced proliferation [8]. The AM abnormalities in sarcoidosis might represent differences in the relative proportions of AM subpopulations, rather than intrinsic differences in the same AM subpopulations. Hance and colleagues [9] have found that AMs from patients with sarcoidosis express antigens that are present on blood monocytes, but AMs from normal subjects do not. Other attributes of sarcoid AMs, such as accessory cell function [10] and spontaneous release of interferon gamma and growth factors [11], are compatible with the influx of less mature AMs in sarcoidosis.

Macrophages are important in the regulation of the immune response, acting as both promoters and suppressors of events that result in immunization or inflammation [12]. In general, AMs that can be obtained with lavage from normal humans contain subpopulations that can increase or suppress lymphocyte proliferation. The result depends on culture conditions—the presence of other accessory cells and the amount of antigen or mitogen present [13–15]. Alterations in accessory cell function might affect pathogenesis. In fact, some studies have shown that AMs from patients with sarcoidosis have accessory cell function more efficient than that in AMs from normal subjects [10,16]. The AMs from asthmatics are less able than AMs from normal subjects to suppress mitogen-induced lymphocyte proliferation [17]. Changes in accessory cell function are gaining acceptance and should be used as markers of environmental exposure and perhaps as markers of susceptibility to disease.

## Interleukin-1

Alveolar macrophages can secrete Il-1, but apparently to a smaller degree than can peripheral blood monocytes [18,19]. The uncertainty surrounding AM Il-1 production and its control mechanisms probably derives from the multiplicity of AM products, some of which antagonize Il-1 production [20]. Interleukin-1 is an important mediator of inflammation [21] and acts as a differentiation signal for several subsets of lymphocytes [3]. In particular, Il-1 promotes differentiation and maturation of helper T cells [22], lymphocytes [23], and natural killer cells [24]. Interleukin-1 is chemotactic for neutrophils, can alter the adherence properties of endothelial cells by stimulating prostaglandin synthesis [25], and can induce fibroblast proliferation [26]. In animals, perturbation of the lung can cause spontaneous release of Il-1 [27]. An analogous situation can occur in sarcoidosis, in that IL-1 is spontaneously released in vitro in the absence of additional signals [28]. The AM Il-1 secretion is influenced by the cytokines, such as interferon gamma. Alveolar macrophages from normal subjects and persons with lung disease might differ in their response to stimuli that increase Il-1 secretion [29].

## Interferon

Alveolar macrophages from normal subjects can release interferon (alpha or gamma) after stimulation by appropriate inducers. Interestingly, cigarette smoking does not influence inducer-stimulated interferon secretion [30]. However, AMs from patients with sarcoidosis spontaneously release interferon, without the necessity of inducers [31].

## Reactive Oxygen Species

Alveolar macrophages release reactive oxygen intermediates, such as $O_2^-$, $H_2O_2$, $OH^-$, and singlet oxygen [3]. Those products have many proinflammatory effects, including inactivation of protective substances in the lung and cytotoxicity. Changes in the signals required for release of those products or changes in the signals controlling the release of protective substances, such as superoxide dismutase, could have important implications for the results of exposure to injurious stimuli. For example, the production of oxygen radicals is increased in AMs from smokers [32] and from subjects with sarcoidosis [33]. Interferon gamma can increase in vitro oxygen radical production [34]. Alveolar macrophages from asthmatics differ from those from normal subjects, in that they have increased numbers of IgE receptors and require only exposure to antigen to secrete reactive oxygen products, whereas AMs from normal persons must first be coated with IgE antibody [35]. Release of other AM products, such as platelet-activating factor and neutrophil and eosinophil chemotactic factors (possibly leukotriene-$B_4$), exhibit the same dependence on surface IgE antibody [36]. The AMs from normal animals do not release reactive oxygen intermediates after stimulation with phorbol esters, unless they are conditioned in vitro with serum; AMs from animals with pulmonary inflammation do not require conditioning to become responsive to phorbol esters [37]. The AM sensitivity to stimuli that cause secretion of toxic oxygen intermediates might be a useful marker of lung damage and inflammation.

## Chemotactic Factors

Alveolar macrophages release chemotactic factors for neutrophils [38], the products of which are important in the pathogenesis of several types of lung injury, such as adult respiratory distress syndrome [39,40] and bronchopulmonary dysplasia. The AMs also release factors that stimulate neutrophil adhesiveness and superoxide generation [40]. Cigarette smokers' AMs, unlike nonsmokers' AMs, spontaneously release a neutrophil chemotactic factor in vitro [41]. Other environmental exposures might similarly change the signals needed to cause release of chemotactic factors.

## Miscellaneous Factors

Pulmonary macrophages can release a factor that causes the release of histamine from human basophils and lung mast cells [42]. Alterations in the mechanisms that control the release could both cause inflammation and be a result of environmental exposure. The AMs release growth factors that influence diverse cell types, such as fibroblasts [43], smooth-muscle cells [44], endothelial cells [45], and type II cells. Although Il-1 is one of these factors, it is clear that additional substances are released by macrophages [46]. The proliferation of the many lung cell types that is a common result of inflammatory stimuli could be caused by AM growth factors, and release of the factors could be an early event in lung changes and provide markers of early lung injury.

# NEUTROPHILS

Neutrophils are not usually found in large numbers in BAL fluid from normal lungs [5,47]. In lavage fluid from the lungs of smokers [48], or from subjects who were exposed to ozone [49,50], an increased percentage and greatly increased absolute numbers of neutrophils are found. As participants in the inflammatory response in lung tissue, neutrophils migrate into the lung from the blood under the influence of one of several chemotactic factors, including complement component C5a and soluble factors secreted by AMs. They might be considered as a secondary line of phagocytic defense for the lungs, which can be recruited into the airspaces in response to exposure to microbial agents or other inhaled materials.

The neutrophil itself contains a host of materials potentially damaging to lung tissue [45,51]. It has the capacity to release oxygen radicals. In addition, it releases several digestive enzymes, especially neutrophil elastase. The release of neutrophil elastase has been associated with significant lung destruction [52], as seen in the adult respiratory distress syndrome, where it has been found in lavage fluid [39]. Neutrophil elastase can cause emphysematous changes in an animal model and has been proposed as a major cause of emphysema. Other less destructive processes have been associated with the release of enzymes from the neutrophil.

## LYMPHOCYTES

The lymphocytes in lung fluid usually are T cells, although natural killer cells and B cells have also been found. The importance of B cells and NK cells in the lung is unclear, inasmuch as these cells are susceptible to the suppressive effects of surfactant and other substances in the lung-lining fluid. T cells have been divided into two classes on the basis of the monoclonal antibodies directed against antigens found on the T-cells' surface. For example, CD4 cells are in the Th/i cell class, and CD8 cells are in the TS/c cell class.

Regardless of the cell type responsible for antigen presentation in the lung, antigen must be presented to stimulate T-cell proliferation. It is not clear whether antigen is presented in the lung, or in the lymph nodes that drain the lung. Results of investigations in dogs and nonhuman primates suggest that antigen deposited in the lung is translocated to draining lymph nodes [53], where proliferation of antigen-specific B cells occurs [54]. The antigen-specific B cells then enter the blood and are recruited by nonspecific means to the lungs [55]. With secondary antigen challenges, results of studies in cynomolgus monkeys suggest that local proliferation of lymphocytes can occur in the lung [56]. Little is known of primary immune responses in the human lung, but mechanisms responsible for the accumulation of T cells in the lungs of patients with sarcoidosis have been the subject of numerous investigations.

It has been shown that T cells in the lower respiratory tract of patients with active sarcoidosis proliferate spontaneously [57] and release Il-2 [58]. The major sources of Il-2 are T cells that express human leukocyte antigens-D region (HLA-DR) on their surface [59]. Interleukin-2 secretion is probably associated with the pulmonary lymphocyte infiltrate in sarcoidosis. When Il-2 release by lung T cells is suppressed by treatment of patients with corticosteroids, the spontaneous proliferation of T cells ceases, and the numbers of T cells and disease activity subside [60–63].

A relatively recent investigation determined that Il-2 gene expression by lung T cells was triggered by enzymes in response to inhaled toxic particles [64].

## PROTEIN AND PROTEIN PRODUCTS

Protein in BAL fluid is measured as a marker of increased permeability of the alveolar–capillary barrier and is a common component of the inflammatory response. Bell and Hook [65] reported that 80% of the soluble protein in human BAL fluid could be accounted for by 19 plasma proteins. The protein content indicated a preferential transfer of smaller proteins across the alveolar–capillary barrier. Immunoglobulin G and IgA constituted a higher fraction of total protein in BAL fluid from smokers than in serum [66]. Transferrin was the only nonimmunoglobulin protein with a higher concentration in lavage fluid than in serum. Serum proteins in BAL fluid from animal studies have proved to be sensitive markers of the inflammatory response [67–71].

The amino acid hydroxyproline is a marker of collagen and has been interpreted as a marker of collagen breakdown. Hydroxyproline content of BAL fluid has been measured as a marker of breakdown or remodeling of pulmonary collagen in ozone-exposed rats [72]. The increase in hydroxyproline in BAL fluid appeared to parallel developing pulmonary fibrosis in hamsters and rats exposed to diesel exhaust [73,74].

Exposure to environmental toxicants can cause damage in single cells at the level of DNA, and that damage can lead to the development of many diseases, including cancer. Toxicant-induced changes in specific (although often unidentified) genes are thought to be the initial events in the development of disease. Identification of genes involved in the development of specific diseases can lead to improved diagnosis, understanding, and treatment, but is not essential. In lieu of disease-specific molecular markers that could be used to study the relation between toxicant exposure and the development of disease, the general interaction between toxicants and DNA can serve as a source of molecular markers of exposure, effect, and susceptibility. The use of molecular markers, defined here as alterations in DNA or RNA, to identify cellular responses or responsiveness to environmental toxicants theoretically can provide information useful in determining the magnitude of exposure, the effects of exposure on human health, and the mechanisms of response.

Molecular markers can be highly sensitive and specific indicators of cell damage or change. Detection of toxicant-induced alterations and use of them as indicators of toxicant exposure, effect, or susceptibility depend on several factors, including the frequency of the alteration which, in turn, can affect the sample size required for its detection; the availability of sufficient material (DNA, RNA, or cells) or analysis; and the ability to culture cells from the respiratory tract. Chapters 26–30 will focus on the monitoring of these cellular and molecular markers in various cells from the respiratory tract, grown in in vitro cultures.

## REFERENCES


1. National Research Council, *Biologic Markers in Pulmonary Toxicology*, National Academy Press, Washington, DC, 1989.
2. National Research Council, *Biologic Markers in Reproductive Toxicology*, National Academy Press, Washington, DC, 1989.
3. National Research Council, *Biologic Markers in Immunotoxicology*, National Academy Press, Washington, DC, 1992.
4. National Research Council, *Biologic Markers in Neurotoxicology*, National Academy Press, Washington, DC, 1992.
5. Henderson, R. F., Use of bronchoalveolar lavage to detect lung damage, *Toxicology of the Lung. Target Organ Toxicology Series* (D. E. Gardner, J. D. Crapo, and E. J. Massaro, eds.), Raven Press, New York, 1988, pp. 239–268.
6. Elias, J., Rossman, M. D., Zurier, R. B., and Daniele, R. P., Human alveolar macrophage inhibition of lung fibroblast growth. A prostaglandin-dependent process, *Am. Rev. Respir. Dis. 131*:94–99 (1985).
7. Elias, J. A., Schreiber, A. D., Gustilo, K., Chien, P., Rossman, M. D., Lammie, P. J., and Daniele, R. P., Differential interleukin 1 elaboration by unfractionated and density fractionated human alveolar macrophages and blood monocytes: Relationship to cell maturity, *J. Immunol. 135*:3198–3204 (1985).
8. Ferro, T. J., Kern, J. A., Elias, J. A., Kamoun, M., Daniele, R. P., and Rossman, M. D., Alveolar macrophages, blood monocytes, and density-fractioned alveolar macrophages differ in their ability to promote lymphocyte proliferation to mitogen and antigen, *Am. Rev. Respir. Dis. 135*:682–687 (1987).
9. Hance, A. J., Douches, S., Winchester, R. J., Ferrans, V. J., and Crystal, R. G., Characterization of mononuclear phagocyte subpopulations in the human lung by using monoclonal antibodies: Changes in alveolar macrophage phenotype associated with pulmonary sarcoidosis, *J. Immunol. 134*:284–292 (1985).
10. Venet, A., Hance, A. J., Saltini, C., Robinson, B. W. S., and Crystal, R. G., Enhanced alveolar macrophage-mediated antigen-induced T-lymphocyte proliferation in sarcoidosis, *J. Clin. Invest. 75*:293–301 (1985).
11. Bitterman, P. B., Adelberg, S., and Crystal, R. G., Mechanisms of pulmonary fibrosis. Spontaneous release of the alveolar macrophage-derived growth factor in the interstitial lung disorders, *J. Clin. Invest. 72*:1801–1813 (1983).
12. Unanue, E. R., Beller, D. I., Lu, C. Y., and Allen, P. M., Antigen presentation: Comments on its regulation and mechanism, *J. Immunol. 132*:1–5 (1984).
13. Liu, M. C., Proud, D., Schleimer, R. P., and Plaut, M., Human lung macrophages enhance and inhibit lymphocyte proliferation, *J. Immunol. 132*:2895–2903 (1984).

14. Toews, G. B., Vial, W. C., Dunn, M. M., Guzzetta, P., Nunez, G., Stastny, P., and Lipscomb, M. F., The accessory cell function of human alveolar macrophages in specific T cell proliferation, *J. Immunol.* *132*:181–186 (1984).
15. Ettensohn, D. B., Lalor, P. A., and Roberts, N. J., Jr., Human alveolar macrophage regulation of lymphocyte proliferation, *Am. Rev. Respir. Dis.* *133*:1091–1096 (1986).
16. Lem, V. M., Lipscomb, M. F., Weissler, J. C., Nunez, G., Ball, E. J., Stastny, P., and Toews, G. B., Bronchoalveolar cells from sarcoid patients demonstrate enhanced antigen presentation, *J. Immunol.* *135*:1766–1771 (1985).
17. Aubas, P., Cosso, B., Godard, P., Michel, F. B., and Clot, J., Decreased suppressor cell activity of alveolar macrophages in bronchial asthma, *Am. Rev. Respir. Dis.* *130*:875–878 (1984).
18. Koretzky, G. A., Elias, J. A., Kay, S. L., Rossman, M. D., Nowell, P. C., and Daniele, R. P., Spontaneous production of interleukin-1 by human alveolar macrophage, *Clin. Immunol. Immunopathol.* *29*:443–450 (1983).
19. Wewers, M. D., Rennard, S. I., Hance, A. J., Bitterman, P. B., and Crystal, R. G., Normal human alveolar macrophages obtained by bronchoalveolar lavage have a limited capacity to release interleukin-1, *J. Clin. Invest.* *74*;2208–2218 (1984).
20. Monick, M., Glazier, J., and Hunninghake, G. W., Human alveolar macrophages suppress interleukin-1 (Il-1) activity via the secretin of prostaglandin $E_2$, *Am. Rev. Respir. Dis.* *135*:72–77 (1987).
21. Dinarello, C. A., Interleukin-1 and the pathogenesis of the acute-phase response, *N. Engl. J. Med.* *311*:1413–1418 (1984).
22. Mizel, S. B., Interleukin 1 and T cell activation, *Immunol. Rev.* *63*:51–72 (1982).
23. Rao, A., Mizel, S. B., and Cantor, H., Disparate functional properties of two interleukin 1-responsive Ly-1 + 2-T cell clones: Distinction of T cell growth factor and T cell-replacing factor activities, *J. Immunol.* *130*:1743–1748 (1983).
24. Dempsey, R. A., Dinarello, C. A., Mier, J. W., Rosenwasser, L. J., Allegretta, M., Brown, T. E., and Parkinson, D. R., The differential effects of human leukocyte pyrogen/lymphocyte-activating factor, T cell growth factor, and interferon on human natural killer activity, *J. Immunol.* *129*:2504–2510 (1982).
25. Bevilacqua, M. P., Prober, J. S., Wheeler, M. E., Cotran, R. S., and Gimbrone, M. A., Jr., Interleukin-1 acts on cultured human vascular endothelium to increase the adhesion of polymorphonuclear leukocytes, monocytes, and related leukocyte cell lines, *J. Clin. Invest.* *76*:2003–2011 (1985).
26. Schmidt, J. A., Mizel, S. B., Cohen, D., and Green, I., Interleukin 1, a potential regulator of fibroblast proliferation, *J. Immunol.* *128*:2177–2182 (1982).
27. Lamontagne, L., Gauldie, J., Stadnyk, A., Richards, C., and Jenkins, E., In vivo initiation of unstimulated in vitro interleukin-1 release by alveolar macrophages, *Am. Rev. Respir. Dis.* *131*:326–330 (1985).
28. Hunninghake, G. W., Release of interleukin-1 by alveolar macrophages of patients with active pulmonary sarcoidosis, *Am. Rev. Respir. Dis.* *129*:569–572 (1984).
29. Eden, E., and Turino, G. M., Interleukin-1 secretion by human alveolar macro-

phages stimulated with endotoxin is augmented by recombinant immune (gamma) interferon, *Am. Rev. Respir. Dis. 133*:455–460 (1986).
30. Nuggent, K. M., Glazier, J., Monick, M. M., and Hunninghake, G. W., Stimulated human alveolar macrophages secrete interferon, *Am. Rev. Respir. Dis. 131*: 714–718 (1985).
31. Robinson, B. W., McLemore, T. L., and Crystal, R. G., gamma Interferon is spontaneously released by alveolar macrophages and lung T lymphocytes in patients with pulmonary sarcoidosis, *J. Clin. Invest. 75*:1488–1495 (1985).
32. Hoidal, J. R., Fox, R. B., Lemarbe, P. A., Perri, R., and Repine, J. E., Altered oxidative metabolic responses in vitro of alveolar macrophages from asymptomatic cigarette smokers, *Am. Rev. Respir. Dis. 123*:85–89 (1981).
33. Greening, A. P., and Lowrie, D. B., Extracellular release of hydrogen peroxide by human alveolar macrophages. The relationship to cigarette smoking and lower respiratory tract infections, *Clin. Sci. 65*:661–664 (1983).
34. Fels, A. O. S., and Cohn, Z. A., The alveolar macrophage, *J. Appl. Physiol. 60*: 353–369 (1986).
35. Joseph, M., Tonnel, A. B., Torpier, G., Capron, A., Arnoux, B., and Benveniste, J., Involvement of immunoglobulin E in the secretory processes of alveolar macrophages from asthmatic patients, *J. Clin. Invest. 71*:221–230 (1983).
36. Gosmset, P., Tonnel, A. B., Joseph, M., Prin, L., Mallart, A., Charon, J., and Capron, A., Secretion of a chemotactic factor for neutrophils and eosinophils by alveolar macropahges from asthmatic patients, *J. Allergy Clin. Immunol. 74*:827–834 (1984).
37. Gerberick, G. H., Jaffe, H. A., Willoughby, J. B., and Willoughby, W. F., Relationships between pulmonary inflammation, plasma transudation, and oxygen metabolite secretion by alveolar macrophages, *J. Immunol. 137*:114–121 (1986).
38. Hunninghake, G. W., Gadek, J. E., Fales, H. M., and Crystal, R. G., Human alveolar macrophage derived chemotactic factors for neutrophils: Stimuli and partial characterization, *J. Clin. Invest. 66*:473–483 (1980).
39. Lee, C. T., Fein, A. M., Lippmann, M., Holtzman, H., Kimbel, P., and Weinbaum, G., Elastolytic activity in pulmonary lavage fluid from patients with adult respiratory distress syndrome, *N. Engl. J. Med. 304*:192–196 (1981).
40. Tate, R. M., and Repine, J. E., Neutrophils and the adult respiratory distress syndrome, *Am. Rev. Respir. Dis. 128*:552–550 (1983).
41. Hunninghake, G. W., and Crystal, R. G., Cigarette smoking and lung destruction: Accumulation of neutrophils in the lungs of cigarette smokers, *Am. Rev. Respir. Dis. 128*:833–838 (1983).
42. Schulman, E., Liu, M., Proud, D., Macglashan, D., Jr., Lichtenstein, L., and Plaut, M., Human lung macrophages induce histamine release from basophils and mast cells, *Am. Rev. Respir. Dis. 131*:230–235 (1985).
43. Chapman, H. A., Jr., Stone, O. L., and Vavrin, Z., Degradation of fibrin and elastin by human alveolar macrophages in vitro. Characterization of a plasminogen activator and its role in matrix degradation, *J. Clin. Invest. 73*:806–815 (1984).

44. Martin, B. M., Gimbrone, M. A., Jr., Unanue, E. R., and Coltran, R. S., Stimulation of nonlymphoid mesenchymal cell proliferation by a macrophage-derived growth factor, *J. Immunol. 126*:1510–1515 (1981).
45. Martin, W. J., 2d, Gadek, J. E., Hunninghake, G. W., and Crystal, R. G., Oxidant injury of lung parenchymal cells, *J. Clin. Invest. 68*:1277–1288 (1981).
46. Leslie, C., McCormick-Shannon, K., Cook, J., and Mason, R., Macrophages stimulate DNA synthesis in rat alveolar type II cells, *Am. Rev. Respir. Dis. 132*: 1246 (1985).
47. Reynolds, H. Y., Bronchoalveolar lavage, *Am. Rev. Respir. Dis. 135*:250–263 (1987).
48. Young, K. R., and Reynolds, H. Y., Bronchoalveolar washings: Proteins and cells from normal lungs, *Immunology of the Lung and Upper Respiratory Tract* (J. Bienenstock, ed.), McGraw-Hill, New York, 1984.
49. Seltzer, J., Bigby, B. G., Stulbarg, M., Holtzman, M. J., Nadel, J. A., Ueki, I. F., Leikauf, G. D., Goetzl, E. J., and Boushey, H. A., $O_3$-induced change in bronchial reactivity to methacholine and airway inflammation in humans, *J. Appl. Physiol. 60*:1321–1326 (1986).
50. Koren, H. S., Devlin, R. B., Graham, D. E., Mann, R., McGee, M. P., Horstman, D. H., Kozumbo, W. J., Becker, S., House, D. E., McDonnell, W. F., and Bromberg, P. A., Ozone-induced inflammation in the lower airways of human subjects, *Am. Rev. Respir. Dis. 139*:407–415 (1989).
51. Henson, P. M., Pathologic mechanisms in neutrophil-mediated injury, *Am. J. Pathol. 68*:593–612 (1972).
52. Janoff, A., Human granulocyte elastase. Further delineation of its role in connective tissue damage, *Am. J. Pathol. 68*:579–592 (1972).
53. Kaltreider, H. B., Caldwell, J. L., and Adam, E., The fate and consequences of an organic particulate antigen instilled into bronchoalveolar spaces of normal canine lungs, *Am. Rev. Respir. Dis. 116*:267–280 (1977).
54. Bice, D. E., Harris, D. L., and Muggenburg, B. A., Regional immunologic responses following localized deposition of antigen in the lung, *Exp. Lung Res. 1*:33–41 (1980).
55. Bice, D. E., Degen, M. A., Harris, D. L., and Muggenburg, B. A., Recruitment of antibody-forming cells in the lungs after local immunization is nonspecific, *Am. Rev. Respir. Dis. 126*:635–639 (1982).
56. Mason, M. J., Bice, D. E., and Muggenburg, B. A., Local pulmonary immune responsiveness after multiple antigenic exposures in the cynomolgus monkey, *Am. Rev. Respir. Dis. 132*:657–660 (1985).
57. Pinkston, P., Bitterman, P. B., and Crystal, R. G., Spontaneous release of interleukin-2 by lung T lymphocytes in active pulmonary sarcoidosis, *N. Engl. J. Med. 308*:793–800 (1983).
58. Humminghake et al., 1983.
59. Saltini, C., Spurzem, J. R., Lee, J. J., Pinkston, P., and Crystal, R. G., Spontaneous release of interleukin 2 by lung T-lymphocytes in active pulmonary

sarcoidosis is primarily from the Leu3$^+$DR$^+$ T cell subset, *J. Clin. Invest* *77*:1962–1970 (1986).
60. Ceuppens, J. L., Lacguet, L. M., Marien, G., Demedts, M., van den Eeckhout, A., and Stevens, E., Alveolar T-cells subsets in pulmonary sarcoidosis. Correlation with disease activity and effect of steroid treatment, *Am. Rev. Respir. Dis.* *129*:563–568 (1984).
61. Pinkston, P., Bitterman, P., Hance, A., Douches, S., Goodman, S., and Crystal, R., In vivo demonstration of the importance of interleukin-2 in the proliferation of human T-lymphocytes [Abstract], *Clin. Res. 32*:355A (1984).
62. Bauer, W., Gorny, M. K., Baumann, H. R., and Morell, A., T-lymphocyte subsets and immunoglobulin concentrations in bronchoalveolar lavage of patients with sarcoidosis and high and low intensity alveolitis, *Am. Rev. Respir. Dis. 132*:1060–1065 (1985).
63. Rossi, G. A., Di Negro, G. B., Balzano, E., Cerri, E., Sacco, O., Balbi, B., Venturini, A., Ramoino, R., and Ravazzoni, C., Suppression of the alveolitis of pulmonary sarcoidosis by oral corticosteroids, *Lung 163*:83–93 (1985).
64. Wolff, R. K., Henderson, R. F., Edison, A. F., Pickrell, J. A., Rothenberg, S. J., and Hahn, F. F., Toxicity of gallium oxide particles following a 4-week inhalation exposure, *J. Appl. Toxicol. 8*:191–199 (1988).
65. Bell, D. Y., and Hook, G. E. R., Pulmonary alveolar proteinosis: Analysis of airway and alveolar proteins, *Am. Rev. Respir. Dis. 119*:979–990 (1979).
66. Bell, D. Y., Haseman, J. A., Spock, A., McLennan, G., and Hook, G. E. R., Plasma proteins of the bronchoalveolar surface of the lungs of smokers and nonsmokers, *Am. Rev. Respir. Dis. 124*:72–79 (1981).
67. Alpert, S. M., Schwartz, B. B., Lee, S. D., and Lewis, T. R., Alveolar protein accumulation, *Arch. Intern. Med. 128*:69–73 (1971).
68. Beck, B. D., Brain, J. D., and Bohannon, D. E., An in vitro hamster bioassay to assess the toxicity of particles for the lungs, *Toxicol. Appl. Pharmacol. 66*:9–29 (1982).
69. Bignon, J. P., Chahinian, G., Feldmann, and Sapin, C., Ultrastructural immunoperoxidase demonstration of autologous albumin in the alveolar capillary membrane and in the alveolar lining material in normal rats, *J. Cell Biol. 64*:503–509 (1975).
70. DeNicola, D. B., Rebar, A. H., and Henderson, R. F., Early damage indicators in the lung. V. Biochemical and cytological response to $NO_2$ inhalation, *Toxicol. Appl. Pharmacol. 60*:301–312 (1981).
71. Lehnert, B. E., London, J. E., Vanderkogel, A. J., Valdez, J. G., and Gurley, L. R., HPLC analyses of lavage fluid constituents following environmental insults [Abstract]. *Toxicologist 6*:138 (1986).
72. Pickrell, J. A., Gregory, R. E., Cole, D. J., Hahn, F. F., and Henderson, R. F., Effect of acute ozone exposure on the proteinase–antiproteinase balance in the rat lung, *Exp. Mol. Pathol. 46*:168–179 (1987).
73. Heinrich, U., Muhle, H., Takenaka, S., Ernst, H., Fuhst, R., Mohr, U., Pott, F.,

and Stöber, W., Chronic effects on the respiratory tract of hamsters, mice and rats after long-term inhalation of high concentrations of filtered and unfiltered diesel engine emissions, *J. Appl. Toxicol. 6*:393–395 (1986).
74. Henderson, R. F., Pickrell, J. A., Jones, R. K., Sun, J. D., Benson, J. M., Mauderly, J. L., and McClellan, R. O., Response of rodents to inhaled diluted diesel exhaust: Biochemical and cytological changes in bronchoalveolar lavage fluid and in lung tissue, *Fundam. Appl. Toxicol. 11*:546–567 (1988).

# 26

# Alternatives to In Vivo Toxicological Testing of Rodent Airway Epithelia

**Leah A. Cohn and Kenneth B. Adler**

*North Carolina State University*
*Raleigh, North Carolina*

The lungs and airways are susceptible to insult from a wide variety of toxicants. Some toxicants are absorbed through nonrespiratory routes, but many are delivered directly into the lung or airways. Respirable toxins may be gases, liquid aerosols, or particulates. The first contact of these toxins is with the epithelium of the airways and the overlying surface liquid layers.

In vivo studies of the effects of respirable toxins in humans are generally limited to low-dose studies of air pollutants or accidental exposures. In vivo studies in animals have proved valuable, but there are drawbacks: inhalation studies in animals are expensive, laborious, and it is often difficult to create appropriately controlled exposures. Direct toxin effects on airways can be difficult to distinguish owing to inflammatory, immunological, neuronal, and hormonal responses that accompany the insult. Some advantages and disadvantages of in vitro systems will be reviewed and a primary cell culture will be described.

In vitro systems lack hormonal, neuronal, inflammatory, immunological, and other systemic influences. This is both the single most important advantage, and disadvantage, of such systems. It is an advantage in that any observed response of the respiratory tissue is due to that tissue itself, without the confounding influence of other body systems; this makes in vitro systems excellent for mechanistic studies. It is a disadvantage in that the response to a particular stimulus in a live animal will be at least partially dependent on these same systemic influences. Ultimately, information gained by an in vitro method will be used to understand responses in a live animal. Since the live animal will have the systemic influences that the in vitro model lacks, observed responses in the model may be at variance with responses in the live animal. Additionally, such isolated in vitro systems can provide no information on how a potential respiratory toxin would affect other organ systems.

Certain chemical compounds exert little effect in their native state, but have profound influence after becoming biologically activated. Although the lung has the capacity for metabolic activation and contains cytochrome P-450 monoxygenases, several compounds are activated outside the lung, often in the liver, and then exert pathological effects on respiratory target tissues. If such a substance were tested in its native state using an in vitro respiratory cell system, a false impression of its toxic potential might result.

In vitro systems allow increased control of experimental variables. Dosages, exposures, temperatures, and many other factors can be controlled more closely in vitro than in vivo. Animal-to-animal variations include not only age, sex, weight, or strain differences, but also many variables that are difficult to assess, such as the presence of subclinical infections or somatic mutations. Many in vitro models allow several treatments to be evaluated on tissue from the same source, keeping variations as small as possible.

Tissues in vitro may be morphologically or functionally different from the same tissue types in vivo. Living animals maintain a remarkable homeostasis; when provided with food, water, and minimal shelter, they closely regulate their own temperature, electrolyte balance, and other vital parameters. In vitro systems depend on the investigator entirely for provision of nutrients, oxygen, temperature regulation, defense from pathogens, and so on. Unless all conditions can be mimicked exactly as they would be in vivo (as of yet a near impossibility), differences between tissue in vitro and in vivo are to be expected. It is always a goal to use tissues as similar as possible to those in situ, but it is our responsibility to acknowledge the differences between the tissue in vitro and in situ, as well as the similarities.

An advantage of in vitro systems that cannot be overemphasized is the minimization of animal distress. An animal that is humanely killed to provide tissue undergoes minimal stress, compared with an animal subjected to repeated handling or application of inhaled gases or particulates. Tissues used in in vitro methods must at some point come from an animal, but many of these methods allow the use of fewer animals than would otherwise be necessary. For example, cell cultures allow tremendous expansion of tissue, and transformed cell lines provide expandable tissue with a virtually unlimited life span.

The in vitro model to be described, involves a primary, biphasic cell culture system. Cell culture techniques can involve primary culture of cells dissociated from the respiratory system or passaged cell lines. Primary culture involves the enzymatic dissociation of cells and their distribution in a nutrient medium on a substrate that allows cellular attachment, spreading, and growth. Subcultures of primary cultures, or virally induced immortalization may give rise to permanent cell lines. Immortalized cells may be quite different from the progenitor cells taken from the animal. Even primary cell cultures may be morphologically and functionally disparate from the same cell types in vivo. A prime feature of epithelial cell culture is that, since respiratory epithelium is not vascularized in vivo, the in vitro need to derive nutrients and oxygen by diffusion is not abnormal for this tissue.

Cell cultures, like all experimental systems, have advantages and disadvantages. Newer techniques allow cells in culture to more closely resemble the in situ differentiated cells. Since a large mass of epithelial tissue is available, several treatments may be applied to cells from one animal, resulting in improved uniformity of results. Cell lines are genetically alike and allow maximum control of experimental variables. Dissociate cells can be enriched for a certain population before culturing. The greatest advantage of cell culture is that only one tissue type (i.e., epithelial cells) is present, ensuring that a response is indeed due to epithelial cells. Primary cultures tend to assure a cause-and-effect relation. On the down side, cells of primary cultures are exposed to proteolytic enzymes during collection, and the relation of epithelial cells to other tissues (such as connective tissues) is lost completely. In many culture systems the epithelia loose their differentiated cellular phenotype, which is critically dependent on the growth medium and on the support surface.

This cell culture system uses guinea pig tracheal epithelium and maintains a remarkably well-differentiated, polarized phenotype. After enzymatic dissociation with pronase, cells are resuspended in Dulbecceo's modified Eagles medium/Ham's nutrient mixture F12 and 5% fetal bovine serum and plated in special chambers atop a collagen-coated nitrocellulose filter. The cells are fed from above and below with the culture medium containing insulin, transferrin, hydrocortisone, epidermal growth factor, cholera toxin, and bovine hypothalamus extract. During the first several days bovine serum is added to the medium, but on about day 4 the serum is no longer used. After these first few days in culture, medium is no longer added above the filter, leaving the top of the cells (the apical surface) exposed to air, as in vivo. After 8 days, the cells form a pseudostratified columnar epithelium, nearly identical morphologically and functionally with mucosal epithelium in vivo.

Morphologically, ciliated, goblet, and basal cells are present in these cultures. The cultures have a volume density for all three major epithelial cell types virtually identical with that found in tracheal epithelium in situ. The volume density of granule content within secretory cell cytoplasm is almost the same as that in situ. The major differences between our cells and guinea pig tracheal epithelium in situ are fewer total cells per area, a slightly reduced height of the pseudostratified columnar epithelium, and fewer basal cells per millimeter basement membrane.

Glycoconjugates released from these cells are mucintype. The hyaluronidase-resistant high molecular weight glycoconjugates have an amino acid composition almost identical with that of mucin, with about 30% serine and threonine residues. There are negligible amounts of chondroitin or dermatan sulfates. The buoyant density of the glycoconjugates are essentially the same as that of respiratory mucins. Most of the high molecular weight glycoconjugates are secreted apically, suggesting a polarity of secretion as well as of morphology. This system retains the advantages of previous cell culture systems while maintaining normal cell orientation, polarity, and epithelial interactions.

Most of the work involving the use of these primary cultures is to characterize effector type functions of the airway epithelium. Classically, tracheal epithelium has been thought of as a tissue that responds to stimuli in a somewhat passive manner (e.g., by secreting more mucus; altering ciliary beat frequency; or undergoing hypertrophy, hyperplasia, or metaplasia). It has become increasingly clear recently that tracheal epithelium can act as an effector; it can act in an autocrine or paracrine manner to alter both its own fate and the fate of neighboring cells. This epithelium produces several mediators that either directly or indirectly change the outcome of an encounter with a stimulant. Examples of substances produced by the tracheal epithelium include eicosanoids (prostaglandins, leukotrienes, and HETES), platelet-activating factor, cytokines, reactive oxygen species, epithelial-derived relaxing factors, and extracellular matrix components.

## CONCLUSIONS

In vitro models of airway epithelium offer both advantages and disadvantages over in vivo models. The elimination of systemic influences and the isolation of one tissue type afforded by cell culture provide an important system for the study of mechanistic reactions in the airway epithelium. This unique, biphasic primary cell culture system using guinea pig tracheal epithelium has many structural and functional similarities to the same tissue in situ. This system has been used extensively in the study of the basic biology of these cell types, and the system offers an excellent opportunity for the study of airway toxins as well.

## ACKNOWLEDGMENTS

This work was supported in part by grants HL 37636, HL 08647, and HL 36982 from the National Institutes of Health, a grant from Hoffmann La Roche, Inc., Nutley, NJ, and a grant from the state of North Carolina. Dr. Adler is an Established Investigator of the American Heart Association.

## BIBLIOGRAPHY

Adler, K. B., Mucin secretion by explants of respiratory tissue in vitro. In vitro *Models of Respiratory Epithelium* (I. J. Schiff, ed.), CRC Press, Boca Raton, FL, 1986, pp. 27–50.

Adler, K. B., Cheng, P. W., and Kim, K. C., Characterization of guinea pig tracheal epithelial cells maintained in biphasic organotypic culture: Cellular composition and biochemical analysis of released glycoconjugates, *Am. J. Respir. Cell Mol. Biol. 2*:145–154 (1990).

Adler, K. B., Schwarz, J. E., Whitcutt, M. J., and Wu, R., A new chamber system for maintaining differentiated guinea pig respiratory epithelial cells between air and liquid phases, *Biotechniques 5*:462–465 (1987).

Alpert, S. E., Kramer, C. M., Brashler, J. R., and Bach, M. K., Generation of lipoxygenase metabolites of arachidonic acid by monolayer cultures of tracheal epithelial cells and intact tracheal segments from rabbits, *Exp. Lung Res. 16*:211–233 (1990).

Cohn, L. A., and Adler, K. B., In vitro mechanisms of lung injury in the rodent, *Toxicol. Pathol. 19*:419–427 (1991).

Cohen, G. M., Pulmonary metabolism of foreign compounds: Its role in metabolic activation, *Environ. Health Perspect. 85*:31–41 (1990).

Emura, M., Riebe, M., Germann, P., Brockmeyer, C., Aufderheide, M., and Mohr, U., Functional culture of hamster and human airway epithelial cells and its application to pulmonary toxicology, *Exp. Pathol. 37*:1–4 (1989).

Fisher, A. B., Furia, L., and Berman, H., Metabolism of rat granular pneumocytes isolated in primary culture, *J. Appl. Physiol. 49*:743–750 (1980).

Hesterberg, T. W., Ririe, D. G., Barrett, J. C., and Nettesheim, P., Mechanism of cytotoxicity of asbestos fibers in rat tracheal epithelial cells in culture, *Toxicol. In Vitro 1*:59–65 (1987).

Kim, K. C., Nassiri, J., and Brody, J. S., Mechanisms of airway goblet cell mucin release: Studies with cultured tracheal surface epithelial cells, *Am. J. Respir. Cell Mol. Biol. 1*:137–143 (1989).

Kim, K. C., Opaskar-Hincman, H., and Ramakrishnan Bhaskar, K. R., Secretions from primary hamster tracheal surface epithelial cells in culture: Mucin-like glycoproteins, proteoglycans, and lipids, *Exp. Lung Res. 15*:299–314 (1989).

Lee, T. C., Wu, R., Brody, A. R., Barrett, J. C., and Nettesheim, P., Growth and differentiation of hamster tracheal epithelial cells in culture, *Exp. Lung. Res.* *6*:27–45 (1984).

Li, A. P., An in vitro lung epithelial cell system for evaluating the potential toxicity of inhalable materials, *Food Chem. Toxicol.* *24*:527–534 (1986).

Li, A. P., Hahn, F. F., Zamora, P. O., Shimizu, R. W., Henderson, R. F., Broods, A. L., and Richards, R., Characterization of a lung epithelial cell strain with potential applications in toxicological studies, *Toxicology* *27*:257–272 (1983).

Lopez, A., Shoji, S., Fajita, J., Robbins, R., and Rennard, S., Bronchioepithelial cells can release hydrogen peroxide in response to inflammatory stimuli [Abstract], *Am. Rev. Respir. Dis.* *137*:81A (1988).

Marchok, A. C., and Wasilenko, W. J., Changes in the responses of rat tracheal epithelial cells to modulators of growth and differentiation during the progression of neoplasia, *In vitro Models of Respiratory Epithelium* (L. J. Schiff, ed.), CRC Press, Boca Raton, FL, 1986, pp. 103–142.

Mass, M. J., and Kaufman, D. G., A comparison between the activation of benzo[*a*]pyrene in organ cultures and microsomes from the tracheal epithelium of rats and hamster, *Carcinogenesis* *4*:297–303 (1983).

Mckinnon, K., Joyce, M., Noah, T., Devlin, R., and Koren, H., BEAS S6 (BEAS) human bronchial epithelial cells produce inflammatory mediators following ozone (O3) exposure [Abstract], *FASEB J.* *5*:A634 (1991).

Menzel, D. B., Vandagriff, R., Ziegler, B., and Rassmussen, R., Intracellular distribution of reactive oxygen species in lung cells in culture [Abstract], *Toxicologist* *11*:226 (1991).

Mossman, B. T., In vitro studies on the biologic effects of fibers: Correlation with in vivo bioassays, *Environ. Health Perspect.* *88*:319–322 (1990).

Mossman, B. T., Ezerman, E. B., Adler, D. B., and Craighead, J. E., Isolation and spontaneous transformation of cloned lines of hamster tracheal epithelial cells, *Cancer Res.* *40*:4403–4406 (1986).

Nikula, K. J., and Wilson, D. W., Response of rat tracheal epithelium to ozone and oxygen exposure in vitro, *Fundam. Appl. Toxicol.* *15*:121–131 (1990).

Oreffo, V. I. C., Morgan, A., and Richards, R. J., Isolation of Clara cells from the mouse lung, *Environ. Health Perspect.* *85*:51–64 (1990).

Richards, R. J., Oreffo, V. I. C., and Lewis, R. W., Clara cell cultures from the mouse and their reaction to bronchiolar toxins, *Environ. Health Perspect.* *85*:119–127 (1990).

Romet, S., Dubreuil, A., Baeza, A., Moreau, A., Schoevaert, D., and Marano, F., Respiratory tract epithelium in primary culture: Effects of ciliotoxic compounds, *Toxicol. In Vitro* *4*:399–402 (1990).

Salari, H., and Wong, A., Generation of platelet activating factor (PAF) by a human lung epithelial cell line, *Eur. J. Pharmacol.* *175*:253–259 (1990).

Schiff, L. J., Effect of nitrogen dioxide on influenza virus infection in hamster trachea organ culture, *Proc. Soc. Exp. Biol. Med.* *156*:546–549 (1977).

Smith, S. M., Lee, D. K. P., Lacy, J., and Coleman, D. L., Rat tracheal epithelial cells produce granulocyte/macrophage colony-stimulating factor, *Am. J. Respir. Cell Mol. Biol. 2*:59–68 (1990).

Steele, V. E., Marchok, A. C., and Nettesheim, P., Establishment of epithelial cell lines following exposure of cultured tracheal epithelium to 12-*O*-tetradecanoyl-phorbol-13-acetate, *Cancer Res. 38*:3563–3565 (1978).

Widdicombe, J. H., Ueki, I. F., Emery, D., Margolskee, D., Yergey, J., and Nadel, J. A., Release of cyclooxygenase products from primary cultures of tracheal epithelial of dog and human, *Am. J. Physiol. 257*:L361–L365.

# 27

# Understanding Mechanisms of Carcinogenesis Using Rat Tracheal Epithelial Cells In Vitro

**David G. Thomassen***

*Inhalation Toxicology Research Institute*
*Albuquerque, New Mexico*

To understand the process of malignant transformation of respiratory epithelial cells, it is necessary to identify mechanisms controlling proliferation and differentiation of normal respiratory cells and to define changes in these mechanisms as cells undergo multistep progression towards malignancy. Cell culture systems and studies with isolated cells are potentially useful experimental models for such investigations. With cell culture systems, cellular and molecular mechanisms involved in carcinogenesis can be investigated under conditions devoid of host-mediated modulating factors. A model system using rat tracheal epithelium as the target tissue has been developed for studies of experimental respiratory carcinogenesis [1]. This model has several useful features. First, the histological structure of the rat trachea resembles that of the human bronchus, a major site of lung cancer in humans [2]. Second, rat tracheas exposed to carcinogens express a range of preneoplastic and neoplastic lesions. [3,4]. Third, the development of preneoplastic and neoplastic rat tracheal cell variants can be quantified in cell culture following exposure of tracheal cells to carcinogens either in vivo or in vitro [5–7]. Fourth, following additional proliferation in vitro, preneoplastic variants can progress to neoplastic cells that are capable of forming squamous cell carcinomas when injected

**Present affiliation*: U.S. Department of Energy, Washington, D.C.

into syngeneic or immunosuppressed hosts [8–10]. Thus, rat tracheal epithelial (RTE) cells, representing a range of phenotypes from normal to neoplastic, can be isolated and used to characterize both individual stages of progression and the mechanism of progression to neoplasia.

However, in vitro model systems that are used to study mechanisms of carcinogenesis or to identify potential carcinogens from exposures to toxicants must accurately reflect in vivo events and risks since the ultimate goal of developing and using an in vitro model system is to understand or predict mechanisms or risks of carcinogenesis in vivo. This chapter describes studies using the RTE cell model for experimental carcinogenesis to address two questions:

1. Are the events observed during the multistep progression of RTE cells to neoplasia in vitro the same as those involved in respiratory carcinogenesis in vivo?
2. Can the in vivo carcinogenic risk for suspected lung carcinogens be estimated from the in vitro response of RTE cells to those carcinogens?

## DOES PROGRESSION IN VITRO EQUAL PROGRESSION IN VIVO?

There are many similarities between changes seen during the neoplastic progression of respiratory cells in vitro and in vivo. For example, a change that may be important at an early stage of transformation in vitro or of carcinogenesis in vivo involves the autocrine production, by altered cells, of transforming growth factor alpha (TGF-α). Normal RTE cells in culture require exogenously added epidermal growth for maximal proliferation. In contrast, preneoplastic and neoplastic RTE cells in culture exhibit maximal proliferation without the addition of epidermal growth factor (EGF) to the culture medium [11,12]. This may be due to production of TGF-α by transformed RTE cells, resulting in an autocrine stimulation of cell proliferation [12]. These results suggest that the increased expression of TGF-α may be a critical, early event in the transformation of RTE cells in vitro. Similarly, hyperplastic and neoplastic lesions induced in the lungs of rats that inhaled $^{239}PuO_2$ express increased levels of TGF-α; 65% of proliferative lesions and 72% of rat lung tumors expressed TGF-α [13]. Increases in TGF-α expression have also been observed in human lung tumors [14] and in preneoplastic and neoplastic lung lesions in dogs [15]. These results suggest that changes in the expression of TGF-α are important at an early stage of respiratory carcinogenesis in vivo. Thus, understanding the role and mechanisms of increased TGF-α expression in RTE cells in vitro may provide information useful in understanding the role and the mechanism of development of this change in carcinogenesis in vivo.

A change that may be important at a later stage of progression to neoplasia is the acquisition of resistance to the proliferation inhibitory effects of transforming growth factor beta (TGF-β). Neoplastic RTE cells in vitro and cells derived from dog and many human lung tumors are resistant to the inhibitory effects of (TGF-β), whereas their normal and preneoplastic counterparts are much more sensitive [14,16]. Understanding the mechanism of resistance to TGF-β in cultured RTE cells is likely to improve our understanding of the role that the genetic change(s) identified by this phenotype play during carcinogenesis in vivo. These and other similarities between specific changes apparently involved in the neoplastic progression of respiratory cells in vivo and in vitro suggest that in vitro models can be used to study specific changes involved in neoplastic progression and their mechanisms of development.

Differences have also been observed in the specific changes that occur during carcinogenesis or transformation of respiratory epithelial cells in vivo and in vitro. Although preneoplastic and neoplastic RTE cells in vitro produce TGF-α, they do not exhibit an increased expression of its receptor, the epidermal growth factor receptor (EGFR) [12]. However, increased expression of the EGFR has been observed in a fraction of preneoplastic and neoplastic lung lesions in rats [15] and dogs [17]. Several possibilities for this difference in the expression of the EGFR in respiratory cells in vitro and in vivo are discussed in the following; however, since changes observed in vivo are not observed in vitro, the in vitro RTE cell model of carcinogenesis may not be a useful model for understanding either the role of increased expression of the EGFR in carcinogenesis or the mechanism by which it is overexpressed.

Up to 40% of the proliferative and neoplastic lesions in the lungs of rats that inhaled $^{239}PuO_2$ have mutations in the first position of the 12th codon of a Ki-*ras* protooncogene [18]. These observations suggested that Ki-*ras* mutations may be important early changes in the development of some lung cancers. To determine if the same changes in Ki-*ras* were induced during the preneoplastic transformation of RTE cells by alpha-radiation, we have examined the Ki-*ras* gene of radiation- and chemical-induced variants for point mutations in the 12th, 13th, or 61st codons. In contrast with the in vivo results, none of 16 total preneoplastic RTE cell variants examined (9 alpha-particle-induced, 4 x-ray-induced, 3 chemical-induced) has Ki-*ras* mutations. These results suggest that the in vitro RTE cell model of carcinogenesis may not be a useful model for understanding either the role of Ki-*ras* mutations in respiratory carcinogenesis or the mechanism by which they develop.

These differing in vivo and in vitro results suggest at least three possible explanations.

1. The in vitro RTE cell transformation assay may detect earlier events than can be identified histologically during carcinogenesis in vivo, and Ki-*ras* mutations and overexpression of the EGFR may not be critical events at this early stage of transformation. If this explanation proves to be true, then the development of these changes may occur at later stages of RTE cell transformation, and an understanding of their development in vitro could still be useful for characterizing their role and development during carcinogenesis in vivo.

2. Pathways of carcinogenesis for proximal airway cells (in vitro results with RTE cells) and peripheral lung cells (in vivo results presumably involving type II cells [19]) are different. If this explanation proves to be true, then results obtained using RTE cells in vitro may be useful only in understanding RTE cell carcinogenesis in vivo. In vitro models using peripheral lung cells may have to be developed to study the development of peripheral lung tumors. Although there are similarities in the types of changes observed for the histologically different types of lung tumors, differences similar to those described here have also been described between histotypes [20,21].

3. Pathways of carcinogenesis for respiratory epithelial cells are different in vivo and in vitro. If this explanation is proved to be true, then in vitro models of respiratory carcinogenesis will not be very useful in understanding carcinogenesis in vivo.

Although one cannot now distinguish between these possibilities, the differing in vivo and in vitro results described in the foregoing demonstrate that in vivo mechanisms of carcinogenesis cannot simply be deduced from an in vitro system without careful validation and interpretation of the model being used. In addition, these results demonstrate a need for in vivo–in vitro comparisons using model systems that are as closely related as possible.

## DOES TRANSFORMATION RISK IN VITRO EQUAL CARCINOGENIC RISK IN VIVO?

The RTE cell model for experimental carcinogenesis enables cells to be exposed to suspected carcinogens either in vivo, before the isolation of RTE cells for analysis in culture, or in vitro, after the cells have been isolated and plated in culture [5,6]. The in vitro transformation responses of RTE cells in vitro following exposure of cells to alpha-particle radiation or ozone, either in vivo or in vitro, have been compared.

The RTE cells were exposed to alpha-particles in vivo by exposing rats to radon progeny [22]. Following exposure, cells were isolated, placed in culture, and the frequency of radiation-induced preneoplastic variants was quantified.

Similarly, RTE cells were exposed to alpha-particles in vitro by exposing cultures to electroplated sources of $^{239}$Pu [23]. The RTE cells isolated from animals given increasing exposures to radon progeny and RTE cells exposed to $^{239}$Pu in culture exhibited exponential decreases in survival and increases in their frequencies of preneoplastic transformation. These results suggest that RTE cells have similar responses to alpha-particles in vitro and in vivo.

The opposite type of result was obtained when RTE cells were exposed to ozone in vivo or in vitro and were examined for preneoplastic transformation. The RTE cells isolated from rats exposed to ozone (maximum total exposure of 1.2 ppm of ozone, 6 hr/day, 5 days/week, for 4 weeks) had no increase in their frequency of preneoplastic variants compared with air-exposed controls [24]. However, ozone exposures in vivo were not without effect, since focal attenuation of the epithelium and loss of cilia were observed in the tracheas of animals exposed to ozone. In contrast, RTE cells exposed to ozone in vitro (0.7 ppm ozone, 40 min, two-times per week for 4.5 weeks) exhibited two- to threefold increases in their frequency of preneoplastic transformation compared with control cultures exposed to air [21]. Although the exact ozone exposures received by RTE cells in these different experiments is not known, these results suggest that RTE cells exposed to ozone in vivo or in vitro do not exhibit the same transformation response. Although it is not known if RTE cells in vivo are protected from ozone by mucus, if the cells are better able to repair ozone-induced damage, or if they simply are not transformed, these results, whatever the reason for the differences in responsiveness to ozone following exposures in vivo or in vitro, demonstrate that the in vivo potency of a toxicant cannot simply be deduced from its behavior in an in vitro system, without careful validation of the model being used.

## CONCLUSIONS

To characterize and understand mechanisms of multistep carcinogenesis in vivo and to define carcinogenic risk from exposure to toxicants,

1. In vitro systems can provide useful information on the mechanisms of carcinogenesis and the carcinogenic risk leading to fewer and more-focused animal studies; however,
2. The relevance of in vitro endpoints or responses used in these analyses must be validated using in vivo determinations.

## ACKNOWLEDGMENTS

Research supported by the U. S. Department of Energy's Office of Health and Environmental Research under contract DE-AC04-76EV01013 in facilities fully accredited by the American Association for Accreditation of Laboratory Animal Care.

## REFERENCES

1. Nettesheim, P., and Barrett, J. C., Tracheal epithelial cell transformation: A model system for studies on neoplastic progression, *CRC Crit. Rev. Toxicol. 12*:215–239 (1984).
2. Kendrick, J., Nettesheim, P., and Hammons, A. S., Tumor induction in tracheal grafts. A new experimental model for respiratory carcinogenesis, *JNCI 52*:1317–1325 (1974).
3. Nettesheim, P., Griesemer, R. A., Martin, D. H., and Canton, J. E., Induction of preneoplastic and neoplastic lesions in grafted rat tracheas continuously exposed to benzo[*a*]pyrene, *Cancer Res. 37*:1272–1278 (1977).
4. Klein-Szanto, A. J. P., Topping, D. C., Keckman, C. A., and Nettesheim, P., Ultrastructural changes in tracheal epithelium, *Am. J. Pathol. 98*:83–100 (1980).
5. Terzaghi, M., and Nettesheim, P., Dynamics of neoplastic development in carcinogen-exposed tracheal mucosa, *Cancer Res. 39*:4003–4010 (1979).
6. Thomassen, D. G., Chen, B. T., Mauderly, J. L., Johnson, N. F., and Griffith, W. C., Inhaled cigarette smoke induces preneoplastic changes in rat tracheal epithelial cells, *Carcinogenesis 10*:2359–2361 (1989).
7. Thomassen, D. G., Kaighn, M. E., and Saffiotti, U., Clonal proliferation of rat tracheal epithelial cells in serum-free medium and their response to hormones, growth factors, and carcinogens, *Carcinogenesis 7*:2033–2039 (1986).
8. Pai, S. B., Steele, V. E., and Nettesheim, P., Neoplastic transformation of primary tracheal epithelial cell cultures, *Carcinogenesis 4*:369–373 (1983).
9. Steele, V. E., Arnold, J. T., and Mass, M. J., In vivo and in vitro characteristics of early carcinogen-induced premalignant phenotypes in cultured rat tracheal epithelial cells, *Carcinogenesis 9*:1121–1127 (1988).
10. Walker, C., and Nettesheim, P., In vitro neoplastic progression of rat tracheal epithelial cells, *Cancer Res. 49*:4427–4430 (1989).
11. Thomassen, D. G., Hubbs, A. F., and Kelly, G., Changes in cellular responses to inducers of differentiation, growth factors, and oncogenes during neoplastic progression of respiratory epithelial cells, *Biology, Toxicology, and Carcinogenesis of Respiratory Epithelium* (D. G. Thomassen and P. Nettesheim, eds.), Hemisphere Publishing, New York, 1990, pp. 205–216.
12. Ferriola, P. C., Walker, C., Robertson, A. T., Earp, H. S., Rusnak, D. W., and Nettesheim, P., Altered growth factor dependence and transforming growth factor gene expression in transformed rat tracheal epithelial cells, *Mol. Carcinog. 2*:336–344 (1989).
13. Stegelmeier, B. L., Schafer, K., Gillett, N. A., Rebar, A. H., Blackstone, K. D., and Kelly, G., Epidermal growth factor receptor and transforming growth factor alpha expression in $^{239}PuO_2$-induced pulmonary neoplasms in rats, 1990 Inhala-

tion Toxicology Research Institute Annual Report, Department of Energy Office of Health and Environmental Research, LMF-129, National Technical Information Service, Springfield, VA, 1990, pp. 152–154.

14. Minna, J. D., Dominant and recessive oncogenes in the pathogenesis of lung cancer, *Biology, Toxicology, and Carcinogenesis of Respiratory Epithelium* (D. G. Thomassen and P. Nettesheim, eds.), Hemisphere Publishing, New York, 1990, pp. 288–305.
15. Gillett, N. A., Stegelmeier, B. L., Chang, I. Y., and Kelly, G., Expression of transforming growth factor α in plutonium-239-induced lung neoplasms in dogs: Investigations of autocrine mechanisms of growth, *Radiat. Res. 126*:289–295 (1991).
16. Hubbs, A. F., Hahn, F. F., and Thomassen, D. G., Increased resistance to transforming growth factor beta accompanies neoplastic transformation of rat tracheal epithelial cells, *Carcinogenesis 10*:1599–1605 (1989).
17. Gillett, N. A., Stegelmeier, B. L., Kelly, G., Haley, P. J., and Hahn, F. F., Expression of epidermal growth factor receptor in plutonium-239-induced lung neoplasms in dogs, *Vet. Pathol. 29*:846–852 (1992).
18. Stegelmeier, B. L., Gillett, N. A., Rebar, A. H., and Kelly, G., The molecular progression of plutonium-239-induced rat lung carcinogenesis: Ki-*ras* expression and activation, *Mol. Carcinog. 4*:43–51 (1991).
19. Masse, R., Histogenesis of lung tumors induced in rats by inhalation of alpha emitters: An overview, *Pulmonary Toxicology of Respirable Particles* (C. L. Sanders, F. T. Cross, G. E. Dagle, and J. A. Mahaffey, eds.), Technical Information Center, U. S. Department of Energy, 1980, pp. 498–521.
20. Viallet, J., and Minna, J. D., Dominant oncogenes and tumor suppressor genes in the pathogenesis of lung cancer, *Am. J. Respir. Cell Mol. Biol. 2*:225–232 (1990).
21. Harris, C. C., Reddel, R., Pfeifer, A. M. A., Amstad, P., Mark, G., Weston, A., Modali, R., Iman, D., McMenamin, M., Kaighn, E., Gabrielson, E., Jones, R., and Trump, B. F., Oncogenes and tumor suppressor genes in human lung carcinogenesis, *Genetic Mechanisms in Carcinogenesis and Tumor Progression* (C. C. Harris and L. A. Liotta, eds.), Wiley–Liss, New York, 1990, pp. 127–152.
22. Thomassen, D. G., Newton, G. J., Guilmette, R. A., and Johnson, N. F., A biodosimetric approach for estimating radiation dose to the respiratory epithelium from inhaled radon progeny, *Radiat. Protect. Dosim. 38*:65–71 (1991).
23. Thomassen, D. G., Seiler, F. A., Shyr, I. J., and Griffith, W. C., Alpha Particles induce preneoplastic transformation of rat tracheal epithelial cells in culture, *Int. J. Radiat. Bio. 57*:395–405 (1989).
24. Thomassen, D. G., Harkema, J. R., Stephens, N. D., and Griffith, W. C., Preneoplastic transformation of rat tracheal epithelial cells by ozone, *Toxicol. Appl. Pharmacol. 109*:137–148 (1991).

# 28

# Development of a Short-Term Bioassay to Assess Pulmonary Toxicity of Inhaled Fibers

**David B. Warheit**

*E. I. du Pont de Nemours & Company*
*Newark, Delaware*

There is a great need for the development of a rapid and reliable short-term bioassay to evaluate the pulmonary toxicity of new and untested fibers. Short-term tests are used to screen potentially toxic fibers for estimates of pulmonary toxicity, evaluations of mechanisms, and in dose-level setting for subchronic or chronic inhalation studies. It is likely that the data generated from short-term inhalation tests could be applied to setting dose levels for 90-day inhalation studies and thus obviate the need for costly and animal-intensive 2-week or 28-day level-setting studies.

The development of our predictive screen is predicated on the hypothesis that the progression of particulate-induced, nononcogenic pulmonary disease is correlated with four interdependent general factors: (a) cytotoxicity, (b) inflammation, (c) alterations in macrophage clearance functions, and (d) persistence of the material in the lung. In addition, we have begun to systematically evaluate mechanistic issues such as in vitro and in vivo fiber durability, fiber deposition, translocation, and clearance. An example of this application can be found in our studies with Kevlar aramid fibrils.

These studies were undertaken to evaluate pulmonary mechanisms and the toxicity of inhaled Kevlar fibers relative to other reference materials. Rats were exposed to ultrafine Kevlar fibers for 5 days at gravimetric concentrations ranging from 2–13 $mg/m^3$ (fiber numbers ranging from 600 to 1300 fibers/

$cm^3$). Cells and fluids from groups of sham and fiber-exposed animals were recovered by bronchoalveolar lavage (BAL). Alkaline phosphatase, lactate dehydrogenase (LDH), protein, and *N*-acetylglucosaminidase (NAG) values were measured in BAL fluids at several time points postexposure. Cells were identified, counted, and evaluated for viability. Alveolar macrophages (AM) were cultured and studied for morphology, chemotaxis, and phagocytosis by scanning electron microscopy. The lungs of additional exposed animals were processed for cell labeling and lung clearance studies, fiber dimensions (from digested lung tissue), histopathology, and transmission electron microscopy.

## MATERIALS AND METHODS

### General Experimental Design

Groups of male Crl:CD BR rats (8 weeks old, Charles River Breeding Laboratories, Kingston, NY) were used to assess the pulmonary effects of short-term, high-dose aerosol exposures to Kevlar fibrils. Rats were exposed 6 hr/day for 5 days to Kevlar fiber concentrations ranging from 600 to 1344 fibers (f)/$cm^3$ (2–13 mg/$m^3$). Following exposures, fiber-exposed animals and aged-matched sham controls were evaluated at several time points postexposure (i.e., 0, 24, or 72 hr, 1 week, 1, 3, or 6 months).

### Fiber Preparations

Ultrafine Kevlar *para*-aramid fibrils were supplied by DuPont Fibers. A special preparation of ultrafine, respirable-sized Kevlar fibers (heretofore referred to as fibrils) that had been prepared for the 2-year inhalation study [1] was used for this study.

### Inhalation Exposure

Dust generation techniques were similar to those described by Lee and colleagues [1] and are described in detail elsewhere [2]. Briefly, atmospheres of ultrafine Kevlar fibrils were generated with a K-tron bin feeder (K-tron Co., Glassboro, NJ) equipped with twin screws. Baffles were inserted into the generation apparatus and served to increase the respirability of the sample. Kevlar fibrils were metered into a plastic funnel connected to a cyclone where high-pressure air transferred the test material into a microjet apparatus (Micro-jet, Fluid Energy Co., Hatfield, PA). This high-pressure air-impingement device was used to separate ultrafine Kevlar fibrils from the larger fiber clumps. Kevlar fibrils were then drawn through another cyclone and then into the chamber. Chamber concentrations of fibrils were maintained by controlling the dust feed rate into the generation apparatus, or by varying the airflow rate. For gravimetric analysis, samples of atmospheric Kevlar were taken from the animal-breathing zone at approximately 30-min intervals by drawing calibrated volumes of chamber atmosphere through preweighed glass fiber filters. Filters were weighed on a Cahn 26 automatic electrobalance. The atmospheric concentrations of Kevlar were determined from filter weight differentials before and after sampling. Particle size measurements of airborne fibers in the test chamber were determined with a Sierra casade impactor and reported as mass median aerodynamic diameter (MMAD). Fiber counts were carried out according to the NIOSH 7400B method.

In addition to gravimetric concentration, fiber numbers in the aerosol, and mass median diameter measurements, fiber size dimensions were analyzed by a scanning electron microscopic (SEM) method. Airborne filters (used for determining fiber concentrations) were prepared for SEM as described previously [3]. The filters were mounted onto carbon stubs, gold-sputtered, and studied in a calibrated JEOL 840 scanning electron microscope (SEM). Fiber dimensions were quantified using a technique that allowed random selection of fibrils by first selecting a field of view at low magnification, to avoid bias. All selected areas were then studied at higher magnification (i.e., 200× or 5000×). Numerous high-magnification SEM micrographs were taken per field of view. The dimensions (lengths and diameters) of at least 160 fibers per filter were then measured on the micrographs with a ruler. For fiber classification purposes in these SEM studies, fibers were defined as having an aspect ratio of 3:1 (length/diameter) and a minimum length of 4 μm. Following tabulation of the data, a count median length (CML) and count median diameter (CMD) for the fiber aerosol were determined.

## Pulmonary Lavage and Biochemical Assays on Bronchoalveolar Lavage Fluid

Bronchoalveolar lavage procedures were conducted according to methods previously described [2]. Lavaged fluids recovered from sham and dust-exposed rats were centrifuged and concentrated for biochemical studies. The cell pellet was resuspended in Eagles minimal essential medium. The methods of quantitation of cell numbers, viabilities, and differential counts have been previously described [2]. All biochemical assays were performed on concentrated BAL fluids at 30°C, using a semiautomated clinical chemical analyzer. Methods for quantifying lactate dehydrogenase (LDH), alkaline phosphatase (ALP), protein, and *N*-acetyl-β-glucosaminidase (NAG) in BAL fluids have been previously reported [2].

## Pulmonary Cell-Labeling Studies

Bromodeoxyuridine (BrdU) cell-labeling methods have been previously described [2]. Groups of Kevlar fiber-exposed and sham-exposed rats were pulsed immediately after 5-day exposure, as well as 1 week or 1 month postexposure with an intraperitoneal injection of BrdU dissolved in a 0.5 *N* sodium bicarbonate buffer solution at a dose of 100 mg/kg body weight. The animals were sacrificed 2 hr later by pentobarbital injection.

## Recovery of Kevlar Fibrils from Digested Lung Tissue

Kevlar fibers were recovered from the lungs of exposed rats according to methods previously reported [2]. Briefly, fixed lung tissue was digested in an 11% KOH solution (in ethanol and water). The numbers of fibers per area of filter were counted by phase contrast light microscopy using the NIOSH 7400B-counting method (*NIOSH Manual of Analytical Methods*). For fiber classification purposes, fibers were defined as particulate materials with an aspect ratio of 3:1 (length/diameter). In addition, only fibers 5 μm or longer were counted. Fiber dimensional analysis was carried out by scanning electron microscopy (SEM); fiber 4 μm or longer were included for dimensional analysis.

# RESULTS AND CONCLUSIONS

## Chamber Atmosphere Analysis

The exposure generation data for aerosolized Kevlar fibrils are summarized in Table 1.

## Analyses of Cellular Constituents in Bronchoalveolar Lavage

Exposures to Kevlar fibrils did not alter the total number of cells recovered by lung lavage. The viability of cells recovered in fiber-exposed or sham-exposed rats was greater than 94% at all time periods.

Cell differential analyses of lavaged cells recovered from Kevlar-exposed rats demonstrated that fiber inhalation produced an early, but transient, pulmonary inflammatory response, as evidenced by increased numbers of neutrophils and biomarkers (i.e., LDH, NAG, and protein) in BAL fluids (Table 2). This effect was measured through a 1-week postexposure period, but was short-lived, since neutrophil numbers were not significantly different from sham

**Table 1** Kevlar Fibril Exposure Data and Inhaled Dose

| Duration | MMAD (δg) | (mg/m³) | Fiber (fiber no./cm³) | Dose (f/g)[d] | CML (μm) | CMD (μm) |
|---|---|---|---|---|---|---|
| Experiment 1[a,b,c] | | | | | | |
| 5 days | 4.5 μm (2.7) | 4.4 | 1073 | $3.2 \times 10^6$ | 9.9 | 0.3 |
| 5 days | 3.4 μm (2.7) | 8.5 | 1344 | $3.5 \times 10^6$ | 9.9 | 0.3 |
| Experiment 2[c] | | | | | | |
| 5 days | 3.2 μm (2.9) | 2.9 | 613 | $1.4 \times 10^6$ | 10.0 | 0.3 |
| 5 days | 4.7 μm (3.2) | 11.1 | 877 | $1.3 \times 10^6$ | 10.0 | 0.3 |

[a]Used for bronchoalveolar lavage studies.
[b]Used for cell-labeling studies.
[c]Used for fiber clearance or retention studies.
MMAD, mass medium aerodynamic diameter; CML, count median length; CMD, count median diameter.
[d]f/g = fibers/gram of lung tissue.

control values at the 1-month–postexposure period, or at any other subsequent time period.

## Cell-Labeling Studies

No significant differences in the labeling index of lung parenchymal cells were detected between Kevlar-exposed rats and their corresponding sham controls at any time period (Table 3). Increased BrdU labeling of terminal bronchiolar cells in Kevlar-exposed rats was measured immediately after exposure. However, no significant differences were measured at 1 week or 1 month postexposure, indicating that this effect was transient. Histopathological analysis at 3 months after Kevlar exposure indicated that no pulmonary lesions were detected in fiber or sham-exposed control rats.

## Fiber Clearance Studies

Clearance studies demonstrated a transient increase in the numbers of retained Kevlar fibrils at 1-week postexposure, with rapid clearance of fibers thereafter (data not shown). The estimated clearance half-time was 90 days. The finding of increased numbers of retained fibers at 1-week postexposure in Kevlar-exposed rats may be related to transverse cleaving of the fibrils. It is likely that fiber shortening during the first week after exposure could account for this apparent increase. Mean fiber lengths recovered from digested lung tissue

**Table 2** Pulmonary Inflammation Responses in Kevlar-Exposed Rats

| | Time after 5-day exposure | | | | |
|---|---|---|---|---|---|
| | 0 hr | 24 hr | 1 wk | 1 mo | 3 mo |
| Numbers of lavaged neutrophils | | | | | |
| Sham | $8.0 \times 10^3$ | $7.8 \times 10^3$ | $7.0 \times 10^3$ | $7.3 \times 10^3$ | $8.6 \times 10^3$ |
| Kevlar | $4.1 \times 10^5$* | $4.4 \times 10^5$* | $3.3 \times 10^5$* | $2.6 \times 10^4$ | $8.7 \times 10^3$ |
| BAL protein, LDH, and NAG values (% of sham control) | | | | | |
| LDH | | | | | |
| Sham | 100 | 100 | 100 | | |
| Kevlar II (1073 f/cc) | 426 | 416 | 111 | | |
| Kevlar III (1344 f/cc) | 375 | 512 | 148 | | |
| Protein | | | | | |
| Sham | 100 | 100 | 100 | | |
| Kevlar II | 193 | 227 | 115 | | |
| Kevlar III | 156 | 304 | 126 | | |
| *N*-Acetyl-β-glucosaminidase | | | | | |
| Sham | 100 | 100 | 100 | | |
| Kevlar II | 300 | 186 | 80 | | |
| Kevlar III | 236 | 189 | 98 | | |

*$p < 0.05$

**Table 3** Pulmonary Cell-Labeling Responses in Kevlar-Exposed Rats

| | Time after 5-day exposure | | |
|---|---|---|---|
| | 0 hr | 1 wk | 1 mo |
| Terminal bronchiolar cells | | | |
| Sham | 0.8 ± 0.2 | 0.9 ± 0.1 | 1.1 ± 0.1 |
| Kevlar | 2.3 ± 0.4* | 1.3 ± 0.1 | 1.2 ± 0.1 |
| Pulmonary parenchymal cells | | | |
| Sham | 1.0 ± 0.1 | 1.1 ± 0.2 | 1.1 ± 0.1 |
| Kevlar | 1.4 ± 0.2 | 1.1 ± 0.6 | 1.0 ± 0.1 |

*$p < 0.05$

decreased from 12.5 to 7.5 μm, measured over a 6-month–postexposure period. Similarly, mean fiber diameters decreased from 0.33 to 0.23 μm over the 6-month period (Table 4). These results suggest that Kevlar fibrils are shortened in the lungs of exposed rats. The pulmonary clearance mechanisms associated with inhaled Kevlar fibrils are different from those associated with chrysotile and crocidolite asbestos, wherein the mean lengths of inhaled fibers are progressively increased over time [4,5].

A short-term inhalation bioassay is being developed to evaluate the potential for inhaled fibers or fibers to produce lung injury (i.e., pulmonary fibrosis). The establishment of this predictive screen is grounded on the detection of a series of biomarkers that may be predictive of the progression of fiber-induced pulmonary injury. In previous studies, we have assessed the efficacy of this short-term inhalation screen by exposing rats to several

**Table 4** Kevlar Fibril Dimensional Data

| | Time after 5-day exposure | | | | |
|---|---|---|---|---|---|
| | 0 hr | 72 hr | 1 mo | 3 mo | 6 mo |
| Mean lengths (μm) | 12.5 ± 2 | 11.1 ± 5 | 9.0 ± 2 | 8.1 ± 2 | 7.45 ± 3 |
| Mean diameters (μm) | 0.33 ± 0.03 | 0.30 ± 0.06 | 0.295 ± 0.04 | 0.27 ± 0.09 | 0.23 ± 0.02 |

concentrations of numerous reference particulate or fibrous materials. These include known fibrogenic dusts, such as α-quartz silica [6] and crocidolite asbestos [7], as well as materials with minimal or moderate biological activity, such as titanium dioxide and carbonyl iron particles [6], or carbon fibers [8]. Short-term exposures to silica or crocidolite asbestos in rats produced persistent pulmonary inflammatory (neutrophilic) responses. In contrast, exposures to Kevlar fibrils caused only transient inflammatory effects. The findings here support the conclusion that acute inhalation of fibers with low durability such as Kevlar are likely to produce only short-term pulmonary effects whereas inhalation of fibrogenic dusts, such as silica, or durable fibers, such as crocidolite, produces sustained pulmonary inflammatory effects along with consistently elevated indicators of cytotoxicity and consequent development of pulmonary fibrosis.

## ACKNOWLEDGMENT

This study was supported by DuPont Fibers.

## REFERENCES

1. Lee, K. P., Kelly, D. P., O'Neal, F. O., Stadler, J. C., and Kennedy, G. L., Jr., Lung response to ultrafine Kevlar aramid synthetic fibrils following 2-year inhalation exposure in rats, *Fundam. Appl. Toxicol. 11*:1–20 (1988).
2. Warheit, D. B., Kellar, K. A., and Hartsky, M. A., Pulmonary cellular effects in rats following aerosol exposures to ultrafine Kevlar aramid fibrils: Evidence for biodegradability of inhaled fibrils, *Toxicol. Appl. Pharmacol. 116*:225–239 (1992).
3. Warheit, D. B., Hwang, H. C., and Achinko, L., Assessments of lung digestion methods for recovery of fibers, *Environ. Res. 54*:183–193 (1991).
4. Roggli, V. L., George, M. H., and Brody, A. R., Clearance and dimensional changes of crocidolite asbestos fibers isolated from lungs of rats following short-term exposure, *Environ. Res. 42*:94–105 (1987).
5. Roggli, V. L., and Brody, A. R., Changes in numbers and dimensions of chrysotile asbestos fibers in lungs of rats following short-term exposure, *Exp. Lung Res. 7*:133–147 (1984).
6. Warheit, D. B., Carakostas, M. C., Hartsky, M. A., and Hansen, J. F., Development of a short-term inhalation bioassay to assess pulmonary toxicity of inhaled particles: Comparisons of pulmonary responses to carbonyl iron and silica, *Toxicol. Appl. Pharmacol. 107*:350–368 (1991).
7. Warheit, D. B., Moore, K. A., Carakostas, M. C., and Hartsky, M. A., Acute pulmonary effects of inhaled wollastonite fibers are dependent on fiber dimensions

and aerosol concentrations, *Mechanisms in Fibre Carcinogenesis* (R. Brown, J. Hoskins, and N. Johnson, eds.), *NATO ASI Ser. A Life Sci. 223*:143–156 (1991).

8. Warheit, D. B., Hansen, J. F., Carakostas, M. C., and Hartsky, M. A., Acute inhalation toxicity studies in rats with a respirable-sized experimental carbon fiber: Pulmonary biochemical and cellular effects, Proceedings of the VII International Symposium on Inhaled Particles, *Ann. Occup. Hyg.* (in press; 1994).

# 29

# In Vitro Toxicity of Refractory Ceramic Fibers to Chinese Hamster Ovary Cells in Culture

**Georgia A. Hart, Mildred M. Newman, and Thomas W. Hesterberg**

*Mountain Technical Center*
*Schuller International, Inc.*
*Littleton, Colorado*

Awareness of the human health hazards associated with asbestos exposure has led to the development of many new man-made and naturally occurring fibers to be used as asbestos substitutes. The importance of testing all fibrous products for their pathogenic potential is generally recognized; however, many fibrous materials currently in use and in the product-development stage have not been evaluated. Animal inhalation testing is currently the only accepted laboratory model for determining the potential human health hazard of respirable particles [1]. Because this form of testing is time-consuming and expensive, short-term tests are needed to screen fibers for their potential toxicity.

Any short-term test system proposed as a means of screening fibers for their toxicological potential must be validated by demonstrating that the effects observed using the screening system correlate with human epidemiological data or animal inhalation data. A recent chronic rodent inhalation study of the toxicity and tumorigenicity of refractory ceramic fibers [2] afforded the opportunity to make these correlations. Conducting in vitro research in parallel with in vivo studies also provides an additional perspective on the toxic mechanisms at the cellular level. Thus, the present study was conducted with these two goals in mind: (a) to develop and validate in vitro assays that could be useful as a part of a battery of short-term fiber toxicology screening tests; and

(b) to contribute to a more complete understanding of the mechanisms of fiber–cell interactions.

# MATERIALS AND METHODS

## Test Fibers

Four different compositions of refractory ceramic test fibers were obtained from the TIMA fiber repository (c/o Manville Sales Corporation, Attn. T. W. Hesterberg, P. O. Box 5108, Denver, CO. 80217-5108): RCF1 (kaolin), RCF2 (zirconia), RCF3 (high purity), and RCF4 (after service). These are size-selected, respirable test fibers that were used in rodent inhalation studies [1–3]. Standard International Union against Cancer (UICC) crocidolite (CD) and UICC chrysotile (CH) asbestos were used. Fiber data (Table 1) were analyzed using scanning electron microscopy (SEM) for RCFs and CD and transmission electron microscopy (TEM) for CH.

## Cell Cultures and In Vitro Assays

Frozen Chinese hamster ovary (CHO)-K1 cells were obtained from American Type Tissue Collection (Rockville, MD) and were thawed and plated in tissue culture flasks with complete medium (CM) consisting of Ham's F12 supplemented with 10% fetal calf serum (FCS), 2 mM L-glutamine, and 1 ml Fungi-Bact solution per 100 ml medium (all from Irvine, Santa Ana, CA). To create a large bank of uniform stock cells for experiments, cultures were incubated (37°C, 5% $CO_2$ and 100% humidity) and allowed to replicate for several days (three to four doublings), then harvested with 0.05% trypsin and 0.02% EDTA

**Table 1** Physical Characterization of Fibers Using Electron Microscopy

| Test Fiber | | | Fiber Dimensions (μm) | |
|---|---|---|---|---|
| Ref. ID | Fiber type | No. fibers/ng | Avg. diameter ± SD | Avg. length ± SD |
| RCF1 | Kaolin | 4.7 | 1.03 ± 0.73 | 21.5 ± 16.12 |
| RCF2 | Zirconia | 4.6 | 1.11 ± 0.82 | 16.7 ± 15.03 |
| RCF3 | High purity | 3.1 | 1.22 ± 0.98 | 24.3 ± 18.82 |
| RCF4 | Aft. serv. | 6.3 | 1.43 ± 0.79 | 9.2 ± 7.08 |
| Asbestos | | | | |
| CD | UICC crocidolite | 2400 | 0.21 ± 0.12 | 1.81 ± 1.94 |
| CH | UICC chyrsotile | 4400 | 0.12 ± 0.07 | 1.65 ± 1.83 |

in Hanks balanced salt solution (Irvine). The stock cells were then frozen at −80°C and later transferred to liquid nitrogen for storage. For experiments, stock cells were thawed and plated in CM and allowed to grow for 2–4 days.

## Inhibition of Cell Proliferation Assay

For the inhibition of cell proliferation (ICP) assays, 60-mm culture dishes were seeded with 100,000 cells in 5 ml CM per dish. After 24-hr incubation, fibers were added to cultures in 1-ml CM per dish. Dishes were gently agitated to disperse fibers over the bottom. Negative control cultures received 1 ml CM per dish. Each exposure group was set up in triplicate. After 3-days exposure, incubated as just described, cells were harvested using trypsin (as described) and counted using a Coulter counter (Coulter Electronics, Hialeah, FL). Relative proliferation was determined by dividing the number of cells present in each exposed culture by the number of cells present in negative control cultures.

## Colony-Forming Efficiency

For the colony-forming efficiency (CFE) assay, 60-mm culture dishes were seeded with 200 cells in 5 ml complete medium and exposed as described for ICP. Each exposure was set up in triplicate. After 5-days exposure, colonies were stained with 0.4% w/v Giemsa in buffered methanol (Sigma). Colonies with more than ten cells were counted, using a stereoscope at low power. The CFE was determined by dividing the number of colonies in each exposed culture by the number of colonies in unexposed cultures.

## Micronuclei and Polynuclei Induction

For the micronuclei (MN) and polynucleus (PN) assays, cultures were prepared as described for ICP. After 2-days exposure, culture dishes were fixed with methanol/acetic acid (3:1, v/v) and stained with 0.01% acridine orange (from Sigma; as described in Clark [4]). With an epifluorescence microscope, the percentage of cells containing MN or PN was determined for each culture dish. An MN was defined as a nucleus that appeared to be less than one-half the size of the normal-sized nuclei; PN was defined as a cell having bi- or multiple nuclei or a lobed nucleus. A minimum of 100 cells per dish and two to three dishes per exposure group were scored for (a) presence of MN, but no other visible nuclear abnormalities; (b) presence of both MN and PN; or (c) PN, but without micronuclei. Calculations included the percentage of cells with MN (including both uninucleate and polynucleate cells), the percentage of PN (with or without MN), and the percentage of cells with either type of nuclear abnormality (NA; having MN or PN).

## RESULTS

Results of the three in vitro assays are shown (Figs. 1–3). These data show the same relative toxicities for the four RCFs when concentration is based on micrograms per square centimeter (μg/cm$^2$). In each case RCF4 is the least toxic, RCF2 is intermediate, and RCF1 and RCF3 are the most toxic. In all three assays, CD was more toxic than the RCFs; in the ICP (see Fig. 1) and MNI (see Fig. 3) assays; CH was more toxic than CD (CH was not included in the CFE assays).

Although some of the error bars (SEM; standard error of the means) overlap, in each individual test performed with these fibers (more than 14 independent tests in all), the same relative toxicities resulted.

When the concentration of the RCFs is expressed as number of fibers per square centimeter (Fig. 4), the similar relative toxicities of the different RCFs is observed, although the spread between RCF1 and RCF3 is increased. Data from the asbestos types are not plotted in this manner, since both the length and diameter are an order of magnitude less than the RCF fibers, making a fiber-with-fiber comparison inappropriate.

The CHO cells in culture exposed to RCF2 are seen in Figure 5. Nuclear

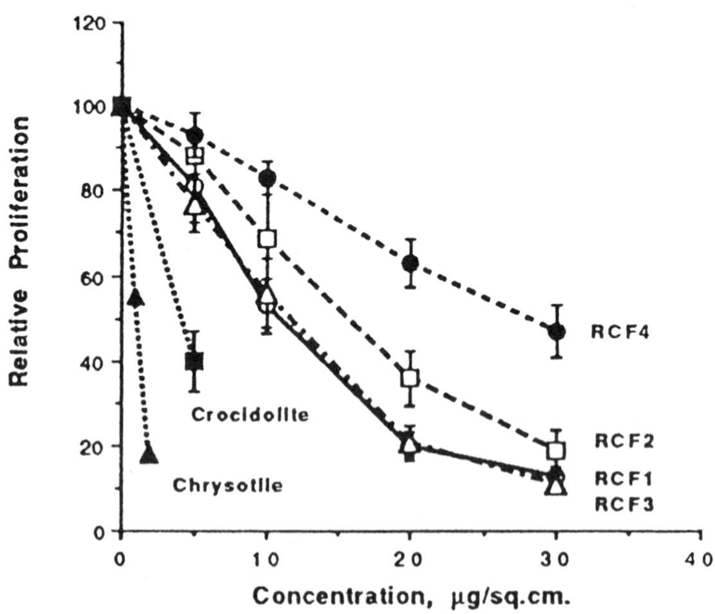


**Figure 1** Inhibition of cell proliferation (concentration based on fiber mass per unit of area).

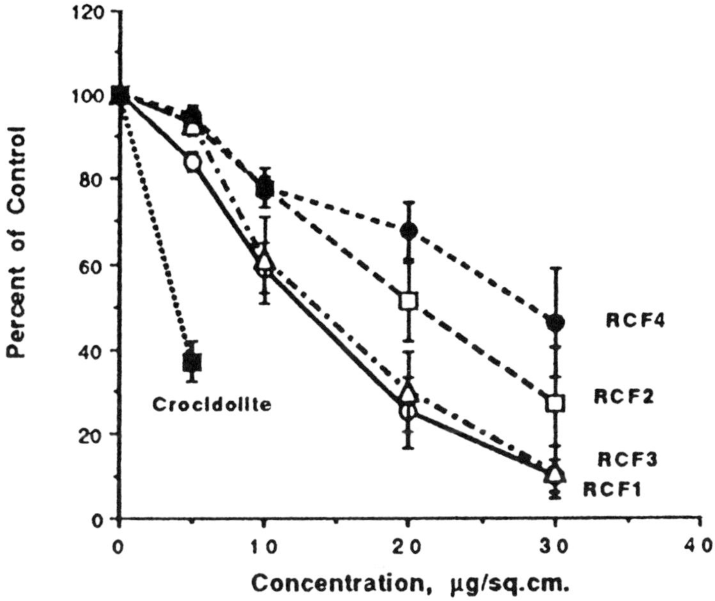


**Figure 2** Colony-forming efficiency.

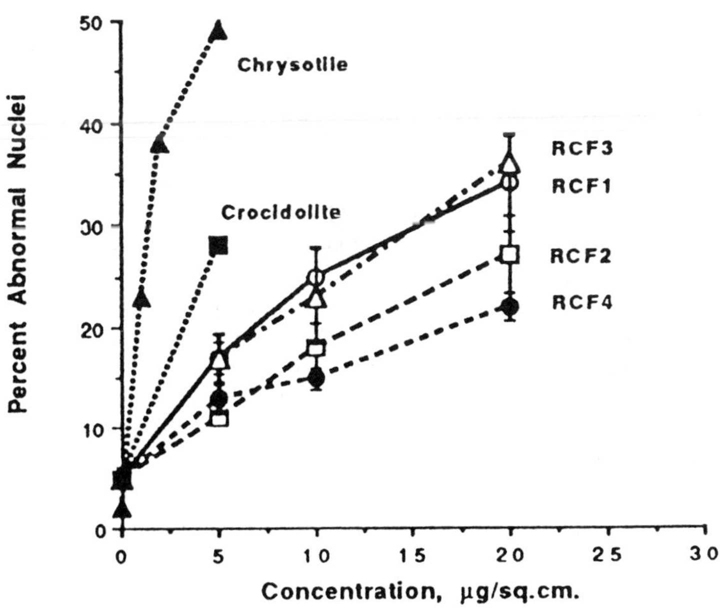


**Figure 3** Induction of nuclear abnormalities (incidence of cells with micronuclei or polynuclei).

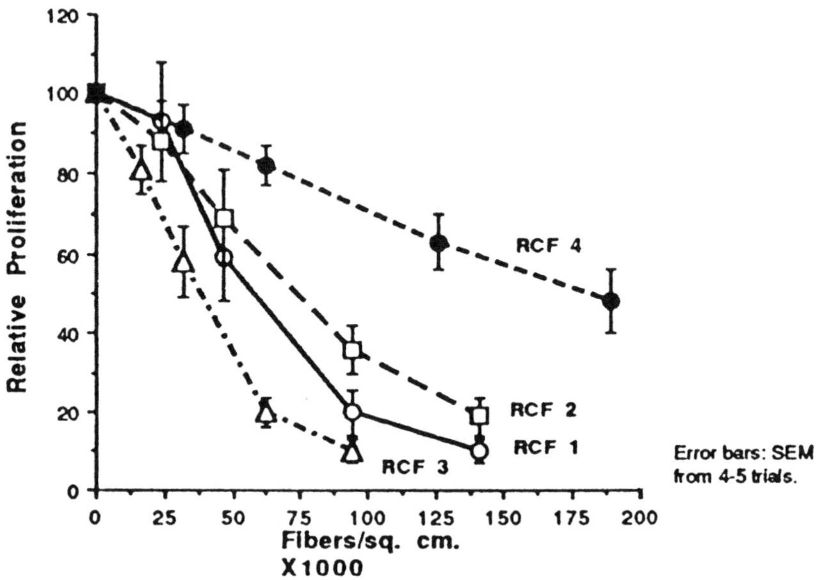


**Figure 4** Inhibition of proliferation (concentration based on number of fibers per square centimeter).

abnormalities (NA) observed in all the fiber-exposed cultures appeared qualitatively similar. Some examples of NA include: PN that appear to be clearly more than one nucleus as well as PN that appear to be bi- or multilobed; micronuclei with one normal-appearing nucleus or with PN. The relation between intracellular fibers and nuclear distortion was unmistakable in all the fiber-exposed cultures and was especially apparent when both PCOM and fluorscent microscopy were used to view the cells. Both wholly and partially internalized fibers were seen.

In the fiber-exposed cultures, the incidence of both types of NA (percentage of cells with MN or PN, or both), was strongly concentration-dependent, whereas the incidence of MN alone, although elevated above background levels, was weakly concentration-dependent. The CH at 5 $\mu g/cm^2$ induced NA in 49% of the cells; CD at 5 $\mu g/cm^2$ induced NA in 28% of the cells. The NA incidence for RCFs at 20 $\mu g/cm^2$ ranged from 20 to 40% (see Fig. 3).

In Table 2 the toxic endpoints of the in vitro studies are compared with those of the animal inhalation studies using the same RCF test fibers. The $LC_{50}$ is the in vitro concentration that resulted in 50% as many cells or colonies in fiber-treated cultures as in unexposed control cultures. The higher the $LC_{50}$, the

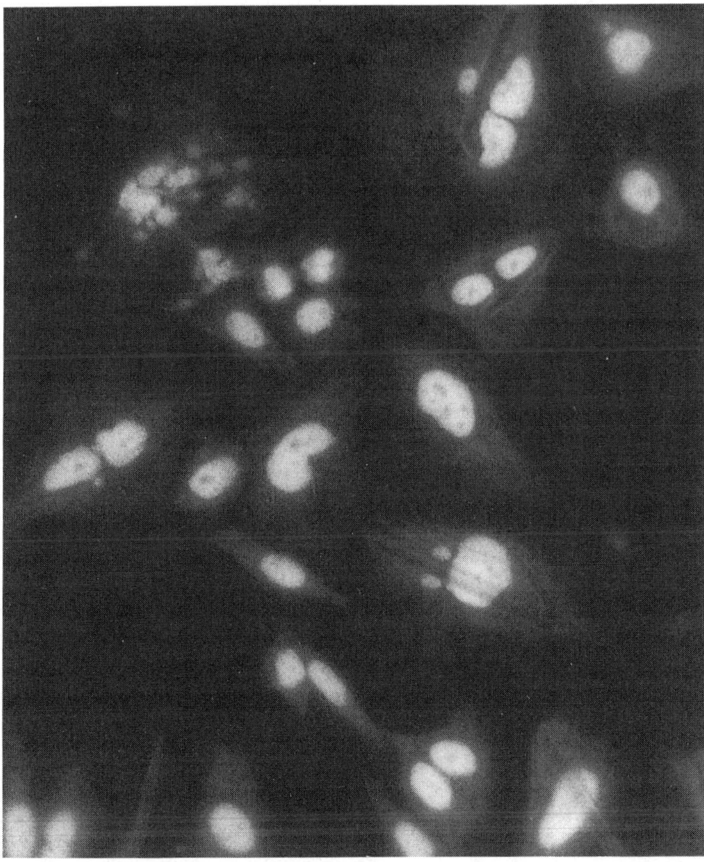

**Figure 5** CHO cells exposed to RCF2. Cells with both normal and abnormal nuclei are visible. Note the micronuclei and polynuclei. (Acridine orange stain, photographed with fluorescent light; 800× magnification).

lower the in vitro toxicity of the fiber. Wagner grades were used to quantitate the lung lesions observed in the chronic inhalation study of RCFs: 1 indicates normal lung; 2 is macrophage infiltration; 3 is inflammation and bronchiolization; 4 is minimal, but irreversible, fibrosis; 5, 6, 7, and 8 are increasingly severe fibrosis [3]. In the animal inhalation studies, three to six animals per group were scored every 3–6 months during the 24-month exposure to fiber aerosol and at the termination of the lifetime study. Only the 6-month scores are given for comparison here, because all fiber-exposed lung scores plateaued near

**Table 2** Comparison of In Vitro and In Vivo Toxic Endpoints

| Exposure group | In vitro $LC_{50}$ ($\mu g/cm^2$) ICP | In vitro $LC_{50}$ ($\mu g/cm^2$) CFE | In vivo Wagner score (6 mo) | In vivo % with lung tumors | In vivo % with mesotheliomas |
|---|---|---|---|---|---|
| Negative control | — | — | 1.0 | 1.6 | 0.0 |
| RCF1 | 13 | 10 | 4.0 | 13.0 | 1.5 |
| RCF2 | 17 | 18 | 3.0 | 6.9 | 2.3 |
| RCF3 | 13 | 12 | 4.0 | 14.6 | 1.5 |
| RCF4 | 30 | 22 | 2.7 | 3.2 | 0.8 |

grade 4 soon after the 6-month time point (air control lungs remained at level 1). The columns on the right in Table 2 indicate the percentage of animals that developed lung tumors and mesotheliomas by the end of the RCF inhalation study. The relative toxicity of the different RCFs seen in vitro correlates with the fibrosis and lung tumors observed in vivo: RCF4 is least toxic, RCF2 is intermediate, and RCF1 and RCF3 are more toxic. Mesothelioma results vary slightly from this pattern in that RCF2 induced more tumors than RCF1 and 3, but RCF4 was still least toxic.

## DISCUSSION

The present study provides several pieces of information that contribute toward the goals of this project: to develop short-term screening tests that can be used to assess the toxicological potential of fibrous dusts and to increase understanding of the mechanisms of fiber toxicity at the cellular level. First, the CHO cell test system is highly responsive to fiber exposure in all three parameters investigated: Concentration-dependent reductions in cell proliferation and cell colony formation as well as increases in nuclear abnormalities were observed following exposure to each of the test fibers. Second, specificity is evident in this model, in the consistent relative response of these cells to the six different fibers. Third, cellular uptake of fibers and alterations of nuclear morphology are clearly major CHO cell responses. The observed fiber-induced nuclear abnormalities fell into two basic categories (MN and PN), and the incidence of each was consistent for the six fibers in three or more separate assays. The mechanism of MN induction may be different from that of PN. Micronuclei are believed to form when chromosomes or fragments are separated from the migrating masses of replicate chromosomes during mitosis. Previous studies have shown a correlation between incidence of micronuclei and aneuploidy in three different cell lines (CHO cells [5], Syrian hamster embryo (SHE) cells [6], human lymphocytes [7]). Polynuclei may result from the failure of cytokinesis following mitosis. And, finally, the incidence of total nuclear abnormalities was approximately the inverse of the proliferation curves, indicating that disruption of mitosis may be the major or immediate cytotoxic effect, rather than the disruption of cytoplasmic metabolic processes. However, validation of this hypothesis would require cell viability testing.

Differences in the length of the four RCFs can probably account for the differences in both the in vitro cytotoxicity and the in vivo lung pathology observed in the animal inhalation study. A comparison of Tables 1 and 2 demonstrates that the fiber length parallels toxicity. Many previous in vitro studies also report a relation between fiber length and cytotoxicity. Brown et al. [8] demonstrated that glass fibers were more toxic to cultured macrophages and to V79-4 cells when average fiber length was greater than 10 μm; Lipkin et al. [9] observed a relation between fiber length and the reduction of cell proliferation in cultures of P388D1 cells exposed to several populations of glass fibers; Tilkes and Beck [10–12] showed a direct relation between fiber length and LDH release (macrophages and ascites) and reduction in proliferation (ascites) in cell cultures exposed to fibrous glass.

A mechanism to explain the relation between fiber length and cytotoxicity could be related to cell diameter. For example, when the flattened, adherent cell becomes spherical as it enters metaphase before cell division, internalized fibers that are longer than the diameter of the metaphase cell might puncture the cell, causing cell death. Fiber length has also been shown to be important in the induction of cytogenetic effects by internalized fibers [13] and in the transformation of cultured cells [14]. In a study that demonstrated direct interaction between intracellular asbestos fibers and migrating chromosomes during anaphase in cultured SHE cells, Hesterberg and Barrett [15] suggested that long fibers are more easily entangled in the migrating chromosomes or spindle apparatus than are short fibers.

## CONCLUSIONS

The concordance between in vitro and in vivo toxic endpoints in the RCF studies is encouraging. However, more research is needed to determine which in vitro test systems and procedures will best correlate with animal inhalation studies and human epidemiology. Again, fibers tested in long-term animal inhalation studies will be needed for validation of the sensitivity and specificity of these short-term assays. Such fibers should be well characterized chemically and physically (length, width, number of fibers per unit mass). Furthermore, the test fibers should include a range of demonstrated in vivo toxicities, from nontoxic to strongly toxic. In vitro testing should include other cell types and other in vitro parameters (e.g., durability, cytotoxicity, and mutagenesis).

Several parameters of in vivo pathogenesis cannot yet be measured in vitro (e.g., lung deposition, lung clearance, and long-term biopersistence). Some work has been done in the third area, by incubating fibers in flowthrough systems using synthetic physiological solutions [16,17]. However, such systems do not contain all of the factors that might be important in the biopersistence of fibers in the lung, such as enzymes, lysosomal secretions, and mechanical removal by phagocytic cells. These limitations might be overcome by including short-term animal inhalation studies in the battery of toxicity-screening tests.

A more detailed report of this study can be found in *Toxicology in Vitro* [18].

## REFERENCES


1. Hesterberg, T. W., Vu, V., McConnell, E. E., Chase, G. R., Bunn, W. B., and Anderson, R., Use of animal models to study man-made fiber carcinogenesis, *Current Communications in Molecular Biology: Molecular Mechanisms of Fiber Carcinogenesis* (B. Brinkley, J. Lechner, and C. Harris, eds.), Cold Spring Harbor Laboratory Press, 1991, pp. 183–205.
2. Hesterberg, T. W., Mast, R., McConnell, E. E., Chevalier, J., Bernstein, D. M., Bunn, W. B., and Anderson, R., Chronic inhalation toxicity of refractory ceramic fibers in Syrian hamsters, *Proceedings of the International Workshop on Mechanisms of Fibre Carcinogenesis* (R. C. Brown, et al., eds.), Plenum Press, New York, 1991, pp. 531–538.
3. Hesterberg, T. W., Mast, R., McConnell, E. E., Vogel, O., Chevalier, J., Bernstein, D. M., and Anderson, R., Chronic inhalation toxicity and oncogenicity study of refractory ceramic fibers in Fisher 344 rats, *Toxicologist 11*:85 (1991).
4. Clark, G., *Staining Procedures*, 4th Ed., Williams & Wilkins, Baltimore, 1981.
5. Sincock, A. M., Delhanty, J. D. A., and Casey, G., A comparison of the cytogenetic response to asbestos and glass fibre in Chinese hamster and human cell lines. Demonstration of growth inhibition in primary human fibroblasts, *Mutat. Res. 101*:257–268 (1982).
6. Oshimura, M., Hesterberg, T. W., and Barrett, J. C., Correlation of asbestos-induced cytogenetic effects with cell transformation of Syrian hamster embryo cells in culture, *Cancer Res. 44*:5017–5022 (1984).

7. Migliore, L., and Nieri, M., Evaluation of twelve potential aneuploidogenic chemicals by the in vitro human lymphocyte micronucleus assay, *Toxicol. In Vitro* *5*:325–336 (1991).
8. Brown, R. C., Chamberlain, M., and Skidmore, J. W., In vitro effects of man-made mineral fibers, *Ann. Occup. Hyg.* *22*:175–179 (1979).
9. Lipkin, L. E., Cellular effects of asbestos and other fibers: Correlations with in vivo induction of pleural sarcoma, *Environ. Health Perspect.* *34*:91–102 (1980).
10. Tilkes, F., and Beck, E. G., Comparison of length-dependent cytotoxicity of inhalable asbestos and man-made mineral fibres, *Biological Effects of Mineral Fibres* (J. C. Wagner, ed.), IARC Sci. Publ., International Agency for Research on Cancer, Lyon, 30:475–483 (1980).
11. Tilkes, F., and Beck, E. G., Macrophage functions after exposure to mineral fibres, *Environ. Health. Perspect.* *51*:67–72 (1983).
12. Tilkes, F., and Beck, E. G., Influence of well-defined mineral fibres of proliferating cells, *Environ. Health Perspect.* *51*:275–279 (1983).
13. Hesterberg, T. W., Butterick, C. J., Oshimura, M., Brody, A. R., and Barrett, J. C., Role of phagocytosis in Syrian hamster cell transformation and cytogenetic effects induced by asbestos and short and long glass fibers, *Cancer Res.* *46*:5795–5802 (1986).
14. Hesterberg, T. W., and Barrett, J. C., Dependence of asbestos- and mineral dust-induced transformation of mammalian cells in culture upon fiber dimension, *Cancer Res.* *44*:2170–2180 (1984).
15. Hesterberg, T. W., and Barrett, J. C., Induction by asbestos fibers of anaphase abnormalities: Mechanism for aneuploidy induction and possibly carcinogenesis, *Carcinogenesis* *6*:473–475 (1985).
16. Law, B., Bunn, W. B., and Hesterberg, T. W., Solubility of polymeric organic fibers and man-made vitreous fibers in gambles solution, *Inhalation Toxicol.* *2*: 321–339 (1990).
17. Law, B., Bunn, W. B., and Hesterberg, T. W., Dissolution of natural mineral and man-made vitreous fibers in Karnovsky's and formalin fixatives, *Inhalation Toxicol.* *3*:309–321 (1991).
18. Hart, G. A., Newman, M. M., Bunn, W. B., and Hesterberg, T. W., *Toxicol. In Vitro* *4*:317–326 (1992).

# 30

# Effects of Phosgene and Perfluoroisobutylene on Permeability of Pulmonary Endothelial Cells in Culture

**Robert J. Werrlein, Janna S. Madren-Whalley, and Stephen Drew Kirby**

*U.S. Army Medical Research Institute of Chemical Defense*
*Aberdeen Proving Ground, Maryland*

Phosgene (CG) and perfluoroisobutylene (PFIB) are toxic, low molecular weight, organohalide gases. Both are practically insoluble in water. Yet, when inhaled into the moist environment of a lung, each of these gases triggers a delayed physiological response that overwhelms the air–blood barrier and floods the alveoli with edematous fluids. It is not certain which cells and mechanisms of the airway and air–blood barrier are targets of these gases. Clearly, the first cells to be exposed during inhalation are epithelial cells lining the airways. Pawlowski and Frosolono [1] have claimed that the initial histopathological effects from CG occur in the bronchioles. Diller et al. [2] have claimed that CG-induced histological lesions begin at the air–blood barrier. From these studies it appears that there is some correlation between increased dose and progressive involvement of more distal portions of the airway. However, a considerable body of evidence suggests that increased permeability following inhalation of toxic gases is a consequence of direct microvascular [3] and endothelial [4,5] injuries. These mixed results on the pathophysiology of toxic gases suggest that the mechanisms effecting inhalation injury and loss of barrier function are complex and elusive subjects, limited by strict adherence to studies with animal models. Cultures of pulmonary tissues provide complementary and powerful alternatives, especially for the study of target cells and mechanisms. Cell cultures have been developed for this purpose and show, in the following study, how endothelial models have helped to investigate endothelial cell (EC)–gas interactions and to understand their involvement in the edemagenic process.

## MATERIALS AND METHODS

Endothelial cells from the distal pulmonary arteries of male sheep were grown to confluence in suspension cultures on Cytodex-3 microcarrier beads (Pharmacia LKB, Piscataway, NJ). The culture medium consisted of HAM's F-12 supplemented with 15% fetal bovine serum (FBS), 50 μg/ml endothelial cell growth factor, 1% L-glutamine, 50 μg/ml gentamicin (Sigma Chemical Co., St. Louis, MO) and 5% Omni serum (Advanced Biotechnologies Inc., Columbia, MD). Under optimized growth conditions, beads seeded with 30,000 cells per square centimeter reached confluence in 7–10 days and produced high-density stationary-phase cell populations in excess of 160,000/$cm^2$. An additional 8 days of growth were required before populations developed cell-to-cell barriers tight enough to expose and test for effects of edemagenic gases on permeability. Dye exclusion assays (Fig. 1) were used to determine whether permeability of EC barriers were altered by direct gas exposure. For these assays erythrocin-B,

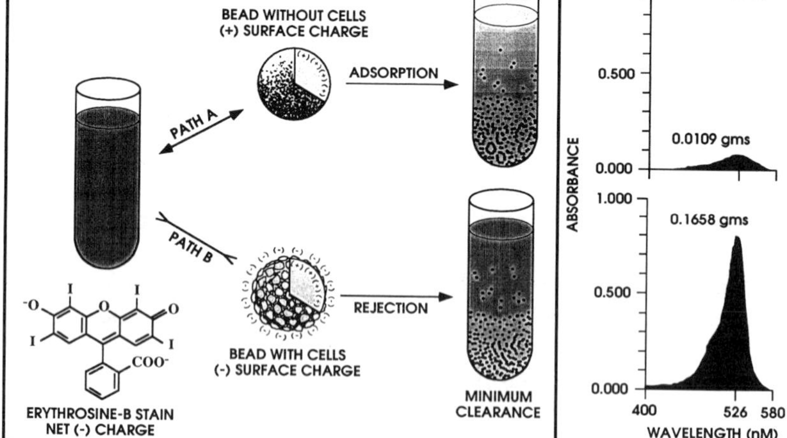

a red viability strain with a negative (−) charge and a peak absorbance of 526 nm was used.

Working solutions contained 8.3 μg of stain in 1 ml of phosphate-buffered saline (PBS). Cytodex-3 beads precoated with 60 $\mu g/cm^2$ of denatured, covalently bound type-I collagen carried a net positive (+) surface charge. When 2 ml of stain were mixed for 1 min with 80,000 beads, without cells (see Fig. 1, path A), roughly 93% of the stain was adsorbed onto the (+) surface of the beads and was cleared from solution as the beads settled. The remaining 7% of the stain could not be adsorbed and remained in solution. When bead surfaces were covered with confluent, untreated, control populations, the cells and their net (−) surface charge blocked adsorption and clearance of the stain (see Fig. 1, path B).

Before exposure, EC populations were washed three times in HAM's F-12 medium without serum, then split into two equal portions for use as controls and experimentals. In dose–response studies, samples from the experimental portion were exposed to CG or PFIB in a humidified (92% RH) gas-tight chamber. Phosgene or PFIB was injected into the chamber at dose equivalents of 0.5–5 $LCt_{50}$ for sheep. Each gas was chased with 150 $cm^3$ of $CO_2$, which raised $pCO_2$ in the chamber to 5% (i.e., 38 mm Hg, and roughly the $CO_2$ tension of a lung). Gas samples were allowed to equilibrate for 2 min before 10-μl samples were drawn for analysis by gas chromatography. Sham-treated control populations were exposed to 5% $CO_2$ and air only. Because CG and PFIB are insoluble, supernatant medium had to be removed from microcarrier cultures before exposure.

Following exposure, microcarrier populations were suspended in working solutions of erythrosin-B and subjected to dye exclusion assays as illustrated in Figure 1. When the beads had settled, supernatants from these erythrocin-B solutions were scanned in an SLM Aminco DW-2C spectrophotometer. The curves generated were copied on bond paper, cut, weighed, and compared with results from untreated and sham-treated controls. Curve weight comparisons were used to quantify changes in permeability of the EC barrier.

## RESULTS

Dose–response studies (Fig. 2) showed that 40-min exposures to phosgene at 0.5, 2, and 5 $LC_{50}$s increased EC permeability and clearance of erythrosin-B by 12, 26, and 57%, respectively. According to Haber's rule ($K = C \times T$), dose-related increases in endothelial permeability (i.e., the biological effect, *K*), were the product of phosgene concentration (*C*) and duration of exposure

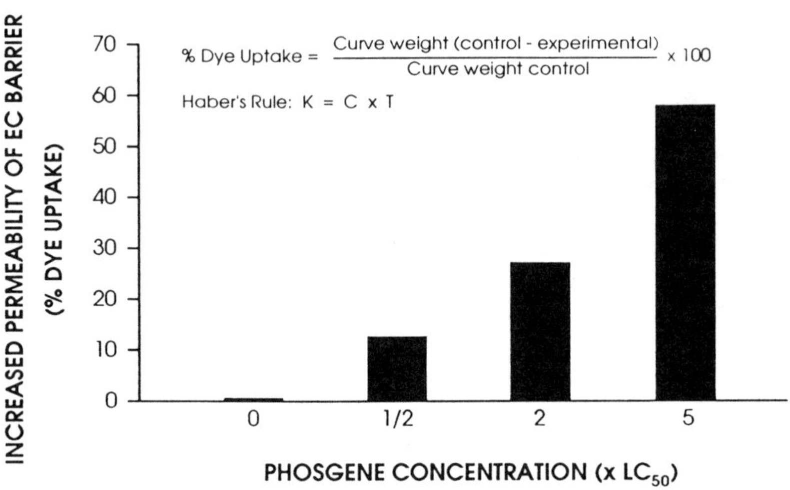


**Figure 2** Dose–response to phosgene.

(*T*). In all subsequent phosgene experiments, exposures were adjusted to 1 $LC_{50}$ for 20 min, making results from in vitro and in vivo studies comparable. Postexposure viability remained greater than 98% throughout each experiment, and cells remained well attached to microcarrier beads.

The onset and duration of phosgene-induced endothelial lesions were determined in a series of postexposure permeability profiles. Results in Figure 3 give the actual and relative amounts of erythrosin-B excluded by EC barriers of untreated controls, sham-treated controls, and exposed populations. Each bar presents information obtained from four separate profiles run at $T = 0, 3, 6, 9$, and 22 hr following sham or CG exposures. Dye exclusion by untreated control populations was consistently high, and variability between assays was low. Differences between sham-treated and untreated controls were barely significant. By contrast, populations exposed to CG responded with an immediate and substantial increase in permeability. Adsorption of stain through the EC barrier of these populations remained between 17.29 and 18.63% during the first 9 hr of the postexposure profile. For technical reasons, sham-treated and exposed populations were maintained in serum-free medium between $T = 0$ and $T = 3$ hr. At $T = 3$ hr, cultures were resuspended in medium containing 15% FBS. Manipulation of the medium did not alter permeability substantially; however, sham-treated and exposed populations both showed a slight, temporary improvement in dye exclusion at $T = 6$ hr, after serum replacement.

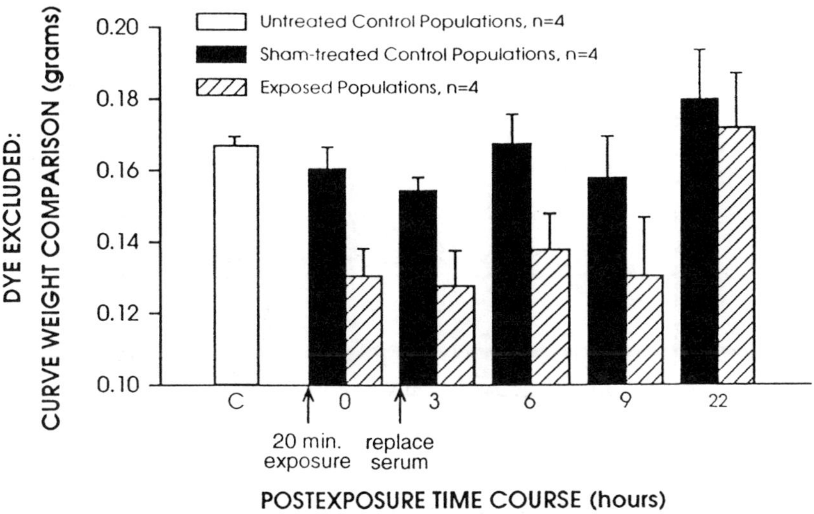


**Figure 3** Phosgene-induced permeability profile. Dose = 3325 ppm·min.

Unlike phosgene, PFIB did not undergo rapid hydrolysis and decay in a moist medium rich environment. In fact, doses of 1–8 $LC_{50}$s remained stable in the exposure chamber, without any detectable changes in GC analyses for periods of up to 4 hr. Under conditions identical with those used with phosgene, results presented in Figure 4 show that PFIB at 1 $LCt_{50}$ had no effect on permeability. In fact, the EC barrier remained unaffected by 20-min exposures to doses as high as 8 $LCt_{50}$, and PFIB produced no delayed effects during 9- and 22-hr postexposure profile periods.

## DISCUSSION

Phosgene and PFIB are lung toxicants that produce pulmonary edema. It is unknown, however, which cells are targets of these gases or which mechanisms they trigger to increase permeability. With cultures of pulmonary EC and tests for permeability it was possible to study endothelial lesions and answer questions that are difficult to address in vivo. The results suggest that the etiology of the lesions producing pulmonary edema differ for the two gases. This is consistent with in vivo results. For example, Keeler and colleagues have shown that postexposure treatment with ibuprofen (an inhibitor of cyclo-oxygenase) protected rats against lethal doses of CG, but was ineffective

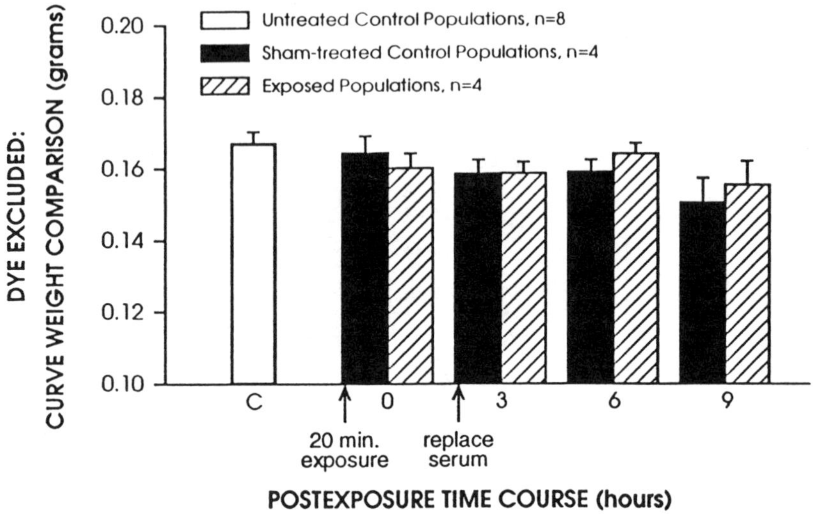


**Figure 4** PFIB-induced permeability profile. Dose = 3230 ppm·min.

against PFIB [personal communication]. There is also evidence, from animal studies, that supports the observation that pathogenesis begins immediately after gas exposure [6,7]. However, with each of these gases, the clinical symptom (i.e., pulmonary edema) is typically delayed by a dose-dependent latent phase of 3–6 hr [2,8]. The fact that exposure to CG caused an immediate increase in EC permeability in vitro does not explain the clinical latent phase. It does underscore that phenomenon, however, as an expression of the lung's complex and overlapping homeostatic mechanisms. Of the eight different cell types recognized in mammalian airways [9], each has a different structure and function. In vitro studies with other chemicals [10] also indicate that organ response to various chemicals can be cell type specific. It is logical to conclude, therefore, that the capacity to develop protective treatments against inhalation injuries produced by these gases and other chemicals can be enhanced by the study of their cell-specific interactions and the use of cultures as models for this purpose.

## CONCLUSIONS

Phosgene causes paracellular leakage in vitro and increases permeability of EC barriers in a dose-dependent fashion. At doses of 1 $LCt_{50}$, CG increased permeability by as much as 18.23% and produced lesions that lasted at least 9 hr. The immediate onset of increased permeability in culture does not explain the clinical latent phase associated with postexposure pulmonary edema. Even at doses of 8 $LCt_{50}$, PFIB did not increase permeability of EC barriers. The data suggest that use of cell cultures can provide complementary and valuable information about target cells and mechanisms by which phosgene and PFIB affect pulmonary edema.

## REFERENCES


1. Pawlowski, R., and Frosolono, M. F., Effect of phosgene on rat lungs after a single high-level exposure. II. Ultra-structural alterations, *Arch. Environ. Health 32*: 278–283 (1977).
2. Diller, W. F., Bils, R. F., Kimmerie, G., and Huth, F., Die Frühphase der Phosgenvergiftung im lichtmikroskopischen, electronmikroskopischen, rötgenologischen und klineschen bild, *Virchows Arch. Pathol. Anat. 348*:230–248 (1969).
3. Hajela, R., Janigan, D. T., Landrigan, P. L., Boudreau, S. F., and Sebastian, A., Fatal pulmonary edema due to nitric acid fume inhalation in three pulp-mill workers, *Chest 97*:487–489 (1990).
4. Hansen-Flaschen, J. H., and Fishman, A., Studies of pulmonary endothelial permeability using tritiated dextrans, *Endothelial Cell Biology in Health and Disease* (N. Simionescu and M. Simionescu, eds.), Plenum Press, New York, 1988, pp. 105–120.
5. Staub, N. C., Pathways for fluid and solute fluxes in pulmonary edema, *Pulmonary Edema* (A. D. Fishman and E. M. Renkin, eds.), American Physiological Society, Washington, DC, 1979, pp. 113–124.
6. Lailey, A. F., Maidment, M. P., and Upshall, D. G., The biochemical toxicology of perfluoroisobutylene, *Proceedings of the 1991 Medical Defense Bioscience Review*, 1991, pp. 299–302.
7. Nold, J. B., Petrali, J. P., Wall, H. G., and Moore, D. H., Progressive pulmonary pathology of two organofluoride compounds in rats, *Inhalation Toxicol. 3*:123–137 (1991).
8. Lehnert, B. E., Kinkead, S. A., Kress, J. D., Kober, E. M., Wood, G. O., Stavert, D. M., and Brainard, J. R., Mechanisms of lung injury caused by perfluoroisobutylene (PFIB) and related agents, *Proceedings of the 1991 Medical Defense Bioscience Review*, 1991, pp. 273–281.
9. Plopper, C. G., St. George, J., Pinkerton, K. E., Tyler, N., Mariassy, A., Wilson, D., Wu, R., Hyde, M., and Evans, M. J., Tracheobronchial epithelium in vivo: Composition, differentiation and response to hormones, *Biology, Toxicology and Carcinogenesis of Respiratory Epithelium* (D. G. Thomassen and P. Nettesheim, eds.), Hemisphere Publishing, New York, 1990, pp. 1–23.
10. Bombick, D. W., Gap junctional communication in various cell types after chemical exposure, *In Vitro Toxicol. 3*:27–39 (1990).

# 31

# In Vitro and Other Alternatives in Cardiovascular Research

**Steven I. Baskin**

*U.S. Army Medical Research Institute of Chemical Defense*
*Aberdeen Proving Ground, Maryland*

**Harry Salem**

*U.S. Army Edgewood Research, Development, and Engineering Center*
*Aberdeen Proving Ground, Maryland*

Implementation of the "3Rs plus 1" (refinement, reduction, replacement, and responsibility) is under consideration for research and safety testing on all organ systems. Refinement of the test methods includes maximizing information obtained from each animal, thereby allowing use of fewer animals and use of those lower on the phylogenetic scale. Reduction in the number of animals employed can be achieved by refinement, and replacement is the ultimate goal, if it can be achieved without compromising human health and safety. In a complex multifunctional organ, such as the heart, consisting of many specialized systems, it may not be possible to replace the intact animal because of many factors influencing its action. However, in vitro tests for some of the individual systems within the heart should be considered. Thus, a battery of in vitro tests may be more applicable for research and safety testing of the heart. It is the responsibility of the scientist and investigator to explore all of these potential alternatives in implementing the 3Rs in cardiovascular research.

The heart is a multifunctional organ incorporating many specialized systems that act to provide electrical, mechanical, bioenergetic, and endocrine

---

The opinions and assertions contained herein are the private views of the authors and are not to be construed as official or as reflecting the views of the Army or the Department of Defense.

responses. For example, the sinoatrial (SA) and atrioventricular (A-V) nodes act in concert with the Purkinje fibers to conduct electrical rhythm within the heart. The actin–myosin system, with specialized invaginations of the T tubule, and the calcium-reuptake systems of the sarcoplasmic reticulum act, in part, to regulate contraction and relaxation. Cardiac mitochondria aid in providing cellular energy for the heart cells, and the right atria produces atrial naturetic factor that aids in the endocrine regulation of sodium in the body. These are only a portion of the individual specialized systems that may be developed in in vitro models to replace the whole-animal experiment.

Because of the complexity of certain irregularities of the heart, such as ventricular arrhythmias, it may involve diffuse areas of the heart and other organs, such as the central nervous system [1] and the adrenal glands [2], in part, which may affect these arrhythmias. It may be necessary to employ multiple in vitro systems to mimic the condition. However, if an underlying site or link for the genesis of the arrhythmia is found [3,4], then a single in vitro model might be applicable.

Development of models to replace or reduce the number of animals used for studies in cardiac toxicity have historically taken place early in the search for cardiac drugs. Myocardial injury [5], metabolic cardiopathies [6], cardiac hypertrophy and failure [7], anatomical pathology [8,9], all represent a variety of cardiovascular toxicologies. Earlier monographs have described alternative models to address the spectrum of toxicities to the heart [10–12].

## ANIMAL MODEL CHOICES

When choosing a source of proteins or enzymes, receptors, cells, tissues, organs, or an intact animal, care must be taken in the species selected. Age, sex, diet, and husbandry factors, as well as the species itself, can influence the qualitative or quantitative responses obtained. For example, the rat heart is quite insensitive to cardiac glycosides and has a negative force–frequency response (whereas other species show a positive curve); has few purinergic receptors; and atypically does not demonstrate a phase 2 plateau in its monophasic electrophysiological ventricular action potential, as do other animals, including humans. The rat characteristically has a quantitatively different response to several drugs than do dogs or humans. For example, a much larger dose of histamine is needed to elicit a response in the rat than in other species such as humans. Animals should be chosen based on their sensitivity to a particular set of criteria under examination [13].

## CARDIAC RECEPTOR MODELS

Many receptors have been isolated in cardiac and neural tissues [14,15]. These includes noradrenergic and cholinergic receptors and transporters for several other modulators that have been isolated or characterized. These isolated or purified receptors offer an opportunity to develop analogues of the substances that act to stimulate the receptors. Specific agonists and antagonists may be examined in isolated receptor models. It may be possible to not only gauge activity, but to obtain an estimate of pharmacokinetics of the product. Lefkowitz and his co-workers [16], for example, have isolated and characterized the noradrenergic receptor in a series of studies. This can lead to further development of noradrenergic agonists and antagonists. Following evaluation of activity in the receptor model, it may be necessary to test the drug in intact animals. However, the information obtained from in vitro studies may result in the use of fewer intact animals. Other factors, such as absorption, distribution, metabolism, and excretion, may influence the in vivo action of the drug; therefore, the data from the in vitro receptor models may not precisely mimic the results observed in the intact animals.

## IN VITRO MODELS

In vitro ligand-binding assays for membrane markers, or proteins that modify ion transport can be used for toxicity testing. Calsequestrin, phospholambam, sodium–potassium ATPase, and myosin ATPase represent only a partial list of proteins or enzyme systems that can be determined as a site for cardiac toxicity. Lipids, carbohydrates, and toxic substances themselves, for example, can also be analyzed for their accumulation and binding to suspected or proposed sites in in vitro toxicity models. Mathematical models for mechanical heart pump function and arrhythmias [17–20] can also provide approaches to define mechanical and electrical phenomena in the heart and the actions of chemicals on these parameters.

## ORGANELLE SUBCELLULAR MODELS

Many subcellular organelles may serve as models for calcium mobilization or calcium overload that could lead to heart failure. These include sarcoplasmic reticulum, sarcolemmal vesicles, myofibrils, or mitochondria, which can be isolated and assayed for enzymic activity. In addition, transport of ions can be measured [21]. In conditions, such as in cardiac ischemia and heart failure, mitochondria are known to be damaged. Intoxiciants that can produce these conditions may be gauged using in vitro estimation of the function of these organelles in the presence of the suspected poison [21]. Abnormalities of myocardial gap junction-mediated cell coupling have been implicated in cardiac arrhythmogenesis. Myocardial gap junctions can be isolated, their properties and numbers determined in vitro, and, thereby, their relation to myocardial ischemia established [22]. It may be possible to quantitate the actions of antiarrhythmic agents that are thought to exhibit their effect at low-resistance gap junctions.

Neurotransmitters can increase the cardiac toxicity under various conditions and after contact with a variety of poisons [12,23]. Several transmitters, such as norepinephrine, are stored in granules in nerve endings. The effect on the rate of release of radiolabeled transmitter or neuromodulator from the granule or nerve ending may provide a measure of cardiac toxicity.

## CELL CULTURE MODELS

With the aid of current methods of cell culture and the use of proteolytic enzymes, it became possible to isolate single cells en masse from tissues and organs to establish myocardial cell cultures [24,25]. The influences of drugs on the growth and survival of normal and tumor cells are quite variable because of the heterogeneity of cell material and conditions of the cell culture [26,27].

## SPECIALIZED TISSUE MODELS

The cardiovascular system includes several tissues that perform particular functions. Studies examining the toxic effects of chemicals on specialized tissues proliferate in the literature. Baskin, as well as others, have shown that Purkinje fibers, the modified conducting muscle, can be isolated from the heart, and then examined with various cardiac poisons that are thought to act at this site [28–30]. Atrial, ventricular and papillary muscle preparations have been used to test the effects of many chemicals on inotropy of these thin and readily oxygen-diffused muscles [31]. Chemicals can exhibit different toxicological effects on atrial, Purkinje, and ventricular animal preparations [32,33].

The clam heart (*Mercenaria mercenaria*) and oyster heart (*Crassostrea virginica*) can serve as a model for cholinergic agents of toxicity. This model employs the use of an animal species lower on the phylogenic scale. Studies by Welsh [34] and others [35], with this model, remain as classic studies for the investigation of cardiac cholinergic receptor function. Quantitative structure–function relationships for other chemicals could be established in these readily available, nonmammalian species. These relationships may provide a fundamental basis for their use in toxicity studies.

Isolated vascular blood vessel models have been employed by many investigators [36] to study the effects of toxic substances on vascular smooth muscle. The roles of chemicals that affect the endothelium, smooth-muscle contraction, and possibly hypertension may be successfully examined in the isolated muscle preparation [37]. An advantage of the use of isolated vascular vessel models is that one animal could provide multiple samples, thereby reducing the numbers of animals employed.

## ISOLATED ORGAN MODELS

The isolated, perfused heart has been used as a useful model of certain cardiovascular parameters. This model has been used, for example, as an index of contractile and electrical activity of the heart. Modifications of the perfused heart have been developed to study the effects of individual toxic substances on atrial and ventricular pacemakers in the same perfused heart. In addition, an index of coronary flow in a constant, pressure-modified, isolated heart, can provide a measure of vasoconstriction to the heart. Neeley et al. [38] have developed a working heart model that may provide a more accurate metabolic model of the heart. This isolated heart model may be used for compound accumulation data, using a radiolabeled compound [39] or oxygen utilization studies. An advantage of an isolated heart model over an intact animal model may be that multiple experiments may be accomplished with the isolated heart versus one experiment in the intact animal. Smaller amounts of drug would be used under greater controlled conditions, without the interference of analgesic or anesthetic agents. Thus, more reliable information may be attained.

## WHOLE-ANIMAL MODELS

Many cardiovascular studies of potentially toxic compounds were traditionally performed in cats or dogs. More recent studies have adapted other animal models that may provide data comparable with what is found in humans. The ferret is considered an excellent cardiovascular model to replace the dog or cat [40]. The pig also could provide a useful alternative to the dog or cat for cardiovascular studies [41]. However, obtaining a blood vessel may be a more difficult task in smaller animal models than in a dog for certain studies.

## CLINICAL MODELS

Libraries of cardiovascular pathology currently exist [9,42,43]. Several large clinical multicenter studies were carried out to demonstrate the effects of aging or disease state on a compound's toxic effect. Reviews by Ahumada [44], Baskin [12], and Wellens et al. [45], provide a list of compounds that can produce clinical cardiac toxicity. In addition, doxorubicin (Adriamycin) [48], digoxin, catecholamines, and cadmium are examples of a few compounds in which retrospective toxicity data have been developed [11,46].

## CONCLUSIONS

Numerous replacements, reductions, and refinements in the use of animals are being developed and used in toxicity testing of many organ systems, including the cardiovascular system. Care needs to be taken in the proper choices of alternatives for specific goals. In vitro tests are being evaluated in the hope of replacing the more elaborate older in vivo ones. The use of newer in vitro validated tests will, we hope, provide similar data with greater reliability and without compromising human health and safety.

## REFERENCES

1. Zavodskaya, I. S., Moreva, E. V., and Novikova, N. A., *Neurogenic Heart Lesions*, Pergamon Press, New York, 1980.
2. Baskin, S. I., Kendrick, Z. V., Goldfarb, A. H., Roberts, J., and Zaydon, A. C., Effect of ouabain on $^3$H-epinephrine washout from the adrenal gland, *Res. Commun. Chem. Pathol. Pharmacol. 29*:365–368 (1980).
3. Noble, D., *The Initiation of the Heartbeat*, 2nd ed., Oxford University Press, Oxford, 1979.
4. Shepherd, J. T., and Vanhoutte, P. M., *The Human Cardiovascular System Facts and Concepts*, Raven Press, New York, 1979.
5. Spitzer, J. J., *Myocardial Injury*, Plenum Press, New York, 1983.
6. Bajusz, E., Experimental "metabolic" cardiopathies and their relationship to human heart diseases, *Ann. N. Y. Acad. Sci. 156*:1–625 (1969).
7. Alpert, N. R., ed., *Myocardial Hypertrophy and Failure*, Raven Press, New York, 1983.
8. Van der Werf, T., *Contractile Behavior of the Heart*, Excerpta Medica, Amsterdam, 1974.
9. Davies, M. J., *Color Atlas of Cardiovascular Pathology*, Oxford University Press, Oxford, 1986.
10. Van Stee, E. W., ed., *Cardiovascular Toxicology*, Raven Press, New York, 1982.
11. Balazs, T., ed., *Cardiac Toxicology*, Vol. 1, CRC Press, Boca Raton, FL, 1981.

12. Baskin, S. I., ed., *Principles of Cardiac Toxicology*, CRC Press, Boca Raton, FL, 1991.
13. Hecht, H. H., and Detweiler, D. K., eds., Comparative cardiology, *Ann. N. Y. Acad. Sci. 127*:1–875 (1965).
14. Hainsworth, R., Kidd, C., and Linden, R. J., *Cardiac Receptors*, Cambridge University Press, Cambridge, 1979.
15. Randall, W. C., ed., *Neural Regulation of the Heart*, Oxford University Press, New York, 1977.
16. Williams, L. T., and Lefkowitz, R. J., *Receptor Binding Studies in Adrenergic Pharmacology*, Raven Press, New York, 1978.
17. Szekeres, L., and Papp, G. J., *Experimental Cardiac Arrhythmias and Antiarrhythmic Drugs*, Akademiai Kiado, Budapest, 1971.
18. Szekeres, L., and Papp, J., *Pharmacological Protection of the Myocardium*, Pergamon Press and Akademiai Kiado, New York, 1985.
19. Garfein, O. B., ed., Clinical pharmacology of cardiac antiarrhythmic agents: Classical and current concepts reevaluated, *Ann. N. Y. Acad. Sci. 432*: (1984).
20. Jalife, J., ed., Mathematical approaches to cardiac arrhythmias, *Ann. N. Y. Acad. Sci. 591*:1–417 (1990).
21. Schwartz, A., ed., *Methods in Pharmacology*, Vol. 5, *Myocardial Biology*, Plenum Press, New York, 1984.
22. Kieval, R. S., Spear, J. F., and Moore, E. N., Gap junctional conductance in ventricular myocyte pairs isolated from post ischemic rabbit myocardium, *Circ. Res. 71*:127–136 (1992).
23. Wilkerson, R. D., ed., *Cardiac Pharmacology*, Academic Press, New York, 1981.
24. Dulbecco, R., Production of plaques in monolayer tissue cultures by single particles of an animal virus, *Proc. Natl. Acad. Sci. USA 38*:747–752 (1952).
25. Rinaldini, L. M., An improved method for the isolation and quantitative cultivation of embryonic cells, *Exp. Cell Res. 16*:477–505 (1959).
26. Yasumura, Y., Tashjian, A. H., Jr., and Sato, G. H., Establishment of four functional tissue culture media, *Ann. N. Y. Acad.Sci. 139*:108–110 (1966).
27. Dawson, M., and Dryden, W. F., Tissue culture in the study of the effects of drugs, *J. Pharm. Sci. 56*:545–561 (1967).
28. Baskin, S. I., Zaydon, P. T., Kendrick, Z. V., Katz, T. C., and Orr, P. L., Pharmacokinetic studies of taurine in bovine Purkinje fibers, *Circ. Res. 47*:673–769 (1980).
29. Dersham, G. H., and Han, J., Actions of verapamil on purkinje fibers from normal and infarcted heart tissues, *J. Pharmacol. Exp. Therp. 216*:261–264 (1981).
30. Davies, M. J., Anderson, R. H., and Becker, A. E., *The Conduction System of the Heart*, Butterworths, London, 1983.
31. Schwartz, A., ed., *Methods in Pharmacology*, Vol. 1, Appleton-Century Crofts, New York, 1971.
32. Sperelakis, N., ed., *Physiology and Pathophysiology of the Heart*, Martinuis Nijhoff, Boston, 1984.

33. Maxwell, D. M., Thomsen, R. H., and Baskin, S. I., Species differences in the negative inotropic action of acetylcholine and soman in rat, guinea pig and rabbit hearts. *Compar. Biochem. Physiol. 100C*:591–595 (1991).
34. Welsh, J. H., Neurohumoral regulation and the pharmacology of a molluscan heart, *Comp. Gen. Pharmacol. 2*:423–432 (1971).
35. Greenberg, M. J., Roop, T., and Painter, S. D., The relative contributions of the receptors and cholinesterases to the effects of acetylcholine on the hearts of bivalve molluscs, *Gen. Pharmacol. 11*:65–74 (1980).
36. Tardos, L., and Rabloczky, G., *Pharmacology of the Vascular System*, Pergamon Press and Akademiai Kaido, New York, 1985.
37. Robinson, C. P., Baskin, S. I., Visnich, N., Jr., and Franz, D. R., The effects of cyanide and its interactions with norepinephrine on isolated aorta strips from the rabbit, dog and ferret, *Toxicology 35*:59–72 (1985).
38. Neely, J. R., Liebermeister, J., Battersby, E. J., and Morgan, H. E., Effect of pressure development on oxygen consumption by isolated rat heart, *Am. J. Physiol. 212*:8804–8814 (1967).
39. Baskin, S. I., Dutta, S., and Marks, B. H., The effects of diphenylhydantoin and potassium on the biological activity of ouabain in the guinea pig heart, *Br. J. Pharmacol. 57*:85–96 (1973).
40. Andrews, P. L. R., Bower, A. J., and Illman, O., Some aspects of the physiology and anatomy of the cardiovascular system of the ferret, *Lab. Anim. 13*:712 (1971).
41. Maduh, E. U., Baskin, S. I., and Nealley, E. W., Cyanide-induced cell energy transients in the brain of Yucatan micropig, *Toxicologist 14*: (in press; 1994).
42. McAllister, H. A., and Fenoglio, J. J., Jr., *Tumors of the Cardiovascular System*, Armed Forces Institute of Pathology, Washington, DC, 1978.
43. Sutton, G. C., and Fox, K. M., eds., *A Color Atlas of Heart Disease*, J. B. Lippincott, Philadelphia, 1990.
44. Ahumada, G. G., ed., *Cardiovascular Pathophysiology*, Oxford Press, New York, 1987.
45. Wellens, H. J. J., Lie, K. I., and Janse, M. J., eds., *The Conduction System of the Heart—Structure, Function and Clinical Implication*, Lea & Febiger, Philadelphia, 1976.
46. Acosta, D., Jr., ed., *Cardiovascular Toxicology*, 2nd ed., Raven Press, New York, 1992.
47. Szekeres, L., ed., *Pharmacology of Antiarrhythmic Agents*, Pergamon Press, New York.
48. Hacker, M. P., Newman, R. A., and Fagan, M. A., The fetal mouse heart: A potential model for anthracycline-induced cardiotoxicity, *Drugs Exp. Clin. Res. 9*: 393–401 (1983).

# 32

# Alternative Tests for Developmental Toxicity

**Thomas J. Flynn**

*U.S. Food and Drug Administration*
*Laurel, Maryland*

Alternative tests for developmental toxicity are unique because of the complexity of the whole-animal systems they attempt to model. Teratogenesis involves complex interactions both within the fetal compartment and between the maternal and fetal compartments. The variety of test systems that have been proposed as alternatives for developmental toxicity testing led to the definition of alternative teratology test systems as "any system that employs test subjects other than the intact pregnant mammal" [1]. Any alternative assay for toxicity testing must have an endpoint relevant to the system being modeled. Endpoints relevant to teratogenesis include (a) differentiation of cells (e.g., biochemical or morphological differentiation, or gene activation or inactivation), (b) pattern formation (e.g., tissue or organ formation), and (c) development of complex organisms from simpler forms (e.g., invertebrates or vertebrate embryos).

Alternative teratogenesis testing systems can be broken down roughly into those that use nonmammalian species and those that use mammalian species. Nonmammalian test systems can be further divided into those that use invertebrates and those that use vertebrates. Mammalian test systems can be divided further on the basis of whether they use isolated cells, organ cultures, or whole-embryo cultures. Table 1 is not comprehensive; it includes only those assays believed to be of historical importance.

**Table 1** Alternative Teratogenesis Test Systems

- I. Nonmammalian
  - A. Invertebrates
    - 1. Drosophila
      - a. Embryos [2,3]
      - b. Embryonic cells [4]
    - 2. Sea urchins [5]
    - 3. Crickets [6]
    - 4. Planaria [7]
    - 5. Hydra [8]
  - B. Vertebrates
    - 1. Fish embryos [9]
    - 2. Amphibian embryos [9]
      - a. Frog embryo teratogenesis assay–*Xenopus* (FETAX) [10]
    - 3. Avian embryos
      - a. Chick embryo [11]
      - b. Chick embryotoxicity screening test (CHEST) [12]
      - c. Chick embryo limb bud cell [13]
      - d. Chick embryo neural crest cell [14]
      - e. Chick embryo neural retina cell [15]
- II. Mammalian
  - A. Isolated cells
    - 1. Established cell lines
      - a. Mouse ovarian tumor (MOT): lectin binding [16]
      - b. Human embryonal palatal mesenchyme (HEPM) [17]
      - c. Chinese hamster lung V79: cell–cell communication [18]
      - d. HEPM: cell–cell communication [19]
      - e. Neuroblastoma [20]
      - f. Teratocarcinoma [21]
      - g. Monkey kidney BSC40: poxvirus morphogenesis [22]
    - 2. Primary embryonic cells
      - a. Rodent limb bud cell micromass [23]
      - b. Rodent limb bud and midbrain cell micromass [24]
  - B. Organ cultures
    - 1. Rodent tooth bud [25]
    - 2. Rodent palatal shelves [26]
    - 3. Rodent salivary glands [27]
    - 4. Rodent limb buds [28]
  - C. Whole-embryo cultures
    - 1. Rodent preimplantation embryos [29]
    - 2. Rodent postimplantation embryos [30]

## NONMAMMALIAN SYSTEMS

### Invertebrates

The fruit fly *Drosophila melanogaster* has been extensively characterized and, therefore, is a desirable model in screening tests for developmental toxicants. Drosophila also has extensive xenobiotic-metabolizing capability. Schuler et al. [2] described a testing protocol and provided data that were obtained with model developmental toxicants. Female flies are mated and allowed to deposit their eggs in medium containing the test compound. The flies complete metamorphosis in approximately 9–10 days. The adult flies, at least 200 per dose level, are then examined extensively for abnormalities. This assay has not been widely accepted, possibly because it requires considerable familiarity with fruit fly morphology. However, a simplified protocol was developed [3] that requires examination of only two easily recognized malformations.

Primary stem cells isolated from drosophila embryos maintain the ability to differentiate into neurons and myoblasts, which can aggregate into discrete structures that resemble ganglia and myotubes. These structures are uniform in size and shape, and they can be quantitated easily with an image analyzer [4]. The basis for the drosophila embryonic cell assay is then the determination of the concentration of test compound in the culture medium that inhibits formation of ganglia or myotubes by 50% relative to controls. There is good correspondence between results from this assay and from whole-animal or human epidemiological studies [31]. A validation study with 100 agents found only two false-positive results (of 45 testing positive) and four false-negative results (of 55 testing negative) [31]. One potential drawback of this assay was the expense of an image analysis system. However, image analyzers are now affordable for most laboratories. New endpoints have also been introduced, including induction of stress (or heat-shock) protein synthesis by teratogenic agents [32].

Sea urchins are a well-established model in developmental biology. Their extensive characterization and morphological simplicity make them a natural choice for screening of environmental teratogens [5]. Although the sea urchin assay was first used for identifying marine pollutants, more recent work [33] suggests attempts to use it as a screen for human developmental toxicants.

The cricket assay [6] evolved from observations of malformed wild crickets near polluted environmental sites. This assay is similar to the drosophila assay, in that crickets deposit eggs in sand contaminated with the test compound. After incubation, the percentage that hatch and the percentage of nymphs with abnormalities are recorded. However, nothing has been published on this assay since its initial development and preliminary validation in the early 1980s. It is, presumably, no longer being pursued actively as an alternative screening test for developmental toxicants.

The planaria assay [7] was developed on the basis of the ability of bisected flatworms to regenerate their missing portions. Abnormal morphogenetic remodeling caused by toxicant exposure is then used to assess the teratogenic potential of the toxicant. No work has been published on this assay in recent years, and it is, presumably, no longer being developed as a teratogen screening assay.

The hydra assay [8], which uses the coelenterate *Hydra attenuata*, is one of the more extensively developed alternative teratogen screening tests and is commercially available. This assay quantitates toxicity in both adult hydra and "artificial embryos" created from reaggregated cells of dissociated adults. The ratio of the concentration of test substance toxic to the adult form (A) to the concentration toxic to the developmental form (D), or A/D, then provides an assessment of teratogenic hazard. Because of their greater toxicity toward the developing organism, agents with a high A/D ratio present more of a hazard than those with a low ratio. Criticisms of the assay include whether the artificial embryo actually represents a developing form and of the concept of the A/D ratio [34,35].

## Vertebrates

Ecotoxicology probably provided the initial impetus for the use of fish or amphibian embryos as biomarkers for identifying environmental contaminants with the potential to cause birth defects in humans. Birge et al. [9] reported testing model toxicants on embryos from a variety of domestic frogs and fish. Endpoints include ability to hatch, larval mortality, and teratogenicity. One of the potential drawbacks is that fish and amphibians tend to be seasonal breeders. This problem was circumvented by using the African clawed frog *Xenopus laevis*. The frog embryo teratogenesis assay–*Xenopus* (FETAX) was developed by Dumont and co-workers [10]. The details of this assay are presented elsewhere in this book (see Chap. 33).

Avian embryos have been used longer than any other system to screen for teratogens, beginning with the pioneering work of Ridgeway and Karnofsky in the 1950s [11]. The basic procedure involves injection of test compounds into the air cell of a chick egg on gestation day 4. The embryos can then be evaluated on day 18 for malformations or incubated further to evaluate their ability to hatch. The chick embryo assay had been criticized for its extreme sensitivity (i.e., too many false-positive results). However, Jelinek [12] refined and standardized the assay into the chick embryotoxicity screening test (CHEST).

More than 130 compounds have been evaluated with CHEST, and the assay appears to be highly predictive of mammalian teratogenicity.

Isolated chick embryo cells have also been used in proposed teratogenicity assays. Mesenchymal cells isolated from limb buds of day-4 chick embryos [13], when plated at high density in "micromass" cultures, will reaggregate and differentiate into chondrocytes. Differentiated foci can be detected and quantitated with stains, such as alcian blue, that are specific for proteoglycans. Test chemicals can be assessed for their ability to inhibit proteoglycan synthesis. Neural crest cells [14] derived from 1-day-old chick embryos will differentiate in culture into either pigment cells, in the presence of fetal bovine serum, or neuronal cells, in the presence of horse serum. Inhibition of differentiation is assessed morphologically. Neural retina cells [15] isolated from day-6 chick embryos will reaggregate into uniformly sized spheres in suspension culture. Cultures can be assessed for cell–cell interactions by measuring reaggregate diameters. The cultures can be induced to differentiate by addition of cortisol, and one can assess differentiation and growth by assaying the astrocyte-specific enzyme glutamine synthetase and total protein, respectively. A preliminary assessment [15] of this assay with 22 chemicals showed a 95% concordance with in vivo mammalian teratogenicity data. This assay is currently being evaluated by an independent laboratory in a blind trial.

## MAMMALIAN SYSTEMS

### Isolated Cells

Isolated mammalian cells in culture, both from established cell lines and primary isolates, are used for many developmental toxicity assays. Cell–cell interaction is an important process in developing embryos. The mouse ovarian tumor (MOT) cell assay [16] measures the effects of agents on the ability of radiolabeled MOT cells to bind to lectin-coated plastic disks. Preliminary validation studies suggest that this assay is not sensitive in identifying teratogens that act by damaging DNA, inhibiting macromolecular synthesis, or binding at specific cell surface receptors. By contrast, the human embryonal palatal mesenchyme (HEPM) cell assay [17] measures growth inhibition and cytotoxicity in a mesenchymal cell line that originated from a day-55 human abortus. Because the MOT and HEPM cell assays are complementary, a validation study on the combined assays was conducted by two contract laboratories, using 44 coded test agents.The results [36] suggested a concordance of about 70% between the combined assay and available human teratogenicity data. Other assays based on cell–cell interactions use gap junction communication as the endpoint. One such assay [18] quantitates the ability of 6-thioguanine (6-TG)-sensitive Chinese hamster lung V79 cells to kill co-cultured 6-TG-resistant cells by transfer of toxic metabolites through gap junctions. A similar assay with HEPM cells [19] uses autoradiography to evaluate the transfer of labeled uridine between cells through gap junctions. In both cases, teratogenicity is assessed by the ability of agents to interfere with gap junction communication. Neuroblastomas and teratocarcinomas are transformed cells that can be induced to differentiate in culture. Assays have been developed [20,21] on the basis of the ability of teratogens to inhibit this differentiation. Virus infection and replication involve a series of steps that can be likened to a developmental pathway. An assay has been developed [22] on the basis of the ability of teratogens to inhibit poxvirus replication in monkey kidney BSC40 cells.

Primary limb bud mesenchymal cells isolated from day-10 mouse embryos can be cultured as high-density micromass cultures, as described earlier for chick embryo limb bud cells. The endpoint for the mouse limb bud cell micromass assay [23] is also the same as for the chick assay. For both assays, inhibition of chondrogenesis (measured by alcian blue staining and [$^{35}$S]sulfate uptake) by test agents is the basis for assessment of the agent as a developmental toxicant. Flint and Orton [24] expanded and refined the micromass assay. They isolated limb bud mesenchymal cells and midbrain neuroepithelial cells from day-12 rat embryos. Development of the limb bud cells is as noted earlier, and the midbrain cells aggregate into ganglionlike foci, which can be visualized and counted after staining with hematoxylin. Cultures of each cell type are also assessed for cytotoxicity by neutral red dye uptake. The Flint micromass assay is currently being validated in a blind, multilaboratory trial.

## Organ Cultures

The suspension culture method developed by Trowell [37] allowed the culture of whole, excised embryonic organs. Cultured embryonic organs have been models in developmental biology for many years, but few organ culture systems have been evaluated as potential teratogen screens. Tooth primordia [26] can be excised from gestation day-14–18 mouse embryos and cultured for up to 14 days. The cultures are then fixed, sectioned, and stained. Sections are evaluated for cell death and for inhibition of growth and differentiation caused by toxicant exposure. Cultured palatal shelves [28] excised from fetal mice (day 13–14) or rats (day 15–16) undergo many of the same developmental processes as palatal shelves in vivo. Palatal shelves are also examined histologically for disruption of normal organogenesis by toxicants. Neither tooth bud nor palatal shelf culture has been developed extensively as a teratogen screening assay, but they have served as model systems for studying mechanisms of normal and abnormal organogenesis.

Fetal mouse salivary glands in culture [25] have been proposed as a screening system for teratogens. Glands excised from day-13 fetuses will undergo the normal process of lobe formation in vitro. A decrease in the number of lobes formed after 48 hr in culture is the basis for classifying agents as developmental toxicants.

Rodent limb bud culture [27] is the most extensively developed organ culture system with potential as a teratogen screen. Limb buds obtained from mouse embryos of 11–13 days in gestation will develop into recognizable paws, with most structures of the long bones and digits present. Limb buds can be cultured as suspension or submersion cultures in either serum-containing or defined media. A variety of morphological and biochemical parameters can be assayed to assess toxicity. In spite of the widespread use of the limb bud assay (over 100 publications through 1991), there is little indication that serious attempts have been made to standardize and validate limb bud culture as a teratogen-screening assay. The present value of this system appears to be for studies in mechanisms of teratogenesis.

## Whole-Embryo Cultures

The procedures for the culture of mouse preimplantation embryos were developed in the early 1960s, and use of this system for screening for developmental toxicants was reported in the mid-1970s [29]. Mice are both time- and superovulated with hormones, and 20–40 two-cell stage embryos are collected per mouse. These embryos develop into blastocysts after 5 days in a defined medium. Blastocysts can be assessed for morphology, total cell count, and chromosomal aberrations. Further developmental potential can be assessed either by propagating blastocysts as explants that can differentiate into embryonic endoderm and ectoderm, or by transferring blastocysts to receptive females, in which they can implant in the uterus and continue to develop. Although agents have been tested on cultured preimplantation embryos to evaluate embryotoxicity, there have been no efforts to standardize and validate this system as a teratogen screen. Because of their rapid rate of cell division and the ease with which metaphase spreads can be prepared from them, preimplantation embryos can be used for evaluating genotoxic mechanisms of embryotoxicity.

New [30] and co-workers established the contemporary method for culture of postimplantation-staged rodent embryos in the 1970s. Neural plate-staged embryos from mouse (day 8) or rat (day 9) continue to develop normally for up to 48 hr when cultured in homologous serum. During this time, they will complete a significant portion of organogenesis. Effects of toxic agents can be assessed easily by gross morphology, but histological and biochemical parameters can be evaluated as well. Several groups (e.g., Govers et al. [38]) are attempting to standardize and validate postimplantation rodent embryo culture as a teratogen screening assay. Because human serum supports growth and development of cultured embryos, Chatot et al. [39] have suggested that cultured embryos may serve as biomarkers for screening the serum of women for factors that may result in adverse reproductive outcomes. Finally, postimplantation embryo culture is an exquisite system for studying mechanisms such as the role of adrenergic agents in situs inversus [40].

## VALIDATION OF ALTERNATIVE TESTS FOR DEVELOPMENTAL TOXICITY

Validations are based on comparisons of assay performance against a standard list of teratogens and nonteratogens [41]. Because these teratogenicity data are drawn from multiple species, protocols, and endpoints, it is difficult to interpret what "validation" against such an assortment means. The standard criteria required for an alternative developmental toxicity assay for assessing human reproductive hazards must be clearly defined.

## ACKNOWLEDGMENT

I thank June Bradlaw for reviewing this manuscript and offering many helpful suggestions.

## REFERENCES

1. Kimmel, G. L., Smith, K., Kochhar, D. M., and Pratt, R. M., *Teratogenesis Carcinog. Mutagen.* 2:221 (1982).
2. Schuler, R. L., Hardin, B. D., and Niemeir, R. W., *Teratogenesis Carcinog. Mutagen.* 2:293 (1982).
3. Lynch, D. W., Schuler, R. L., Hood, R. D., and Davis, D. G., *Teratogenesis Carcinog. Mutagen.* *11*:147 (1991).
4. Bournias-Vardiabasis, N., and Teplitz, R. L., *Teratogenesis Carcinog. Mutagen.* 2:333 (1982).
5. Hagström, B. E., and Lönning, S., *Acta Pharmacol. Toxicol.* *32*(Suppl. 1) (1973).
6. Walton, B. T., *Environ. Entomol.* *9*:18 (1980).

7. Best, J. B., and Morita, M., *Teratogenesis Carcinog. Mutagen. 2*:277 (1982).
8. Johnson, E. M., *J. Environ. Pathol. Toxicol. 4*:153 (1980).
9. Birge, W. J., Black, J. A., Westerman, A. G., and Ramey, B. A., *Fundam. Appl. Toxicol. 3*:237 (1983).
10. Dumont, J. N., Schultz, T. W., and Newman, S. M., *Teratology 25*:37A (1982).
11. Ridgeway, L. P., and Karnofsky, D. A., *Ann. N. Y. Acad. Sci. 55*:203 (1952).
12. Jelinek, R., *Teratogenesis Carcinog. Mutagen. 2*:255 (1982).
13. Wilk, A. L., et al., *In Vitro 16*:269 (1980).
14. Greenberg, J. H., *Teratogenesis Carcinog. Mutagen. 2*:319 (1982).
15. Daston, G. P., Baines, D., and Yonker, J. E., *Toxicol. Appl. Pharmacol. 109*:352 (1991).
16. Braun, A. G., Emerson, D. J., and Nichinson, B. B., *Nature 282*:507 (1982).
17. Pratt, R. M., Grove, R. I., and Willis, W. D., *Teratogenesis Carcinog. Mutagen. 2*: 313 (1982).
18. Trosko, J. E., Chang, C. C., and Netzloff, M., *Teratogenesis Carcinog. Mutagen. 2*:31 (1982).
19. Welsch, F., and Stedman, D. B., *Environ. Health Perspect. 57*:125 (1984).
20. Mummery, C. L., van den Brink, C. E., van der Saag, P. T., and de Laat, S. W., *Teratology 29*:271 (1984).
21. Piersma, A. H., Willemse, A., Mummery, C. L., and de Laat, S. W., *Teratology 34*:420 (1986).
22. Keller, S. J., and Smith, M. K., *Teratogenesis Carcinog. Mutagen. 2*:361 (1982).
23. Hassell, J. R., and Horigan, E. A., *Teratogenesis Carcinog. Mutagen. 2*:325 (1982).
24. Flint, O. P., and Orton, T. C., *Toxicol. Appl. Pharmacol. 76*:383 (1984).
25. Kollar, E. J., *Tests of Teratogenicity In Vitro* (J. D. Ebert and M. Marois, eds.), North Holland, Amsterdam, 1976, pp. 303–333.
26. Lahti, A., and Saxen, L., *Nature 215*:1217 (1967).
27. Lyng, R. D., *Teratology 39*:591 (1989).
28. Aydelotte, M. B., and Kochhar, D. M., *Dev. Biol. 28*:191 (1972).
29. Spielman, H., and Eibs, H.-G., *Methods in Prenatal Toxicology* (D. Neubert, H.-J. Merker, and T. E. Kwasigroch, eds.), Georg Thieme, Stuttgart, 1977, pp. 421–433.
30. New, D. A. T., *Biol. Rev. 53*:81 (1978).
31. Bournias-Vardiabasis, N., Teplitz, R. L., Chernoff, G. F., and Seecof, R. L., *Teratology 28*:109 (1983).
32. Bournias-Vardiabasis, N., and Buzin, C. H., *Developmental Toxicology: Mechanisms and Risk, Banbury Report 26* (J. M. McLachlan, R. M. Pratt, and C. L. Markert, eds.), Cold Spring Harbor Laboratory, Cold Spring Harbor, NY, 1987, pp. 3–16.
33. Kahn, T. A., Blumer, J., Silverman, R. A., and Bickers, D. R., *Fundam. Appl. Toxicol. 11*:511 (1988).
34. Daston, G. P., et al., *Fundam. Appl. Toxicol. 17*:696 (1991).

35. Setzer, R. W., and Rogers, J. M., *Teratology 44*:653 (1991).
36. Steele, V. E., et al., *Fundam. Appl. Toxicol. 11*:673 (1988).
37. Trowell, O. A., *La Culture Organotypique*, Editions du Centre National de la Recherche Scientifique, Paris, 1961, pp. 237–249.
38. Govers, M. J. A. P., et al., *Teratology 39*:456 (1989).
39. Chatot, C. L., Klein, N. W., Piatek, J., and Pierro, L. J., *Science 207*:1471 (1986).
40. Flynn, T. J., and Gibson, R. R., *Teratology 43*:447 (1991).
41. Smith, M. K., et al., *Teratogenesis Carcinog. Mutagen. 3*:461 (1983).

# 33

# Frog Embryo Teratogenesis Assay–*Xenopus*: A Nonmammalian Method for Developmental Toxicity Assessment

**Robert A. Finch and Henry S. Gardner, Jr.**

*U.S. Army Biomedical Research and Development Laboratory*
*Fort Detrick, Frederick, Maryland*

**John A. Bantle**

*Oklahoma State University*
*Stillwater, Oklahoma*

The frog embryo teratogenesis assay–*Xenopus* (FETAX) is a 96-hr, whole-embryo, nonmammalian developmental toxicity screening test that utilizes the embryos of the South African clawed frog *Xenopus laevis*. FETAX was first developed and standardized as a definitive assay for developmental toxicants by Dr. James Dumont and his co-workers [1] at the Oak Ridge National Laboratory in 1983. The assay originally was developed to assess the potential developmental toxicity of complex mixtures derived from the synthetic fuels program at Oak Ridge National Laboratory. The assay is based on a large body of information generated by various researchers in studies on normal embryonic development in which the developing xenopus embryo was used as the model system. The assay, as it is currently performed, is useful in screening for the potential developmental toxicity of both single chemicals, such as pharmaceuticals or commodity chemicals, and complex chemical mixtures, such as environmental samples.

---

The views of the author do not purport to reflect the positions of the Department of the Army or the Department of Defense (Para. 4-3, AR 360-5).

## ADVANTAGES AND DISADVANTAGES OF *XENOPUS*

There are a number of advantages in using *Xenopus laevis* as a model system to study developmental toxicity. The adult frogs used for breeding can be raised in the laboratory and are available commercially. The adults are easy to maintain in the laboratory, since they are very hardy and do not require live food. The most important advantage of *Xenopus laevis* is that the adults can be induced to breed throughout the year by the injection of human chorionic gonadotropin to induce ovulation and mating. This is in contrast with native frog species, which have defined breeding seasons. Adult xenopus are very fecund. A normal mating of a single pair of adults can produce up to several thousand fertile eggs and, thus, can easily provide an adequate number of embryos for testing.

One of the disadvantages of using *Xenopus laevis* is that it is not a native species, so care must be taken not to release the adults, the fertile eggs, or the embryos to the environment. However, there is a positive aspect to this, since the use of xenopus eliminates the need to use native amphibian species, which appear to be in decline.

## ADVANTAGES AND DISADVANTAGES OF FETAX

The FETAX method has several advantages as a screening test for developmental toxicants. It is a relatively inexpensive short-term assay. Embryos are exposed continuously to the test samples over a 96-hr period, from midblastula–early gastrula (stage 8–11) to stage 46 [2]. At the end of this period, the assay is terminated, and data are collected. The assay uses a whole embryo in which all of the developmental processes are intact. Since the period of exposure to test materials coincides with the period of primary organogenesis in the embryo, the influence of developmental toxicants on the processes occurring during this period can be assessed. The stage 46 [2] larvae are transparent. This permits the easy identification of malformations in internal organs, such as the heart and the gut. The use of FETAX as a developmental toxicity-screening test permits a reduction in the number of laboratory mammals required for classic teratogenesis assays.

There are several disadvantages associated with FETAX. The assay is an aqueous-based system which, although an advantage for the testing of aqueous-based environmental samples, such as groundwater or wastewater, presents problems for the testing of hydrophobic materials. This incompatibility requires the use of carrier solvents, such as acetone, dimethyl sulfoxide (DMSO), or triethylene glycol [3–6] to allow the testing of these hydrophobic materials. The use of these solvents introduces the possibility of solvent–test material interactions [4,5]; thus, the results of any assays in which a carrier solvent is used should be viewed with caution. The developing embryos lack a metabolic activation system for xenobiotics. Because of this, an exogenous metabolic activation system must be included in the assay if proteratogens are to be tested in FETAX. *Proteratogens* are compounds that require metabolic activation to become active developmental toxicants. The addition of the metabolic activation system is especially important if FETAX is to be used to assess the potential developmental toxicity hazard of a test material for mammals or humans. Finally, an obvious disadvantage of FETAX is that the assay fails to duplicate the placental function or the maternal–embryo interactions found in mammals.

## USES

The FETAX method can be used as a screening test in ecotoxicology or in the identification of potential mammalian and human health hazards. In ecotoxicology applications, it can provide information on the developmental toxicity of individual chemicals or complex mixtures of chemicals that may affect amphibians and other species found in the environment. It has been used to assess the potential developmental toxicity of complex chemical mixtures found in surface waters [7], groundwater [8], and sediments [9]. With the addition of an exogenous metabolic activation system [3,10–12], FETAX can be used to evaluate the potential developmental toxicity of pure chemicals or complex chemical mixtures for mammals and humans, and it can be useful in toxicity reduction assays. In these assays, complex mixtures of chemicals are fractionated, and each fraction is tested for developmental toxicity with FETAX. Once the developmentally toxic fraction is identified, the components of that fraction can be identified, and steps can be taken to reduce or eliminate the toxic components, thereby reducing the toxicity of the complex mixture. An example of this approach to toxicity reduction is the work of Dr. Mendel Friedman at the U.S. Department of Agriculture in Albany, California [13,14]. Friedman is using genetic-engineering methods in an effort to reduce or eliminate developmentally toxic alkaloids found in potatoes and tomatoes. FETAX is being used as a screening test to evaluate the success of this effort. Schultz and Dawson have been using FETAX as a short-term assay for developmental toxicity in an attempt to relate the molecular structure of chemicals to their toxic activity [15–17]. FETAX can be used to evaluate the effects of metabolism on the developmental toxicity of xenobiotics. Chemicals can be subjected to metabolism in vitro through the use of an exogenous metabolic activation system incorporated in FETAX or known metabolites of the chemicals can be tested directly in

FETAX. Another use of FETAX is in the investigation of the cellular and molecular mechanisms of developmental toxicity [18–20].

## TOXICITY ENDPOINTS

The primary toxicity endpoints used in FETAX are embryotoxicity, as indicated by the 96-hr median lethal concentration ($LC_{50}$), teratogenicity as indicated by the 96-hr median effective concentration ($EC_{50}$; malformation), and growth inhibition as indicated by the minimum concentration to inhibit growth (MCIG). The MCIG is the minimum concentration of test material that significantly inhibits growth, as determined by measurement of head–tail length of the 96-hr larvae exposed to the test material. The 96-hr $LC_{50}$ and the 96-hr $EC_{50}$ are determined using probit analysis. The MCIG is determined by the *t*-test for grouped observations at $p < 0.05$. In addition, developmental retardation is assessed by determination of the stage of development attained by the exposed embryos by the end of the 96-hr exposure period. The teratogenic index (TI) also is calculated [TI = 96-hr $LC_{50}$/96-hr $EC_{50}$ (malformation)]. Additional toxicity endpoints that have been used occasionally include changes in normal locomotion or swimming behavior, abnormal pigmentation, and failure or delay in hatching into free-swimming larvae.

## TEST METHODS

A detailed description of the procedures used in the performance of FETAX can be found in the American Society for Testing and Materials (ASTM) *Standard Guide for Conducting the Frog Embryo Teratogenesis Assay–Xenopus* (FETAX) [21]. Sexually mature adult South African clawed frogs, *X. laevis*, are obtained from a commercial breeder (Xenopus I, Ann Arbor, MI). The breeding stock is held in aquaria connected to a flow-through well water system to maintain water cleanliness. It also is possible to use a static-holding system in which the water in the aquaria is periodically replaced with dechlorinated tap water. Since the frogs are extremely sensitive to the toxic effects of chlorine, all traces of chlorine must be removed before the frogs are exposed to the tap water. The water temperature for the adult frogs must be maintained at 23°±3°C. Adult frogs are fed beef liver, supplemented with liquid multiple vitamins. Breeding pairs are selected from the breeding stock. The day before the start of the test, both the males and the females of two to five breeding pairs are injected subcutaneously with an appropriate amount of human chorionic gonadotropin to induce ovulation and mating. Each of the injected breeding pairs is placed into a separate mating aquarium containing aerated FETAX solution [22], a balanced salt solution. Ovulation, mating, and egg deposition occur during the night. The following morning, the adult frogs are removed from the mating aquaria and returned to the breeding stock to be used again for mating after approximately 6–8 weeks. The eggs deposited in the bottom of the mating aquarium are removed and inspected immediately for fertility and general quality. After completion of this inspection, the jelly coat on the eggs is removed by treating the eggs with a solution of L-cysteine prepared in FETAX solution. Removal of the jelly coat enhances the susceptibility of the embryos to developmental toxicants by removing this potentially protective layer. Removal also reduces the adhesiveness of the eggs, thereby facilitating their manipulation. After the jelly coat has been removed, the embryos are transferred to dishes containing FETAX solution and again inspected for damage or abnormal development. Normally developing embryos in midblastula to early gastrula (stage 8 to stage 11) [2] are selected for use in the test. These embryos are randomly assigned to control or treated groups. Embryos from only one breeding pair are used in each test. Embryos from more than one breeding pair are never pooled. In the standard FETAX renewal exposure regimen, the embryos are exposed to the test material or the controls for 96 hr. During this period, the solutions to which the embryos are exposed are renewed every 24 hr. The temperature is maintained at 24°±2°C in a controlled-temperature environmental chamber. At each renewal, embryos are examined for mortality, and dead embryos are removed from the exposure dishes. At the end of the 96-hr exposure period, a final mortality count is performed and the remaining live embryos are fixed, examined for malformations, and head–tail length measurements are made with a computerized image-analysis system. The collected data are used to determine the 96-hr $LC_{50}$, the 96-hr $EC_{50}$ (malformation), the MCIG, and the TI. In general, an initial range-finding test is recommended. The range-finding test should consist of a series of at least seven concentrations of the test material that differ by a factor of 10. This test should be adequate to delineate the concentration range needed to establish the 96-hr $LC_{50}$ and the 96-hr $EC_{50}$ (malformation). Three definitive tests should be conducted on each test material. Each test should consist of at least five concentrations, which should be adjusted to permit the determination of the 96-hr $LC_{50}$, the 96-hr $EC_{50}$ (malformation), and the MCIG.

## Special Applications Methods

Currently, methods are being developed [R. A. Finch, unpublished] to permit the use of FETAX in a flow-through exposure system in a mobile biomonitoring laboratory that is to be used for on-site environmental toxicity assessment.

FETAX is performed according to the *ASTM Standard Guide for FETAX* [21], but the embryos are placed on the bottom of a mesh-bottomed beaker (Fig. 1). The mesh-bottomed beaker is made by removing the glass bottom of the beaker with a glass-cutting saw and replacing it with a section of plastic mesh that is cemented to the beaker with silicone sealant. The embryos are exposed to the test material in the flow-through system by suspending the mesh-bottomed beakers in aquaria through which different concentrations of the test material are flowing (Fig. 2). The different concentrations are produced through the use of a solenoid-controlled dilutor apparatus. The developing embryos on the bottom of the beakers are bathed in the test material as it exchanges through the plastic mesh bottom. After 96 hr of exposure in the flow-through system, the embryos are removed from the mesh-bottomed beakers and processed according to standard FETAX procedures. This application of FETAX can provide on-site information about the potential developmental toxicity of chemicals found in surface waters, groundwaters, or wastewater effluents.

Linder and co-workers [23] have been developing methods for in situ applications of FETAX at contaminated waste sites. They have developed a plastic mesh exposure cage (Figs. 3A and B) to allow in situ exposure of the developing xenopus embryos. The exposure cage is placed into the test matrix (e.g., sediment or water column) at on-site locations and then secured with

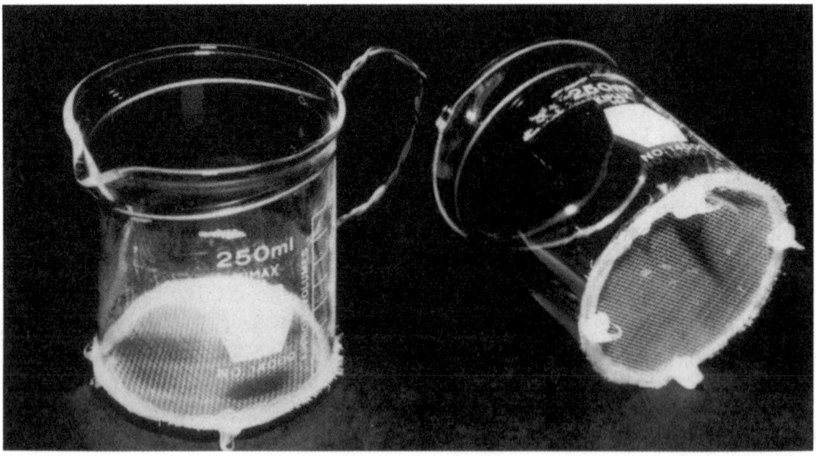

**Figure 1** Mesh-bottomed beakers used to expose developing xenopus embryos to test material in FETAX performed in a flow-through exposure system.

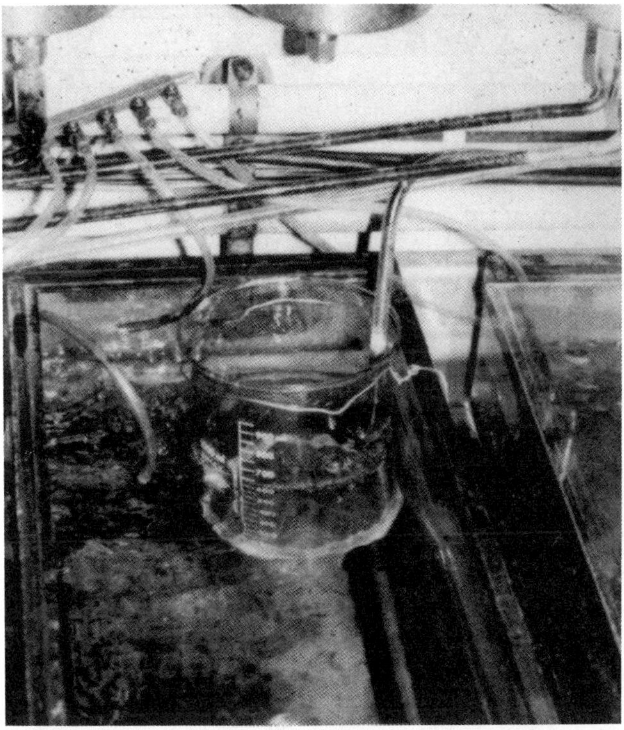

**Figure 2** A mesh-bottomed beaker suspended in an aquarium through which test material is flowing.

stainless steel stakes or other restraints. This in situ exposure method has recently been used at an abandoned mine site in Montana [23].

## REPRESENTATIVE RESULTS

One of the primary FETAX toxicity endpoints is the presence of morphological malformations in embryos exposed to developmentally toxic materials. These malformations can vary from single minor abnormalities, to multiple severe malformations. Figure 4 illustrates the severe malformations produced in a stage 46 [2] embryo exposed to 0.05 mg/L *trans*-retinoic acid for 96 hr in FETAX. The neural tube has failed to fuse during neurulation. Con-

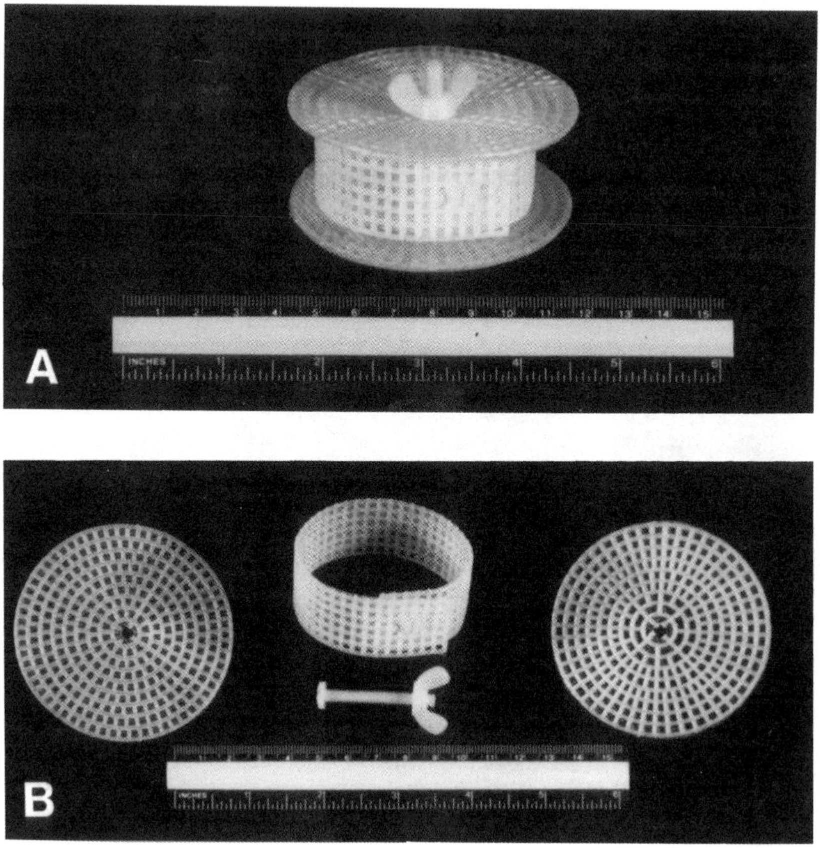


**Figure 3** (A) An assembled plastic mesh exposure cage used in in situ applications of FETAX. (B) Component parts of the plastic mesh exposure cage.

sequently, this embryo has no brain or eyes but was alive at the conclusion of the test.

Typical FETAX results for a nonteratogen (saccharin), a weak to moderate teratogen (caffeine), and a strong teratogen (5-fluorouracil) are shown in Figures 5 and 6. Figure 5 presents concentration–response curves for mortality and malformation, as well as the TI, for each of these compounds. The TI values represent the degree of separation between the malformation and mortality curves. The greater the TI value, the greater the potential developmental toxicity hazard of the compound. For example, there is a large separa-

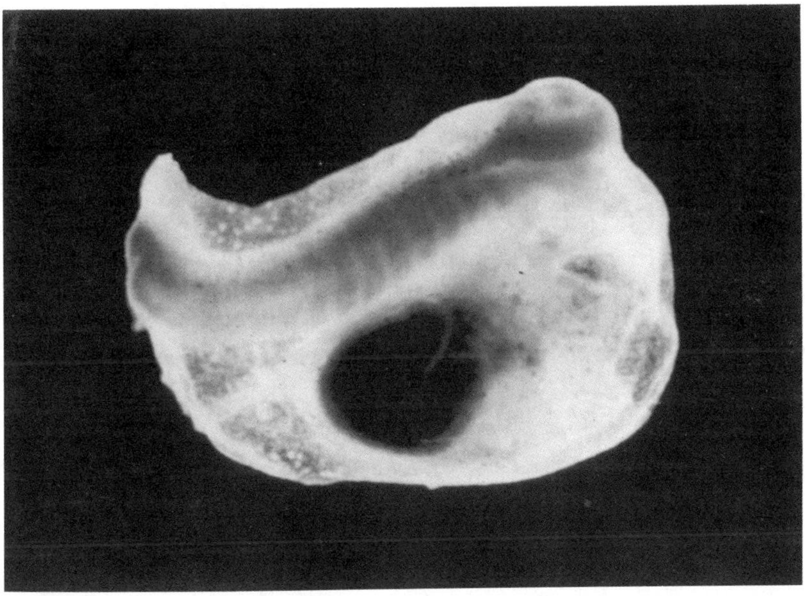

**Figure 4** An example of the severe malformations produced in a stage 46 embryo that had been exposed to 0.05 mg/L *trans*-retinoic acid during the 96-hr exposure period of FETAX. The neural tube has not fused and the embryo has no brain or eyes, however the embryo was alive at the conclusion of the test. All other organs are malformed as well. (Modified after Ref. 26.)

tion between the malformation and mortality curves for 5-fluorouracil, indicating teratogenic effects are observed at far lower concentrations than embryo lethality. At concentrations where most or all of the surviving embryos are malformed, there is little if any mortality. The probability of an embryo being alive and malformed at these concentrations is very high. By comparison, the malformation and mortality curves for saccharin cover the same concentrations and crossover. This gives saccharin a TI equal to 1. This indicates that the probability of an embryo being alive and malformed is equal to or less than the probability of being dead. Generally, in FETAX, TI values less than 1.5 indicate low teratogenic hazard [3,10,18,22,24]; however, the degree of teratogenic hazard indicated by the TI cannot be used alone to determine if a compound is a teratogen or not. Courchesne and Bantle [18] tested dactinomycin (actinomycin D) on xenopus embryos and observed a crossover of the malformation of mortality dose–response curves similar to that noted for

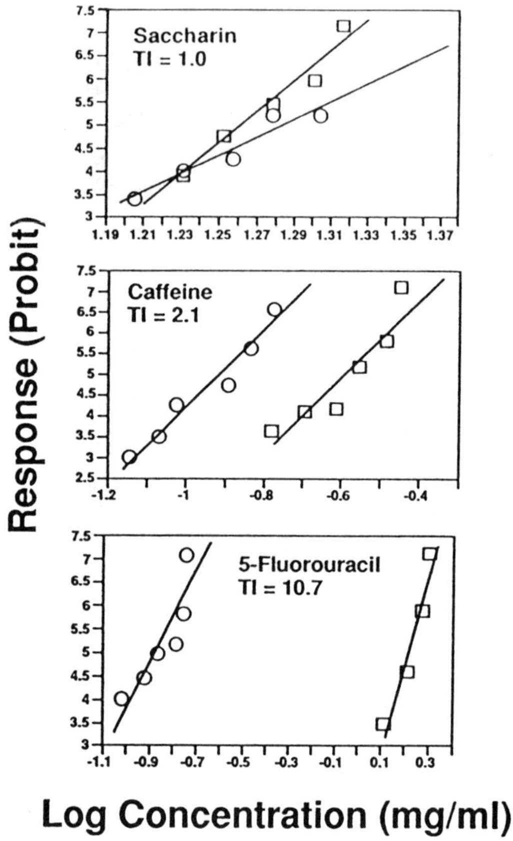


**Figure 5** Concentration–response curves for morality (open squares) and malformation (open circles) for xenopus embryos exposed to saccharin, caffeine, and 5-fluorouracil. (Modified after Ref. 22.)

saccharin (see Fig. 5). The calculated TI for actinomycin D would have been less than 1, indicating a low teratogenic hazard. However, the malformations induced in embryos exposed to this compound were so severe that it was considered teratogenic [18]. Growth inhibition appears to be a very sensitive toxicity endpoint that can be used to aid in the determination of the degree of teratogenicity of a compound. Figure 6 relates the concentration of the test compound as a percentage of its 96-hr $LC_{50}$ to its inhibitory effect on growth as indicated by the percentage of mean control length attained by the exposed

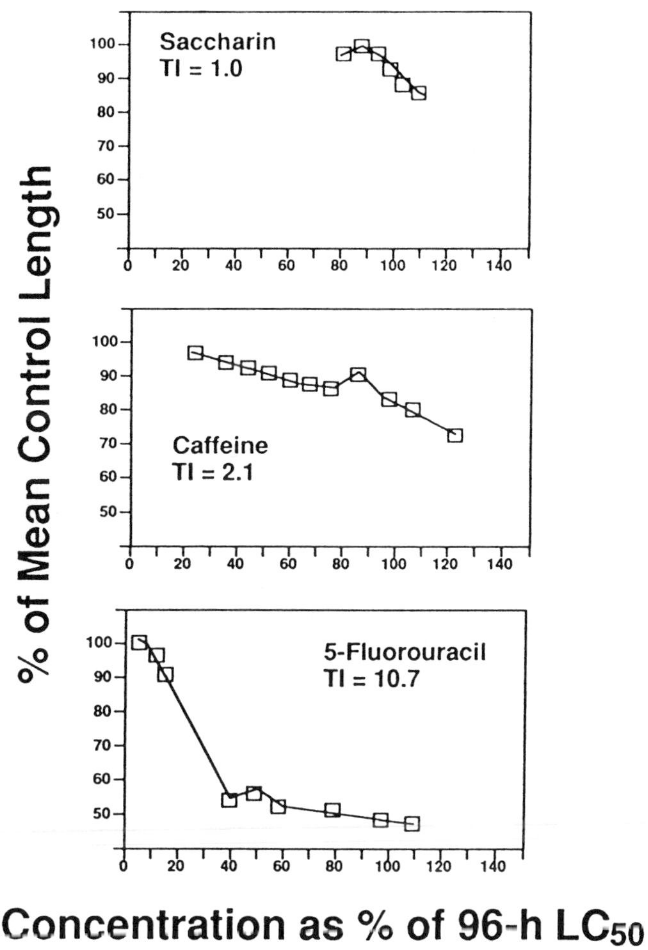


**Figure 6** Representative growth curves for xenopus embryos after 96 hr of exposure to saccharin, caffeine, and 5-fluorouracil. (Modified after Ref. 22.)

embryos. The MCIG for saccharin, caffeine, and 5-fluorouracil were 92.5, 38.9, and 7.4%, respectively, of their respective 96-hr $LC_{50}$ values. As the teratogenicity of the compound increases, the rate of growth inhibition and the total reduction in length increase proportionally. Dawson et al. [24] suggest that compounds with significant teratogenic potential generally inhibit growth at concentrations less than 30% of the respective 96-hr $LC_{50}$ values.

By adding an exogenous metabolic activation system, based on purified rat liver microsomes, to FETAX, it is possible to use FETAX as a test to screen for potential mammalian and human proteratogens [3,10–12]. Figure 7 illustrates the effect of the incorporation of a metabolic activation system in FETAX in which the proteratogen cyclophosphamide was tested. Metabolic activation shifted both the dose–response curves for malformation and for mortality to the left (i.e., to lower effective concentrations). In addition, activation increased the distance between the two dose–response curves, which resulted in a greater TI (TI = 3.5 vs. 1.3). In effect, metabolic activation increased the developmental toxicity of cyclophosphamide.

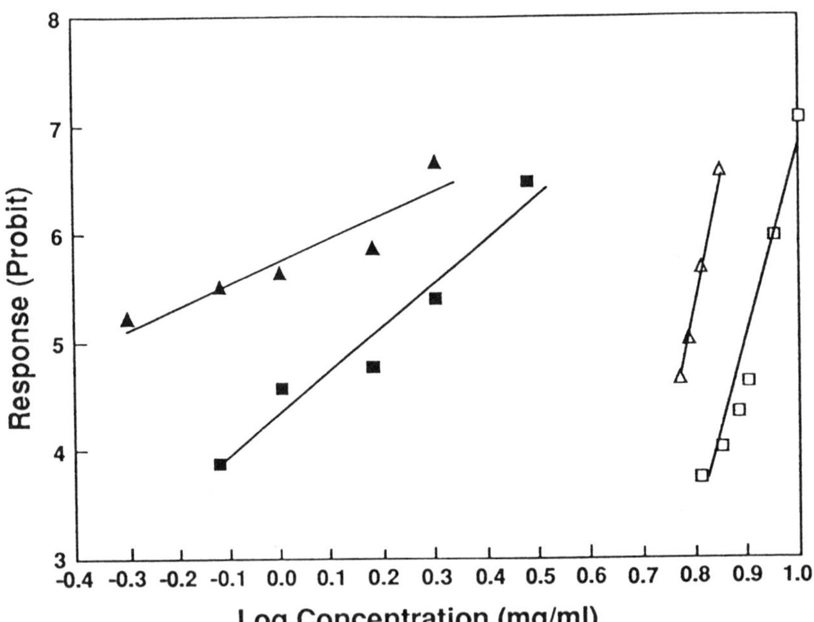


**Figure 7** Concentration–response curves for mortality and malformation for xenopus embryos exposed to the proteratogen cyclophosphamide, with or without the incorporation of a metabolic activation system in FETAX: (solid squares) activated, mortality; (solid triangles) activated, malformations; (open squares) unactivated, mortality; (open triangles) unactivated, malformations.

## CURRENT STATUS

Currently, the U.S. Army Biomedical Research and Development Laboratory at Fort Detrick is sponsoring, under contract to Dr. John Bantle of Oklahoma State University, an extensive interlaboratory validation study of FETAX. This study involves the participation of six independent laboratories. The purpose of the study is to determine the accuracy and reproducibility of the results of FETAX tests performed on coded samples of known mammalian teratogens and nonteratogens. The preliminary results of this study indicate that FETAX is very accurate in distinguishing between teratogens and nonteratogens, and that the results for a given test compound are very reproducible among the participating laboratories. Approximately 100 compounds have been tested in FETAX in earlier intralaboratory validation studies. The results of some of these tests are summarized in Table 1. In these studies, the overall predictive accuracy of FETAX for the identification of mammalian teratogens and non-teratogens was approximately 90% when the test incorporated an exogenous metabolic activation system. Intralaboratory validation studies of FETAX with new test compounds are continuing [25]. Studies to identify appropriate carrier solvents for use with hydrophobic test materials in FETAX, as well as their potential interactions with the test materials, are in progress [4,5]. Further development of the exogenous metabolic activation system is continuing [12]. To facilitate FETAX's further validation and establishment in new laboratories, an American Society for Testing and Materials (ASTM) standard guide for FETAX has been published [21]. The *Atlas of Abnormalities: A Guide for the Performance of FETAX* [26] also has been published. This atlas is intended as a companion to the ASTM standard guide for FETAX. The information and illustrations contained in the atlas provide a basis for the initial establishment of FETAX in a laboratory, as well as acting as a guide in the identification and interpretation of developmental abnormalities observed in embryos during the performance of the test. Copies of the ASTM standard guide for FETAX, as well as the *Atlas of Abnormalities* are available free from Dr. Robert A. Finch or Dr. John A. Bantle.

## CONCLUSIONS

The increasing costs of performing mammalian-based developmental toxicity tests, as well as the desire to reduce the number of laboratory mammals used in these tests has stimulated the development and validation of short-term non-mammalian screening tests to assess the developmental toxicity of a wide

**Table 1** Validation of FETAX

| Compound | Teratogen (TI[a] > 1.5) | Nonteratogen (TI[a] < 1.5) |
|---|---|---|
| Direct acting | | |
| Acetaminophen | | 1.1 |
| Actinomycin D (dactinomycin) | | 0.9 |
| Amaranth | | 1.0 |
| 5-Aminonicotinamide | 560.0 | |
| Aspartame | | NT[b] |
| 5-Azacytidine | 42.0 | |
| Caffeine | 2.1 | |
| Cotinine | 6.0 | |
| Cycloheximide | | 1.3 |
| Cytosine arabinoside (cytarabine) | 7.1 | |
| Diphenhydramine | 10.3 | |
| Ethanol | 1.6 | |
| 5-Fluorouracil | 10.7 | |
| Hydroxyurea | 4.7 | |
| Isoniazid | 37.9 | |
| Naphthalene | | 1.1 |
| Nitroloacetate | | 1.1 |
| Penicillin G | | 1.1 |
| Propylene glycol | | 1.2 |
| *d*-Pseudophedrine | | 1.4 |
| Puromycin | | 1.0 |
| *trans*-Retinoic acid | 10.5 | |
| Saccharin | | 1.0 |
| Sodium cyclamate | | 1.1 |
| Triton X-100 | | 1.1 |
| Urethane | 3.3 | |
| Methotrexate | 23.0 | |

[a]The TI is the leading indicator of teratogenic hazard. However, the severity of the malformations and the MCIG relative to the 96-hr $LC_{50}$ also must be considered in assessing the hazard.

[b]NT, Nontoxic. An $LC_{50}$ could not be determined and therefore a TI could not be calculated.

variety of pure chemicals and complex mixtures. The Frog Embryo Teratogenesis Assay—*Xenopus* (FETAX) provides a rapid, low-cost alternative to the mammalian developmental toxicity tests currently in use. When combined with an exogenous metabolic activation system, FETAX is approximately 90% accurate in distinguishing between mammalian teratogens and nonteratogens. In addition, the results obtained with FETAX are very reproducible both on an intra- and interlaboratory basis. Final validation of FETAX will allow it to be used to screen and rank compounds for further developmental toxicity testing, to assess complex environmental mixtures for human and ecological effects, and as a tool for studying the basic mechanisms of teratogenesis.

## ACKNOWLEDGMENT

The authors would like to thank Jeannine von Loewe for her assistance in the typing and formatting of this manuscript. A portion of the research described in this report was supported by U. S. Army Medical Research and Development Command contracts DAMD17-88-C-8031 and DAMD17-91-C-1048 to Oklahoma State University. The views, opinions, or findings contained in this report are those of the authors and should not be construed as official Department of the Army position, policy, or decision, unless so designated by other official documentation. Citations of commercial organizations or trade names in this report do not constitute an official Department of the Army endorsement or approval of the products or services of these organizations. Research was conducted in compliance with the Animal Welfare Act and other federal statutes and regulations relating to animals and experiments involving animals and adheres to principles stated in the *Guide for the Care and Use of Laboratory Animals*, NIH publication 86-23, 1985 edition.

## REFERENCES

1. Dumont, J., Schultz, T., Buchanan, M., and Kao, G., Frog embryo teratogenesis assay: *Xenopus* (FETAX)—a short-term assay applicable to complex environmental mixtures, *Short-term Bioassays in the Analysis of Complex Environmental Mixtures*, Vol. 3 (M. Waters, S. Sandhu, J. Lewtas, L. Claxton, N. Chernoff, and S. Nesnow, eds.), Plenum Publishing, New York, 1983, pp. 393–405.
2. Nieuwkoop, P., and Faber, J., *Normal Table of* Xenopus laevis (Daudin), 2nd ed., North Holland, Amsterdam, 1975.
3. Fort, D., James, B., and Bantle, J., Evaluation of the developmental toxicity of five compounds with the frog embryo teratogenesis assay: *Xenopus* (FETAX) and a metabolic activation system, *J. Appl. Toxicol. 9*:377–388 (1989).

4. Rayburn, J., Fort, D., McNew, R., and Bantle, J., Synergism and antagonism induced by three carrier solvents with *t*-retinoic acid and 6-aminonicotinamide using FETAX, *Bull. Environ. Contam. Toxicol. 46*:625–632 (1991).
5. Rayburn, J., DeYoung, D., and Bantle, J., Altered developmental toxicity caused by three carrier solvents, *J. Appl. Toxicol. 11*:253–260 (1991).
6. Dresser, T., Rivera, E., Hoffmann, F., and Finch, R., Teratogenic assessment of four solvents using the frog embryo teratogenesis assay–*Xenopus* (FETAX), *J. Appl. Toxicol. 12*:49–56 (1992).
7. Dawson, D., McCormick, C., and Bantle, J., Detection of teratogenic substances in acidic mine water samples using the frog embryo teratogenesis assay—*Xenopus* (FETAX), *J. Appl. Toxicol. 5*:234–244 (1985).
8. Bantle, J., Fort, D., and James, B., Identification of developmental toxicants using the frog embryo teratogenesis assay—*Xenopus* (FETAX), *Hydrobiologia 188/189*: 577–585 (1989).
9. Dawson, D., Stebler, E., Burks, S., and Bantle, J., Evaluation of the developmental toxicity of metal-contaminated sediments using short-term fathead minnow and frog embryo–larval assays, *Environ. Toxicol. Chem. 7*:27–34 (1988).
10. Fort, D., Dawson, D., and Bantle, J., Development of a metabolic activation system for the frog embryo teratogenesis assay: *Xenopus* (FETAX), *Teratogenesis Carcinogen. Mutagen. 8*:251–263 (1988).
11. Fort, D., and Bantle, J., Use of frog embryo teratogenesis assay–*Xenopus* and an exogenous metabolic activation system to evaluate the developmental toxicity of diphenylhydantoin, *Fundam. Appl. Toxicol. 14*:720–733 (1990).
12. Fort, J., Rayburn, J., DeYoung, D., and Bantle, J., Assessing the efficacy of an Aroclor 1254-induced exogenous metabolic activation system for FETAX, *Drug Chem. Toxicol. 14*:143–160 (1991).
13. Friedman, M., Rayburn, J., and Bantle, J., Developmental toxicology of potato alkaloids in frog embryo teratogenesis assay—*Xenopus* (FETAX), *Food Chem. Toxicol. 29*:537–547 (1991).
14. Friedman, M., Rayburn, J., and Bantle, J., Structural relationships and developmental toxicity of solanum alkaloids in the frog embryo teratogenesis assay—*Xenopus* (FETAX), *J. Agric. Food Chem. 40*:1617–1624 (1992).
15. Dawson, D., Schultz, T., Baker, L., and Mannar, A., Structure–activity relationships for osteolathyrism: III. Substituted thiosemicarbazides, *J. Appl. Toxicol. 10*:59–64 (1990).
16. Schultz, T., and Dawson, D., Structure–activity relationships for teratogenicity and developmental toxicity, *Practical Applications of Quantitative Structure–Activity Relationships (QSAR) in Environmental Chemistry and Toxicology* (W. Karcher and J. Devillers, eds.), Kluwer Academic, Dordrecht, 1990, pp. 389–409.
17. Dawson, D., Schultz, T., and Baker, L., Structure–activity relationships for osteolathyrism: IV. *para*-Substituted benzoic acid hydrazides and alkyl carbazates, *Environ. Toxicol. Chem. 10*:455–461 (1991).
18. Courchesne, C., and Bantle, J., Analysis of the activity of DNA, RNA, and protein

synthesis inhibitors on *Xenopus* embryo development, *Teratogenesis Carcinog. Mutagen.* *5*:177–193 (1985).

19. Bantle, J., Fort, D., and Dawson, D., Bridging the gap from short-term teratogenesis assays to human health hazard assessment by understanding common modes of teratogenic action, *Proceedings of the 12th Aquatic Toxicology Symposium*, (STP-1027) ASTM, 1989, pp. 46–58.
20. Fort, D., and Bantle, J., Analysis of the mechanism of isoniazid-induced developmental toxicity with frog embryo teratogenesis assay: *Xenopus* (FETAX), *Teratogenesis Carcinog. Mutagen.* *10*:463–476 (1990).
21. Standard Guide for Conducting the Frog Embryo Teratogenesis Assay–*Xenopus* (FETAX), *Annual Book of ASTM Standards*, Designation: E 1439-91, American Society for Testing and Materials, Philadelphia, 1991, pp. 1–11.
22. Dawson, D., and Bantle, J., Development of a reconstituted water medium and preliminary validation of the frog embryo teratogenesis assay—*Xenopus* (FETAX), *J. Appl. Toxicol.* *7*:237–244 (1987).
23. Linder, G., Wyant, J., Meganck, R., and Williams, B., Evaluating amphibian responses in wetlands impacted by mining activities in the western United States, *Issues and Technology in the Management of Impacted Wildlife*, Proceedings of a National Symposium, Snowmass Resort, Colorado, April 8–10, 1991, Thorne Ecological Institute, 1991, pp. 17–25.
24. Dawson, D., Fort, D., Newell, D., and Bantle, J., Developmental toxicity testing with FETAX: Evaluation of five validation compounds, *Drug. Chem. Toxicol.* *12*: 67–75 (1989).
25. Young, D., Bantle, J., and Fort, D., Assessment of the developmental toxicity of ascorbic acid, sodium selenate, coumarin, serotonin, and 13-*cis*-retinoic acid using FETAX, *Drug. Chem. Toxicol.* *14*:127–141 (1991).
26. Bantle, J., Dumont, J., Finch, R., and Linder, G., *Atlas of Abnormalities: A Guide for the Performance of FETAX*. Oklahoma State University Publications Department, January, 1991.

# 34

# Earthworms as Substitutes for Rodents in Metal Toxicity

**Arthur Furst and Paul K. Chien**

*University of San Francisco*
*San Francisco, California*

Certain segments of the public are pressuring the scientific community to find alternatives to animals for toxicology research. As a result of this, many investigators are using quite different systems as substitutes for experimental animals [1,2]. Among these are cell cultures; however, these cell systems are limited; they do not mimic a metabolizing rodent, as an example. Other test species mentioned in the literature are *Drosophila melanogaster*, nematodes [3], planaria [4], and even *Daphnia magna* [5]. Not all of these species will give reproducible results in toxicological testing.

For toxicological investigations of metals, a species is required with a metabolizing system containing some of the "detoxification" enzymes that are found in the mouse liver. The common earthworm, *Lumbricus terrestris*, does possess catalase [6] and glutathione-*S*-transferase (GSH-S-t) [7]. It can be assumed that these worms also contain superoxide dismutase [8]. Thus, this worm was selected to study the acute toxicity of some metal compounds. Worms are cold-blooded animals, and their anatomy and physiology are well documented [9].

Our results on metal ion toxicity were compared with known values from rodents. Also, these worms were compared with mice for the actions of metal ions on enzymatic activity of catalase and glutathione-*S*-transferase.

The following investigations have been carried out:

1. The acute toxicity of selenite versus selenate
2. The relative toxicity of three nickel salts: the acetate, chloride, and sulfate
3. The induction of metallothionein after a challenge with cadmium chloride
4. The inhibition action of a variety of metals on the catalase activity of the earthworm
5. A comparison of two substrates, 1-chloro-2,4-dinitrobenzene (CDB) and 4-nitroquinoline-*N*-oxide (4NQO) on the activity of GSH-S-t
6. The effect of two dimercapto compounds as arsenic antidotes, *meso*-dimercaptosuccinic acid (DMSA), and 2,3-dimercapto-1-propanesulfonic acid (DMPS). Comparisons were made with dimercaprol (British anti-lewisite; BAL) on three arsenic compounds: arsenate, arsenite, and phenyl dichloroarsine (PDA).

## EXPERIMENTAL

### Procedure for Acute Toxicity

The worms can be obtained from local bait shops. They are kept in potting soil fortified with moist corn meal. The holding cages were kept in a cold room at 15°C. When used, the worms are sorted by length and weight (usually 8–10 cm in length and between 4 and 5 g in weight). The actual ages are unknown.

Each worm is rinsed with distilled water, gently pressed to remove the soil, blotted dry, and weighed to calculate the appropriate injection dosages. The worms are then placed on a plastic bag containing ice to minimize activity. The measured amount of saline solution, either as a vehicle control or containing a known concentration of the metal compound is injected into the coelom, directly behind the cliteium, at an acute angle to avoid entering the alimentary tract. They are then placed in a beaker, and allowed to warm to room temperature. Mortality is recorded at 2, 4, 8, 24, and 48 hr. Between 5 and 10 worms were tested at every dose level. Each level is repeated three to five times. Lethality versus dose is plotted on probit paper. The median lethal dose ($LD_{50}$) is estimated from the graph.

### Procedure for Other Tests

For the determination of metallothionein (MT), the worms are first treated with a sublethal dose of cadmium chloride (with tracer amounts of $^{109}Cd$) and, after a time lapse, the worms are homogenized, and the supernatant separated on Sephacryl S-200 columns. The metallothionein is determined indirectly by analyzing the cadmium by atomic absorption spectroscopy and by liquid scintillation counting.

For the arsenic antidote studies, the worms are first injected with the specific arsenic compound to determine its relative $LD_{50}$ graphically using probit paper. The experiment is repeated for each of the three antidotes, by injecting the arsenic compound and then the antidote. The new $LD_{50}$ values are determined.

Catalase activity was determine either volumetrically, by noting the rate of evolution of oxygen after hydrogen peroxide was added to a worm homogenate; or spectrophotometrically by measuring the rate of decomposition of the $H_2O_2$.

The substrates 1-chloro-2,4-dinitrobenzene and 4-nitroquinoline-*N*-oxide are tested spectrophotometrically for the activity of glutathione-*S*-transferase. The effectiveness of the two is compared.

The protective action of sodium selenate against acute toxicity of cadmium chloride is determined by pretreating the worms with 1% of the $LD_{50}$ for the selenate and then injecting the cadmium salt.

## RESULTS

### Acute Toxicity of Selenium Compounds

The $LD_{50}$ value of selenite and selenate in earthworms are 31 and 60 mg/kg, respectively; for the rat the values are 3 and 5. The rats are about ten times more sensitive to the selenium compounds. However, the ratios of toxicity of selenate to selenite are similar in both species [10].

### Acute Toxicity of Three Nickel Compounds

The acute toxicities of nickel acetate, chloride, and sulfate were determined in the earthworms. The values are 69, 52, and 54 mg/kg. For the mice the values are 32, 26, and 21 mg/kg. The ratios of the toxicity of the chloride and sulfate to the acetate are similar in both species [11].

### Antidotes to Arsenic Toxicity

The values of the acute toxicities of the arsenic compounds are given. These are followed by the values after the respective antidotes were injected.

Arsenite, alone = 191.0; after DMPS = 254.6; after DMSA = 222.3; and after BAL = 211.9

Arsenate, alone = 519.41; after DMPS = 841.0; after DMSA = 607.6; and after BAL = 531.2

Dichlorophenyl phosphine (DCP), alone = 189.5; after DMPS = 287.7; after DMSA = 225.0; and after BAL = 317.4.

Dimercaprol is more active against the organic arsenic compound. The dimercapto derivatives are effective against inorganic arsenic. The latter results are not as spectacular as those reported by Aposhian [12] in mice for inorganic arsenic compounds; however, the earthworms do show similar protection. The DCP values were not found in the literature for rodents [13].

### Induction of Metallothionein

Following the injection of cadmium chloride and determining its acute toxicity [14], the induction of metallothionein was determined. Results indicated that two peaks were formed by measuring the cadmium content of the supernatant with atomic absorption spectroscopy. The experiment was duplicated using $^{109}Cd$. Similar to the action in rodents, cadmium induced two isoforms of metallothionein in the worms [15].

### Effect of Two Substrates for Glutathione-*S*-Transferase

1-Chloro-2,4-dinitrobenzine was superior to 4NQO as substrate for GSH-S-t. The 4NQO reacted with the earthworm cytosol noncatalytically, which negated its effectiveness as a substrate [16].

### Effect of Heavy Metals on Catalase

The divalent ions of Cu, Zn, Fe, Pb, Cd, and Ni at levels up to 1000 ppm, mixed with worm extract, had a moderate inhibitory effect on the catalase activity. Lead and copper were the most effective inhibitors. Either Cd or Ni, when injected into the whole worm 48 hr before the catalase determination, exhibited enzyme enhancement above the base line [17].

## DISCUSSION

From the variety of experiments conducted with the earthworm (*Lumbricus terrestris*) it does appear that useful preliminary information can be obtained on the toxicity of metal compounds. Although the worms are less sensitive than rodents to the acute toxicity of the metal compounds, the ratios of toxicity of one compound to another are almost the same in these two disparate species. The effect of metal ions on the enzymes of the worms are the same as that of the rodents.

Others have used worms for studying the toxicity of pesticides with results similar to those of rodents. For example, Karr et al. [18] used an earthworm to evaluate *d*-limonene, a monocyclic monoterpenoid for toxicity and neurotoxic effects.

The European Economic Community (EEC) now recommends that earthworms be tested for the toxicity of chemicals, especially organic compounds [19]. They suggest that an artificial soil be used, and that the species, *Eisenia foetida* be the test subjects. Heimbach [20] compared the activities of both *E. foetida* and *L. terrestris* against a variety of toxic agents and found a correlation of 0.81; no inorganic agents were in their test compounds.

## CONCLUSIONS

Worms are readily available, are cheap, and can be maintained easily. Following a series of experiments on toxic aspects of inorganic ions, it is concluded that this species shows promise for obtaining comparative information on the toxicity of inorganic compounds.

## REFERENCES

1. Dawe, C. J., Invertebrate animals in cancer research, *Gann Monogr. 5* (O. Muhlbock and T. Nomura, eds.), Maruzen Co., Tokyo, 1989.
2. Chengelis, C., Examples of alternative use in toxicology for common species, *J. Am. Coll. Toxicol. 9*:319–342 (1990).
3. Williams, P. L., and Dusenbery, D. B., Use of the nematode, *Caenorhabditis elegans*, to predict mammalian acute lethality to metallic salts, *Toxicol. Ind. Health 4*:469–478 (1988).
4. Cowgill, U. M., Why round-robin testing with zooplankten often fails to provide acceptable results, *Aquatic Toxicology and Environmental Fate*, Vol 6 (T. M. Poston and R. Purdy, eds.), ASTM STP 921. American Society for Testing Materials, Philadelphia, 1986, pp. 349–356.
5. Best, J. B., and Morita, M., Toxicology of planarians, *Hydrobiologia 227*:275–383 (1991).
6. Fischer, E., and Horvath, I., Evidence for the presence of extraperoxisomal catalase in chloragogen cells of the earthworm, *Lumbricus terrestris*, *Histochemistry 56*:165–171 (1978).
7. Stenersen, J., Guthenberg, C., and Manner, B., Glutathione-*S*-transferase in earthworms (Lumbricidae), *Biochem. J. 181*:47–50 (1979).
8. Moment, G. B., Tolmasoff, J. M., and Cutler, R. G., Superoxide dismutase thermal respiratory acclimation and growth in an earthworm *Eisenia foetida*, *Growth 44*:230–234 (1980).
9. Mill, P. J., *Physiology of Annelids*, Academic Press, New York, 1978.

10. Serda, S., and Furst, A., Acute toxicity of selenium to earthworms, *Proc. West. Pharmacol. Soc. 30*:277–278 (1987).
11. Furst, A., Chien, Y., and Chien, P. K., Worms as substitute for rodents in toxicology: Acute toxicity of three nickel compounds, *Toxicol. Meth. 3*(1):19–23 (1993).
12. Aposhian, H. V., Carter, D. E., Hoover, T. D., Hsu, C.-A., Maiorion, R. M., and Stine, E., DMSA, DMPS, and DMPA-as arsenic antidotes, *Fundam. Appl. Toxicol. 4*:558–570 (1984).
13. Li, W., Chien, P. K., and Furst, A., The response of worms to arsenic antidotes, in press, *J. Appl. Toxicol.*, 1994.
14. Nguyen, Q., and Furst, A., Acute toxicity of cadmium and zinc in earthworms (*Lumbricus terrestris*), *Biol. Trace Element Res. 18*:81–83 (1988).
15. Furst, A., and Nguyan, Q., Cadmium induced metallothionein in earthworms (*Lumbricus terrestris*), *Biol. Trace Element Res. 21*:81–85 (1989).
16. Chen, S., Chien, P. K., and Furst, A., Comparison of two substrates for earthworm GSH-S-transferase (in preparation; 1994).
17. Furst, A., Chien, Y. S., Xie, W., and Chien, P. K., Effect of heavy metal ions on activities of earthworm catalase, *J. Am. Coll. Toxicol. 10*:620 (1991).
18. Karr, L. L., Drewes, C. D., and Coats, J. R., Toxic effects of *d*-limonene in earthworm *Eisenia foetida* (Savigny), *Pestic. Biochem. Physiol. 36*:175–186 (1990).
19. Edwards, C. A., and Neuhauser, F. F., eds., *Earthworms in Waste and Environmental Management*, SPB Academic, The Hague, 1988.
20. Hembach, F., Comparison of laboratory methods using *Eisenia foetida* and *Lumbricus terrestris*, for assessment of the hazard of chemicals to earthworms, *J. Plant. Dis. Protec. 92*:186–193 (1985).

# 35

# The Development and Validation of the Miniature Swine, Mouse, and Rabbit Models as Alternatives to the Use of the Dog in Drug Testing

**James A. Vick**

*U.S. Food and Drug Administration*
*Washington, D.C.*

Calcium channel blocking agents have been extensively used in Europe for the treatment of cardiomyopathies, arrhythmias, and angiospastic angina [1]. The use of β-adrenergic blocking agents in similar cardiovascular disorders is likewise common throughout the world. It is not surprising, therefore, that several recent reports have dealt with the possible synergistic toxicity of the two drug types when used together [2–5]. This study was designed to investigate the possible toxic interaction of propranolol and one of the calcium channel blocking agents, verapamil, and to attempt to develop an antidote for the treatment of this type of reaction. Furthermore, these studies represent an attempt to validate the use of research animals lower on the phylogenetic scale to replace the previous use of companion animals in drug testing.

## MATERIALS AND METHODS

The experiments were performed in rabbits, minipigs, and mice. All studies were conducted using *dl*-propranolol (P) and verapamil (V).

*Rabbits* (2.5–4.5 kg)—control group: Each of 30 rabbits received propranolol, 1 mg/kg IV, followed by verapamil, 0.1 mg $kg^{-1}$ $min^{-1}$ infused IV over a period of 15 min. Treated group: Each of 12 rabbits received propranolol and verapamil as previously described, followed by IV $CaCl_2$ 300 mg $kg^{-1}$. All rabbits were observed for 72 hr or until death.

*Minipigs* (24–28 kg)—control group: Eight minipigs were given propranolol, 0.5 mg $kg^{-1}$ IV, followed by verapamil, 0.5 mg $kg^{-1}$ IV, and observed for 3 hr, or until death. Treated group: Five minipigs received propranolol and verapamil as in the control group, but were treated with $CaCl_2$ 150 mg $kg^{-1}$ IV at the time at which both the heart rate and the blood pressure had decreased to 50% of control and the electrocardiogram (ECG) showed signs of auriculoventricular (A-V) blockage. Blood pressure, heart rate, ECG and respiration were measured in all animals using standard recording apparatus and techniques. All animals were observed for 72 hr, or until death.

*Mice* (20–24 g)—control group: 30 mice received propranolol, 13.63 mg $kg^{-1}$ IP, followed immediately by verapamil, 6.36 mg $kg^{-1}$ IP. All animals were observed for 3 hr, or until death. Pretreated group: 30 mice received $CaCl_2$ 600 mg $kg^{-1}$ IV before being given the combination of propranolol and verapamil. All animals were observed for 72 hr, or until death.

## RESULTS

*Rabbits*, control group: All animals in this group died within 50 min after having received the propranolol and verapamil combination. Treated group: All animals treated with $CaCl_2$ survived.

*Minipigs*, control group: All animals died within 20 min, exhibiting marked A-V blockade, profound bradycardia and severe hypotension. Respiration remained normal until shortly before cardiac arrest, at which time there was hyperpnea followed by apnea. Treated group: All animals treated with $CaCl_2$ survived. Blood pressure and heart rate decreased following propranolol and verapamil, as in the control group. Following $CaCl_2$, the blood pressure increased markedly with a significant widening of pulse pressure. The heart rate remained slow, however, following $CaCl_2$ administration and did not return to normal for at least 12–24 hr. This prolonged bradycardia did not affect the ultimate survival of the treated minipigs. The A-V block produced by propranolol and verapamil was likewise reversed by $CaCl_2$.

*Mice*, control group: All animals died within 15 min after having received propranolol and verapamil. Pretreated group: All animals pretreated with $CaCl_2$ survived the challenge with propranolol and verapamil. The results are shown in Table 1.

**Table 1** The Effect of Calcium Chloride on Propranolol–Verapamil Toxicity in Experimental Animals

| Animals | Propranolol | Verapamil | Calcium chloride | Animal survival (%) |
|---|---|---|---|---|
| Rabbits | | | | |
| Model ($n = 30$) | 1.0 mg kg$^{-1}$ IV | 1.5 mg kg$^{-1}$ IV | — | 0 |
| Treated ($n = 12$) | 1.0 mg kg$^{-1}$ IV | 1.5 mg kg$^{-1}$ IV | 300 mg kg$^{-1}$ IV | 100 |
| Minipigs | | | | |
| Model ($n = 8$) | 0.5 mg kg$^{-1}$ IV | 0.5 mg kg$^{-1}$ IV | — | 0 |
| Treated ($n = 5$) | 0.5 mg kg$^{-1}$ IV | 0.5 mg kg$^{-1}$ IV | 150 mg kg$^{-1}$ IV | 100 |
| Mice | | | | |
| Model ($n = 30$) | 13.6 mg kg$^{-1}$ IP | 6.4 mg kg$^{-1}$ IP | — | 0 |
| Treated ($n = 30$) | 13.6 mg kg$^{-1}$ IP | 6.4 mg kg$^{-1}$ IP | 600 mg kg$^{-1}$ IV | 100 |

## DISCUSSION

A successful model for the study of a possible toxic interaction between propranolol and verapamil was developed in rabbits, minipigs, and mice. The reaction was characterized by marked hypotension, bradycardia, and A-V block, representing an exaggerated pharmacological action of both drugs. Simultaneous administration of the two drugs produced death at doses well below the reported toxic range. Administration of $CaCl_2$ successfully antagonized the development of the "cardiogenic shock" in all animal models and resulted in the survival of all treated animals.

Calcium plays an important role in the excitation–contraction-coupling process of cardiac muscle. Depolarization of the cardiac sarcolemma is associated with an influx of calcium into the cell that initiates the release of more calcium from the sarcoplasmic reticulum to activate the myofibrils [6]. Recent studies confirmed that the calcium that participates directly in myocardial contraction is derived from intracellular and extracellular sources [7–9]. Therefore, it is not surprising that an ample extracellular supply of calcium is essential for normal cardiac function or to reverse the effects of calcium blockade.

Both propranolol and verapamil induce direct cardiac depression by β-adrenergic receptor blockade and inhibition of calcium transport, respectively, which may explain their synergistic toxic interaction [10]. Administration of calcium restores extracellular–intracellular balance through the remaining patent channels, antagonizing the observed "toxic effect of propranolol and verapamil." Therefore, calcium can be considered as an appropriate antidote against the toxic interaction of β-adrenergic blockers and calcium channel blockers.

Previous studies in dogs have shown a similar synergistic toxicity between the calcium antagonists and the β-adrenergic blocking drugs [11]. Both the decreases in blood pressure and in the heart rate observed in the dog appear to be similar to if not identical with the changes seen in these studies. Not only was the toxic interaction seen in the dogs quite like that observed in the minipig, but the ability of calcium to reverse these otherwise lethal effects was also the same [12].

## CONCLUSIONS

Results of these studies strongly support the validation of alternative animal species lower on the animal chain to replace the dog or monkey in drug research.

## REFERENCES

1. Braunwald, E., Calcium channel blockers, *Am. J. Cardiol. 46*:1045 (1980).
2. Benaim, M. E., Asystole after verapamil, *Br. Med. J. 2*:169 (1972).
3. Krikler, D., and Spurrell, R., Asystole after verapamil, *Br. Med. J. 2*:405 (1972).
4. Rumboldt, Z., Bakovic, Z., and Bagatin, J., Verapamil–propranolol: A dangerous cardiodepressive interaction, *Lijec. Vjesn. 101*:430 (1979).
5. Hughes, J., Barraclough, B. M., Opie, L. H., and White, D. W., Adverse interaction between nifedipine and blockade, *Br. Med. J. 281*:1462 (1980).
6. Adams, R. J., and Schwartz, A., Comparative mechanisms for contraction of cardiac and skeletal muscle, *Chest 78*(Suppl.):123 (1980).
7. Langer, G. A., Events at the cardiac sarcolemma: Localization and movement of contractile-dependent calcium, *Fed. Proc. 35*:1274 (1976).
8. Fabiato, A., and Fabiato, F., Calcium release from the sarcoplasmic reticulum: A brief review, *Circ. Res. 40*:119 (1977).
9. Fabiato, A., and Fabiato, F., Calcium and cardiac excitation–contraction coupling, *Annu. Rev. Physiol. 41*:473 (1979).

10. Kohlhardt, M., Bauer, B., Krause, H., and Fleckenstein, A., Differentiation of transmembrane Na and Ca channels in mammalian cardiac fibres by the use of specific inhibitors, *Pflugers Arch. 335*:309 (1972).
11. Vick, J. A., Lethal effects of beta blockers and calcium antagonists, *Physiologist 32*:216 (1989).
12. Vick, J. A., Kendil, A., Herman, E. H., and Balazs, T., Reversal of propranolol and verapamil toxicity by calcium, *Vet. Hum. Toxicol. 25*:8 (1983).

# 36

# Regulatory Requirements for Validation of In Vitro Alternative Tests

**Sidney Green**

*U.S. Food and Drug Administration*
*Laurel, Maryland*

The U.S. Food and Drug Administration (FDA) cannot precisely prescribe what is required for validation of a test or method, because there are a variety of purposes for which tests are used. However, there are some criteria that must be met before an alternative test can be accepted as valid by the general scientific community and used for regulatory decisions [1].

When developing an alternative test, one must determine whether the alternative assay is to be used as a screen, or as a replacement for the animal test. An assay used as a screen will require less stringent acceptance criteria than are required for a replacement test. A screen is a preliminary test for the assessment of a toxicological effect. It is used for making preliminary decisions, or for establishing the direction for further testing. Screens answer fewer and less complex questions than do replacement tests, and the results from screens must be confirmed by more definitive testing. A replacement test, however, must provide the same answers as in vivo methods for the assessment of toxicity. It must provide data that allow a definitive assessment of toxicity.

One must also have a thorough knowledge of the in vivo assay to be replaced. This knowledge should include the test procedures and the regulatory information the test provides (e.g., eye irritation responses of the cornea, conjunctiva, and the iris, and the methods of scoring). In certain instances, such as for eye irritation responses, one must also be familiar with how these data are used to classify test substances.

An investigator should clearly state the rationale for recommending the use of a particular in vitro test in relation to the other available tests. A few alternatives to the use of live animals for the eye irritation test include the use of chick chorioallantoic membrane [2], agarose diffusion [3], Eyetex (Ropak, Irvine, CA) [4], uridine uptake inhibition [5], neutral red uptake [5], rabbit corneal cells [6], and the tetrahymena motility assay [7]. Each test has its own strengths and weaknesses, and new tests must contribute additional needed information, or they must be much improved (e.g., easier to use or less expensive) over the more traditional methods that are already available. The endpoint for the in vitro test should be similar to the in vivo endpoint. For example, if the in vivo assay provides information about ocular irritation, the endpoint of the alternative test should also address ocular irritation.

Occasionally validation studies focus on the correlation between positive and negative results in the new assay, compared with the animal method, whereas the relevance of the endpoint is not addressed. A good correlation between test results for the alternative and animal assays should not be the major criterion for assessing the validity of alternative tests.

For example, the human embryonic palatal mesenchyme (HEPM) cell line has been suggested as a screen for teratogens. The endpoint for this test is inhibition of cellular proliferation. Cellular proliferation is essential to normal development, and its interruption could lead to depressed growth and possibly to teratogenesis. However, how would a cytotoxic agent be differentiated from a teratogen? The correlation between known teratogens and nonteratogens in the assay must be established, but without the basic scientific foundation underlying the endpoint, the conclusions reached are suspect. It is far simpler to determine and express correlation in terms of an endpoint (e.g., to say, "this teratogen inhibits cellular proliferation" than to state "because this agent inhibits cellular proliferation it is a suspected teratogen.") The latter statement is more scientifically sound, because it implies some relationship between the endpoint in the alternative assay and the effect that is found in the animal. Correlations are only as good as the biological or physiological endpoint. How relevant the endpoint is to the effect being measured strengthens or weakens the interpretations and conclusions drawn from such an alternative assay and, consequently, its regulatory usefulness.

Validation studies can be undertaken only after a biologically or physiologically relevant endpoint has been established and a standardized protocol has been developed. As the validation studies proceed, the protocol must be modified to enable identification of all agents that are to be tested. Validation, therefore, is a dynamic process and cannot be described by a rigid set of guidelines or rules.

Once the protocol is developed, the next step is to conduct intralaboratory and interlaboratory testing. These investigations should include as many agents and laboratories as practically possible, given the type and quantity of information the test will provide.

Qualitative data, which provide information on whether the assay responds in the same manner as the more traditional assays, are also needed. A broad spectrum of chemical classes must be tested with the new assay before any correlation between results from the alternative and traditional tests can be objectively evaluated.

Quantitative data, which could involve determining whether the rank-order of positive agents in terms of toxicity or other measured effect is the same or similar in both assays, are also needed. Should an alternative test closely match the animal test that it is to replace in predicting which agents are the most and least hazardous? If there are wide and disparate results, how much confidence can be placed in the alterative test? These questions must be addressed before these new technologies are used.

An in vitro alternative test should not be discarded simply because it does not function with universal perfection. Indeed, it is unlikely that any single alternative will be completely acceptable in all circumstances. For example, some tests may be highly effective only for water-soluble materials. Others may be effective for solids, granular substances, or pastes. It is likely that certain types of products may prove amenable to particular tests, whereas other products may need a different battery of tests. For instance, it may be possible to develop tests that reliably indicate the eye irritation potential of surfactants in soaps and detergents. However, these tests may be less useful for assessing the irritation potential of halogenated organic solvents used in dry cleaning and in spot removers. The critical consideration is that the usefulness of any test depends on the understanding of the capabilities and limitations of the in vitro test for the types of materials to be tested.

When one is selecting the products to be used for validation studies, John Frazier [8] suggests the following: The test substances used should include chemicals from all the general classes of toxicity (nontoxic, mildly toxic, and highly toxic) and from all defined chemical classes (so structure–activity relationships can be developed). Some of the products used should provide data on mixture interactions. Finally, in vivo data on each of the products should be provided so that the test results can be compared between assays.

## REFERENCES


1. Green, S., Regulatory issues associated with the use of alternative tests, *Advances in Modern Environmental Toxicology*, Vol. 10, *Safety Evaluation: Toxicology, Methods, Concepts and Risk Assessment* (M. A. Mehlman, ed.), Princeton Scientific Publishing, Princeton, NJ, 1987.
2. Leighton, J., Nasauer, J., and Tchao, R., *Food Chem. Toxicol. 23*:293 (1986).
3. Wallin, D., and Jackson, E., *J. Toxicol. Cutan. Ocul. Toxicol. 6*:239 (1987).
4. Gordon, V. C., and Bergman, H. C., EYETEX: An in vitro method for evaluation of ocular irritancy, *In Vitro Toxicology—Approaches to Validation*, Vol. 5, *Alternative Methods in Toxicology*, Mary Ann Liebert, New York, 1987, pp. 293–302.
5. Shopsis, C., Validation study: Ocular irritancy prediction with the total cell protein, uridine uptake, and neutral red assays applied to human epidermal keratinocytes and mouse 3T3 cells, *In Vitro Toxicology—New Directions*, Vol. 7, *Alternative Methods in Toxicology* (A. M. Goldberg, ed.), Mary Ann Liebert, New York, 1989, pp. 273–287.
6. North-Root, H., Yakovich, Demetrulias, F. J., Gugula, N., and Heinze, J. E., *Toxicol. Lett. 14*:207 (1982).
7. Silverman, J., and Pennisi, S., *J. Toxicol. Cutan. Ocul. Toxicol. 6*:33 (1987).
8. Frazier, J. M., The validation approach of the Johns Hopkins Center for Alternatives to Animal Testing, *In Vitro Toxicology, Approaches to Validation*, Vol. 5, *Alternative Methods in Toxicology* (A. M. Goldberg, ed.), Mary Ann Liebert, New York, 1987.

# 37

# Consumer Safety, Harmonization of Test Methods and Classification Systems, and Validation of Alternatives

**Kailash C. Gupta**

*U.S. Consumer Product Safety Commission*
*Bethesda, Maryland*

The consumer population includes susceptible individuals, such as children, the elderly, and the infirm, whereas the worker population involved in manufacture, transport, and distribution, generally includes healthy adults. The workers are generally exposed to a few hazardous substances, whereas the consumer can be exposed to multiple hazardous substances that are present in and around residences, transportation, and public places. Exposure of multiple hazardous substances to a susceptible population may pose a greater risk of injury. For these and other reasons, consumer safety has always been given special consideration in the regulatory process. It is crucial that this special consideration continue during the current efforts to harmonize test methods and classification systems. Concerns about trade and animal welfare should not overwhelm concerns about consumer safety.

International harmonization of test procedures, classification systems, and communication systems for hazards to humans and the environment should improve protection of humans and environment, facilitate international trade, and reduce the number of animals used in testing. International acceptance of test results from harmonized test protocols that are designed to generate data

The opinions expressed in this article are those of the author and do not necessarily represent the views of the commission.

suitable for use in a harmonized classification system will reduce the number of tests currently conducted to meet various national and regional classification systems. The harmonization of test protocols and classification systems should also facilitate and possibly accelerate the development and acceptance of alternative test methods, since the alternative methods need to meet the requirements of a harmonized system, rather than multiple national and regional systems. However, the interlaboratory validation phase of an alternative test will need to include laboratories in different countries. Involvement of an appropriate harmonizing organization early in the validation process should facilitate international interlaboratory validation and adoption of alternative methods. However, during the development and validation of alternative methods, one should assure that the alternative methods generate data suitable for the evaluation of risk of injury, with an adequate margin of safety for consumers and, especially, for susceptible populations.

The Organization for Economic Cooperation and Development (OECD) through its Chemicals Group and Management Committee has been involved in the development of test protocols for hazards to humans and environment posed by chemicals. The data generated using these protocols are accepted by the member nations. The membership of OECD includes most of the industrialized nations. These include the United States, Canada, Germany, Australia, Austria, Belgium, Denmark, Spain, Finland, France, Greece, Iceland, Ireland, Italy, Japan, Luxembourg, Norway, New Zealand, Portugal, Turkey, United Kingdom, Sweden, Switzerland, the Netherlands, and the European Economic Community. The OECD will continue to be the primary forum for the harmonization of test methods.

At the United Nations Conference on Environment and Development (UNCED, June 3–13, 1992) held in Rio de Janeiro, Brazil, it was agreed to develop an internationally harmonized system for testing, classification, and communication of hazards posed by chemicals [1]. This responsibility was assigned to the International Programme for Chemical Safety (IPCS). The IPCS is composed of World Health Organization (WHO), International Labour Office (ILO), and United Nations Environment Programme (UNEP). The IPCS has formed a Coordinating Group for the Harmonization of Chemical Classification Systems (IPCS/CG.HCCS). This group has drafted a Terms of Reference and Work Programme [2] and has decided that the focal point for the harmonization of physical hazards posed by chemicals will be ILO, and the focal point for the harmonization for all health effects will be OECD.

## REFERENCES

1. Final Text of Agreement Negotiated by Governments at the United Nations Conference on Environment and Development, June 3–13, 1992, Rio de Janeiro, Brazil, Agenda 21: Programme of Action for Sustained Development, Chapter 11.
2. Provisional Record of the 3rd Consultation of the IPCS Coordinating Group for the Harmonization of Chemical Classification Systems, November 16–17, 1992, WHO Headquarters, Geneva (IPCS/CG.HCCS/93.1).

# 38

# Epilogue

**Alan M. Goldberg**

*Johns Hopkins University*
*Baltimore, Maryland*

The number of papers that have been published dealing with in vitro science is markedly increasing. This can now be quantified from the National Library of Medicine's *Annotated Bibliography on Alternatives to the Use of Live Vertebrates in Biomedical Research and Testing*. This has been published for at least 5 years. The current bibliography (Summer, 1993) has 52 pages, which includes 124 abstracts. When it started, there was only 1 or 2 pages. In addition, a number of companies who have never done in vitro testing are now doing it, and those who have been doing in vitro testing are increasing these activities. A new industry devoted specifically to in vitro sciences has emerged. Companies who are not doing in vitro tests themselves are contracting to have them done. There are now also up to at least five journals devoted to in vitro sciences. The first Conference on Teratogenesis held in 1982 was devoted to in vitro studies, thus initiating the approximate 10-year history of in vitro alternatives.

It was pointed out earlier in this book that many of the major advances in medicine and science could not have happened without the use of animals. That is only part of the story. These advances could not have occurred without the use of in vitro studies as well. It is not only animals, and not only in vitro, it is a combination of both. It is important to make this point over and over again because that is the reality.

This book, written to highlight new approaches and concepts as well as to review the advances made for alternatives to animal testing, has successfully accomplished its mission.

The chapters in this book clearly demonstrate that progress has been made in the in vitro sciences. The media and press always ask, and sometimes in a hostile way, if the use of animals in research and testing has been replaced or reduced. A prime example of replacement is in pregnancy testing. At one time it was animal-based; now, it is exclusively in vitro. In addition, for pyrogen testing, rabbits were used routinely for measuring bacterial endotoxin; the limulus lysate assay for this measurement is now the standard. Previously most hormone assays were bioassays; now they are being done by high-performance liquid chromatography (HPLC). These point to some specific examples of significant progress made in this field over the span of a 10-year history in alternatives, which is more than could have been hoped for.

In the chapter by Tyson, it is pointed out that, at the Society of Toxicology meetings, approximately 25% of the papers are now devoted to in vitro studies. Gerry Zbinden did much the same kind of calculation for the FASEB meeting and found that, up to 1989, about 50% of the papers were in vitro and 50% in vivo. In vitro is not something that is adjunct or supplemental, it is coincidental and important. Proctor and Gamble has published that, over the last 5 years, they have reduced animals in household and consumer goods development and testing by 89%. An 89% decrease in terms of animal use is highly significant. The fact that the regulatory agencies in Europe and the United States have all begun to harmonize the requirements for toxicity testing is yet another example of future direction. Although incredible progress has been made, the growth phase of the exponential curve has not yet begun.

Highlights of some of the other chapters in this book are as follows. Mel Anderson, among others, calls for the increased use of human tissue for in vitro studies in the hope that the uncertainty introduced by extrapolation to humans can be decreased. The answer is not known but it is testable, and that is the important part, to determine if human cells offer unique opportunities in this area. Howard Maibach reported that "skin is not skin is not skin." He showed that there are differences in sensitivities from the wrist to the elbow. This changing sensitivity can be shown for skin all over the body. Skin is not skin is not skin. If this is true, how can there be hope for, or even search for, a single or simple assay? There are no simple assays. There will be complex assays, and many of them will include mechanistically based tests.

Correlative tests may have a place, although a very limited one. Correlative tests allow one to only predict the chemicals that were in the battery that were used to establish the test. The weekly news magazine, *The Economist*, once indicated that correlative tests lead to trivial understandings. The example they used was on the noise associated with the firing of a gun. There is no relevance of that noise to the functioning of the gun. This may be true for some of the current correlative tests.

The fact is that in vitro alternatives in toxicology are just beginning. Some of the tests described in this book may lead to more mechanistically based tests. There may be hundreds or thousands of mechanistically based tests, which for some may be a depressing thought. However, this should be seen as a unique opportunity that might really make a difference.

The chapters from the regulatory sectors describe the role that they are beginning to take and point out the major advances and refinement to date in the Draize eye test. It has also been emphasized that the need for chronic eye irritation tests should not be forgotten. All of the communities that are involved are concerned that the tests used adequately protect the health of the public, and that is the mission for all of us. These chapters also point out that progress has been made. The critics and cynics among us may agree with Sheldon Kopp, who stated from his eschatological list that "progress is an illusion." Not true, in this area of in vitro alternatives.

The issue of validation has come up many times and is misunderstood. Early in this field, validation was considered something that just happened. Clearly, this is not true. In a recent novel by John Kenneth Galbraith—*A Tenured Professor*—he describes that tenure may actually destroy what it is meant to protect. Instead of protecting dissenting opinion, it may restrain early outspokenness of the faculty. The definitions and restraints that are put on validation, especially on some of the earlier, inadequately developed methods, one hopes, will not destroy what we hope to accomplish.

Some expect that in vitro methods must give the same information as an in vivo test. That is not fully correct. Replacement methods should do better and provide more specific information, and examples of this have already occurred. In vitro methods are much more specific. It is difficult to compare a hormone assay in an animal with the quality of information in an HPLC printout. The HPLC clearly provides a better set of information. However, the necessity for mechanistically based tests in the regulatory process is essential.

Advances that have been made are described throughout this book. In conclusion and as a cautionary note: in the book *Zen and the Art of Motorcycle Maintenance*, Robert Pirsig describes the assembly of a barbecue. There is a very vivid description of putting a barbecue together, with no deviations permitted from taking part A to part B, and number 1 to number 2. A rigid structure is necessary. It is essential that we do not evolve into a very rigid structure and limit our thinking in alternatives. Flexibility is necessary to continue to develop mechanistically based tests. This book indicates that this is being done.

# Index